The **Brilliance** of **Bioenergy**

In Business and in Practice

The **Brilliance** of **Bioenergy**

In Business and in Practice

Ralph E H Sims

Routledge
Taylor & Francis Group

LONDON AND NEW YORK

First published 2002 by James & James (Science Publishers) Ltd

Published 2016 by Routledge
4 Park Square, Milton Park, Abingdon, Oxon OX14 4RN
605 Third Avenue, New York, NY 10017

Routledge is an imprint of the Taylor & Francis Group, an informa business

First issued in paperback 2015

This edition published in 2009 by Earthscan

A catalogue record for this book is available from the British Library.

ISBN 978-1-84971-069-5 (hbk)
ISBN 978-1-138-98884-2 (pbk)

Cover photos courtesy of Ralph Sims and Dr Jorg Gigler.

Earthscan publishes in association with the International Institute for Environment and Development.

Contents

Preface   x

1   What is biomass and what is bioenergy?   **1**
So what is biomass?   1
   Fuels from biomass   5
   Sustainability issues from using biomass   8
So what is bioenergy?   10
Terms and units used in biomass production   10
   Physical structure   10
   Weight   12
   Volume   12
   Density   12
   Bulk density   13
   Densification   14
   Moisture content   15
   Heat value   16
   Energy density   16
Terms and units relating to conversion of biomass to bioenergy   16
   Capacity   17
   Cogeneration   17
   Efficiency   17
Biomass fuel characterization   18
   Proximate analysis   19
   Ultimate analysis   19
   Energy (or heat) values   20
   Bulk density and volume   20
   Ash   21
Carbon sinks and greenhouse gas mitigation from the use of biomass   23
   Role of bioenergy systems for greenhouse gas reduction   24
   Forest carbon sinks   25

2   The woody biomass resource   **28**
Forest residues   29
Short-rotation energy forest plantations   32
Municipal green waste   34
Environmental implications of the use of woody biomass   35
Woody biomass resource assessments   39
Characteristics of woody biomass   42
Security of fuelwood supply   43

Cost of the biomass   45
Case study 1. Orbost Power Co-operative, forest arisings, Victoria, Australia   46
Case study 2. Auspine sawmill drying kiln, Tarpeena, South Australia   50

3   The non-woody biomass r   esour ce   **52**
Agricultural crop residues   53
   Rice husks   53
   Bagasse   53
   Cereal straw   57
Energy crops   59
Animal wastes   61
   Sewage sludge   62
Municipal wastes   62
   Industrial wastes   64
   Landfill gas   65
Case study 3. Straw-fired heating boiler, Woburn Abbey, UK   67
Case study 4. Bagasse and MGW cogeneration plant, Rocky Point Sugar Mill,
   Australia   69

4   The supply chain: harvesting, transport and pr   ocessing   **75**
Forest harvesting systems   75
   Harvesting   76
   Extraction   76
   Short-rotation forest harvesting   80
Chipping and chunking   81
System options   85
Drying the biomass fuel   87
The overall system of delivery to the conversion plant   89
Case study 5. Harvesting and processing systems for short-rotation willow
   coppice in the UK   89
   Concepts and scale of operation   90
   Comparison of alternative SRC biomass fuel supply systems   93
   System descriptions of the eight scenarios compared in the model   94
   Recommendations resulting from the study analysis   99
Case study 6. Collection and transport of *P. radiata* forest arisings in New Zealand 101
   Systems of transport and handling compared   101

5   Ther mochemical conversion by combustion and the steam cycle   **104**
The combustion process   106
   Designs of combustion systems   108
   Chemistry of combustion   112
   Emissions   115
Co-firing of biomass   116
Novel biomass combustion and waste-to-energy systems   119
   Waterwide close-coupled gasifier   119
   Offset direct-fired biomass combustion   120

Sawdust   120
Municipal solid waste to RDF   120
Biorefinery   121
The steam cycle   122
Feedwater treatment   122
Steam distribution system   124
Steam boilers   125
Steam turbines   126
Combustion of wastes and MSW   129
Case study 7. The Carter Holt Harvey 39 MW$_e$ plant using wood process residues,
Kinleith, New Zealand   133
Boiler plant   137
Turbine   139
Generator   140
Case study 8. MSW waste-to-energy plant, London, UK   141

6   Ther mochemical conversion by gasification and pyr   olysis   143
Biomass gasification   144
The gasification process   146
Biomass gasifier design   149
Gas turbine systems   151
Emissions   152
Examples of demonstration and commercial gasifier plants   154
Case study 9. Arable Biomass Renewable Energy project 10 MW$_e$ BIGCC,
Yorkshire, UK   157
Pyrolysis   160
Case study 10. Pyrolytic oil from sewage sludge, Western Australian
Water Corporation, Perth   164

7   Biochemical conversion of wet biomass   168
Anaerobic digestion   170
Digester design   174
Biogas plant operation   176
Biogas utilization   178
Environmental benefits   179
International experience   180
Economic analyses   182
Case study 11. Anaerobic digestion of pig waste, Masterton, New Zealand 184
Case study 12. Woodman Point sewage treatment and anaerobic digestion plant,
Perth, Western Australia   185
Landfill gas   190
Municipal wastes and integration of treatments   193
Case study 13. Nova Gas landfill plant, Porirua, New Zealand   194

8   Cogeneration of combined heat and power   196
Cogeneration technology   197

Environmental benefits   201
Process energy   201
Turbine operation   202
Assessing performance of a cogeneration plant   204
Case study 14. Wood-fired district heating scheme and power plant, Vaxjo, Sweden 207
Biomass fuel contracts   209
The boiler system   209
Steam turbines   210
Emissions   210
Controls   212
District heating   213

9   Biofuels for transport fr   om the biochemcial conversion of biomass   214
Biodiesel   218
Case study 15. The potential for biodiesel in New Zealand   225
Oil and fat resource assessment   226
Raw vegetable oil fuels   227
Ester production and blending   227
Fuel property evaluation   228
Engine performance tests and fleet demonstrations   229
Economic analysis   231
Bioethanol   233
Hydrolysis of ligno-cellulose   236
Case study 16. The Brazilian ethanol fuel programme   238
Other biofuel production options   240
Biomethanol   240
Biogas   241
Bio-oil   241
Syngas conversion to Fischer-Tropsch liquids   242
Pollution reduction potential   242
Air pollution   242
Water pollution   243
Greenhouse gas emissions   244
Tropospheric ozone formation   244
Acid rain   245

10 Small-scale bioener   gy systems: pr   esent and futur   e   246
Primary energy conversion technologies   247
Combustion   247
Gasification   249
Pyrolysis   252
Secondary energy conversion technologies   253
Internal combustion engines   253
Steam turbines   256
Steam engines   256
Stirling engines   257

Indirect-fired gas turbines   260
Direct-fired pressurized gas turbines   261
Micro-turbines   262
Fuel cells   263
Electricity generation system analysis at the small scale   272
Case study 17. Small-scale bioenergy project, Taveuni Island, Fiji   276
Power demand   279
Power prices   281
Capital investment costs   282

11 The futur e potential for bioener   gy and the barriers   **284**
Greenhouse gas balances for bioenergy systems   285
Life cycle analysis techniques   287
Social and other environmental issues   290
Distributed generation systems   292
Barriers to implementation   295
Economic barriers   297
Technical barriers   298
Environmental barriers   299
Social barriers   299
Policy barriers   300
Perception barriers   300
Overcoming the barriers   300
Economic instruments   300
Non-economic instruments   301
Case study 18. Western Power's integrated oil mallee eucalyptus project,
Western Australia   302
Conclusions   305

References and bibliography   308

Index   315

Preface

The time for modern biomass has come. It has long languished in the shadows of the other more 'sexy' renewable energy technologies such as wind, solar and hydro, and still retains a poor image relating to smoking chimneys, rubbish tips, industrial factories and poverty. Its carbon dioxide neutrality with regard to mitigation of greenhouse gas emissions remains a mystery to many policy makers. Misleading links made between biomass and the cutting and burning of native forests just to produce cheap energy are still the target of various environmental groups, even though this rarely occurs and is certainly not advocated in this book. The potential for biomass to act as a store of solar energy and yet be able to be converted efficiently when required into heat, power, transport fuels and even substitutes for plastics and petrochemicals is not clear to many. However, the recent abundance of well-designed, successful bioenergy projects around the world has created a renewed interest in this renewable, sustainable and low-emission-producing source of energy.

To meet the challenges of the rapidly increasing atmospheric levels of carbon dioxide (one of the six major greenhouse gases) resulting mainly from the continued burning of fossil fuels, the use of alternative cleaner energy sources will be necessary, together with energy efficiency and conservation measures. There will not be a single technical or social solution to this hugely complex environmental problem. We shall need to utilize all possible means of mitigation at our disposal. The use of biomass in a wise, clean and sustainable manner will be one of them.

Biomass is a broad term used for many resources. Bioenergy covers a wide range of energy conversion technologies. An attempt has been made in this book to cover all the main resources and technologies, giving examples from around the world in both developed and developing countries. Both small- and large-scale projects are included, as are the social and environmental issues along with the more commonly understood economic and technical ones.

This book is for those wishing to learn more about biomass and the business opportunities available from developing bioenergy projects. Having worked on biomass research since the oil shocks of the 1970s, I find it gratifying to see many of the ideas reaching commercial reality. Developments are now happening so rapidly that the book is already out of date, even before it has even been sent to the publishers. However, the implementation of a biomass project remains a challenge for the resource owner and developer compared with the use of many other energy technologies. In order to learn from practical experiences, both good and bad, details of real-world case studies are provided.

For simplicity, US 2000 dollars have been used when discussing costs throughout the book except where stated. All costs quoted should be taken as indicative only, as operating projects and procuring biomass fuels are very site specific. In addition, when costs have been quoted from the literature it was not always clear exactly what was included, or what discount rate was used in the economic analysis.

The idea of writing this book arose following my role as the Western Power Visitor in Renewable Energy Engineering to Murdoch University, Western Australia, in 2000/2001, when I developed teaching resource materials on bioenergy. This included input into several postgraduate courses taught over the internet by the Australian Co-operative Research Centre for Renewable Energy (ACRE). Contributions to the information contained in the book and the courses were provided in one form or another by Jorg Gigler, John Gifford, Ian Bywater, Kingiri Senelwa, Julije Domac, Paul Mitchell, Vesna Harrison, Lanbin Guo, Peter Read, Katrin Jakob, Keith Richards, Damian Culshaw, Drusilla Riddell-Black, Brian Cox, Ralph Overend, Martina Calais, Phil Calais, Louis Arnoux, Peter Billins, John Adams, Roberto Moreira, Fiona Weightman, Nike Pogacnik and all those others whose work appears in the bibliography. Any errors or inaccuracies are however all mine. The late Professor David Hall, the guru of the modern biomass movement, was an inspiration over the years, and his book review comments will be sadly missed.

The continued support from Cathy, my wife, is acknowledged as always, and grateful thanks also go to Bettie and Muriel, my two pet Kunekune pigs. Given to me by my children Adam and Miranda, they were a never-ending source of amusement during the many happy hours spent writing the biogas section!

Chapter 1

What is biomass and what is bioenergy?

The extremely varied nature of biomass, and the many routes possible for converting the biomass resource to bioenergy in the form of useful energy products and services, make this whole topic a complex subject. For other renewable energy sources, such as wind, solar and hydro, the energy conversion technology employed is the key component. For biomass it is the whole system that is critical, and this entails gaining an understanding of:

- the nature of biomass and its greenhouse gas mitigation potential (Chapter 1)

- the range of diverse biomass resources, including wastes (Chapters 2 and 3)

- the processing and delivery of these resources to the energy conversion plant (Chapter 4)

- the thermochemical conversion of dry biomass fuels (Chapters 5 and 6)

- the biochemical conversion of wet biomass fuels (Chapter 7).

Biomass can be transformed into both heat and electricity simultaneously through cogeneration (Chapter 8), into transport fuels (Chapter 9), and even into petrochemical substitutes. At the smaller domestic and village community scale, biomass also has good potential as a fuel for micro-turbines and fuel cells (Chapter 10). In all cases environmental issues are key factors, and can be dealt with by developing good practice guidelines and understanding life cycle analysis. A range of socio-economic benefits will also result, particularly at the smaller scale. There seems little doubt that biomass will provide an increasing share of the global primary energy supply, and that in the future new and more efficient bioenergy technologies will continue to be developed (Chapter 11).

In the words of Professor David Hall, one of the pioneers of modern bioenergy who sadly died in late 1999 prior to his vision being fulfilled: 'Biomass is forever.'

So what is biomass?

From a renewable energy perspective, biomass can be defined as:

recent organic matter originally derived from plants as a result of the photosynthetic conversion process, or from animals, and which is destined to be utilized as a store of chemical energy to provide heat, electricity, or transport fuels.

Biomass resources include wood from sustainably grown plantation forests, residues from agricultural or forest production and organic waste by-products from food and fibre industries, domesticated animals and human activities.

The chemical energy contained in the biomass is derived from solar energy using the process of photosynthesis (based on the Greek *photo*, implying to do with light, and *synthesis*, the linking together of several parts). Photosynthesis is the process by which plants take in carbon dioxide and water from their surroundings and, using energy from sunlight, convert them into sugars, starches, cellulose, lignin etc. These components of vegetable matter are loosely termed carbohydrates, shown chemically for simplicity as $[CH_2O]$. Oxygen is produced and emitted into the atmosphere during the process:

$$CO_2 + 2H_2O \xrightarrow[\text{Heat}]{\text{Light}} ([CH_2O] + H_2O) + O_2$$

It is of course not that simple, and many specialist books exist on the very complex photosynthetic process for anyone wanting more information.

All plant material on Earth, both terrestrial and marine, is formed using this process. Further down the food chain animals that graze plant material as well as carnivorous species all indirectly depend on photosynthesis. Animal products and organic wastes can therefore be classified as forms of biomass when used for energy purposes.

Only a small proportion (0.02%) of the solar energy reaching the Earth each year is fixed and stored by terrestrial biomass, and more by the algae, plankton, aquatic plants, etc. found in the oceans and waterways. The small amount of solar energy captured is equivalent to perhaps seven or eight times the global anthropogenic primary energy consumption, which currently exceeds 400 EJ/year (details of energy units are given later in the chapter).[1]

Depending on the time of year and specific location, the global irradiance is around 600–1000 W/m^2 of the Earth's surface on a clear sky summer day, reducing to 200–500 W/m^2 on a cloudy day to perhaps 50–100 W/m^2 on an overcast winter day. Since the sky conditions vary day by day, season by season, and with latitude, the solar energy reaching the Earth's surface is very variable but can be assumed to average around 1000 kWh (or 3600 MJ)/m^2 per year.

Taking an example, the total annual solar energy reaching a one hectare (1 ha = 10,000 m^2) wheat field in say the Canadian Prairies, a temperate climate in a location receiving about the average solar irradiation of 1000 kWh/m^2 per year, is therefore

- annual energy received = 36,000 GJ

- but only one third arrives during the growing period of the plants = 12,000 GJ

- of which only 20% reaches the leaves of the growing plants = 2400 GJ

- 20% loss occurs from reflection of some of the light = 2000 GJ

- 50% is lost as photosynthetically active heat radiation = 1000 GJ

- 30% of the remaining energy is converted into stored energy = 300 GJ

- of which 40% is consumed in sustaining the plant = 180 GJ.

So, typically, 1 ha of land can store approximately 180 GJ/year in the growing crops.

If in the example the field is planted in say *Miscanthus*, a vegetative grass grown for energy purposes in Europe which yields around 9 t dry matter/ha per year, then, since biomass dry matter contains around 20 GJ/t, this equates to around 180 GJ of available energy contained in the above-ground harvestable biomass. This is approximately only 0.5% of the 36,000 GJ solar input received, which is therefore the conversion efficiency percentage of solar energy to stored chemical energy in the crop. It results from a range of factors and therefore varies widely.

A plant grows by photosynthesis. It also continually respires, during which process the carbohydrates are oxidized to produce carbon dioxide and water. When the plant, or components of it, die (as in the leaf fall of deciduous trees), the material decays. Oxygen is used and heat is released, as is the case for combustion, although the latter occurs over a far shorter period. Thus combustion of biomass can be considered to be simply a more rapid process of either of the natural processes of decay or respiration. In essence the interception of biomass from the natural or agricultural cyclical process and use for energy purposes is the reversal of photosynthesis.

Globally around 55 EJ/year of biomass is currently used for energy purposes, mainly for cooking and heating in developing countries, but also for running a growing number of large-scale modern biomass energy plants. By comparison the world population consumes around 10 EJ/year of energy in the form of food, which of course is a biomass energy resource in itself. (One bowl of cornflakes costing say 10 cents contains enough energy for the consumer of it to cycle 10 km as opposed to driving a car the same distance and consuming around 1 litre of petrol costing around $1 – never mind the additional externality costs involved from the greenhouse gas emissions!)

When the biomass, with its store of chemical energy, can be used usefully it becomes a fuel. In simple terms, the atoms making up the molecules of fuel are linked by electrical forces. Combustion can be considered as a process that converts this stored electrical energy into heat, since the energy in the carbon dioxide and water molecules remaining at the end is much lower. Provided that biomass consumption does not exceed the natural level of plant growth occurring to replace the volume of biomass consumed, then, when converting biomass into useful bioenergy, no more total heat is generated or carbon dioxide created than would have been produced by the natural decomposition processes. So in theory biomass is an energy source that, when managed and used sustainably, has few if any adverse effects on the environment.

In addition to the aesthetic value of the planet's flora and its wonderful biodiversity, biomass represents a useful and valuable resource. The value of biomass is related to the chemical and physical properties of the large molecules of which it is made. Humans have long exploited biomass by burning it for heat and eating it for the nutritional energy from its oil, sugar and starch content. More recently, particularly in the last 150 years, humans have exploited fossilized biomass in the form of coal, oil and gas. This **fossil fuel** is the result of very slow chemical transformations that converted the sugar polymer fraction of the biomass into a chemical composition that resembles the lignin fraction. The additional chemical bonds formed in the fossil fuels represent a more concentrated source of energy. Since it takes millions of years to convert biomass into fossil fuels, they are not considered to be renewable, based on the short time frame over which we are consuming them.

The chemical composition of biomass varies among plant species, but it generally consists of approximately 50% carbon, 44% oxygen and 6% hydrogen (plus water), in the form of 25% lignin and 75% cellulose and hemicellulose (carbohydrates). Most species also contain about 5% of smaller molecular fragments such as resins, collectively known as *extractives*. Typically 1 t of biomass contains around 5–20 GJ of energy depending on the type of material and moisture content. The amount of energy contained in any given source of biomass is valuable information to have in order to enable the size of conversion plant to be designed around the amount of fuel available.

When a crop is to be grown to provide a fuel for bioenergy, it is also useful to know in advance the amount of biomass energy that can be obtained from a particular area of land in terms of GJ/ha per year. In practice this depends upon the climate, weather, location, soil type, soil nutrient levels, water supplies and of course the species grown and crop yield. For example, the genus *Eucalyptus* has over 500 species which vary enormously in their optimum growth requirements. Where one species grows well another will die. Even within a species there is considerable variation, depending on the provenance (or origin) of the seed. Plant breeding, species selection and, in the future, genetically modified varieties all have a major impact on potential biomass yields for a given site. Thus yields can vary over a very wide range between 1 and 30 t/ha, giving between 15 and 500 GJ/ha per year, which has a major impact on the cost of the biomass (in terms of $/GJ) when delivered to the conversion plant gate.

Biomass, in the form of wood, is the oldest form of energy: it was used by humans even before the Stone Age to make life more comfortable. Traditionally biomass has been used to provide heat for warmth and cooking, mainly through direct combustion on open fires, when 90% of the energy in the firewood or dung is wasted. It is still widely used in this way by many of the 2 billion people living in developing countries without access to electricity or other fuels.

In Europe until the 17th century biomass was virtually the only source of energy used for heating, apart from natural hot water springs and a little coal sourced from seams found near the surface of the ground or washed up on the beach. Also, for lighting, tallow from animal fat was used to make candles and lamps long before coal gas was used. The conversion of wood to charcoal was the first improvement in biomass technology. It enabled metals to be extracted from ores requiring high temperatures not easily obtained by wood alone. Brazil still uses charcoal today for its steel industry.

At the beginning of the Industrial Revolution in Great Britain and other parts of Europe, coal began to replace wood and charcoal for several reasons:

- There was an increasing shortage of wood for firewood as forests were increasingly cleared for agriculture or to meet the non-sustainable growth in demand for energy (the first energy crisis!).

- Scientific inventiveness led to technological change, with coal displacing not only wood but also traditional windmills and water wheel power.

- Growing wealth encouraged further technology innovation, such as steam engines, for which coal was the preferred fuel.

- The rising price of wood led to deep coal mining techniques becoming economic, and also to the development of mechanically powered pumps to empty flooded mines.

The fossil fuel era had begun.

Many developing countries today, such as Ethiopia and Kenya, are in a position similar to that Britain was in then, as their scarce firewood supplies continue to decrease from overuse. Developing countries annually consume around 140 EJ of primary energy per year (being 36 GJ/person), of which biomass provides 35% of the total. This equates to a consumption of between 0.5 and 2 t of air-dry biomass per person per year. By contrast, industrialized countries consume over 250 EJ/year (210 GJ/person per year), of which biomass provides around 3% (0.25–0.3 t air-dry/person per year). In wealthier countries such as Switzerland, Finland, Sweden, Austria and the USA technologies for converting biomass, particularly from residues and wastes, are well advanced. The contribution from biomass to their total primary energy supply is far higher than for most other OECD countries. Hence globally around 14% of total primary energy supply is thought to come from biomass, but exact figures are hard to obtain, and biomass is even at times left out of global energy statistics entirely!

Fuels from biomass

Biomass fuels[2] are defined as any solid, liquid or gaseous product derived from a wide range of organic raw materials, either directly from plants or indirectly from industrial, commercial, domestic, forest or agricultural wastes, and produced in a variety of ways. These cover a very wide range of energy sources and scales, from simple firewood for small domestic fires to half a million tonnes of bagasse a year used to fire a 50 MW cogeneration plant. Table 1.1 gives an indication of the relative scale of bioenergy projects.

Table 1.1 An indication of the relative sizes of energy conversion plants with approximate volumes of oven dry tonnes (odt) of biomass fuels consumed and a comparison with typical fossil fuel power plants

Plant capacity	Properties served	Annual fuel demand	Vehicle movements	Conversion technology	Physical size	Investment cost
Domestic heating (15 kW$_{th}$)	Family dwelling	3–5 odt firewood	2–3 trailer loads/yr	Wood burner or boiler	Large suitcase	$100s
Small business heating (350 kW$_{th}$)	School or a small factory	80–120 odt wood or straw	30–40 small truck loads/yr	Straw/wood burner/boiler and fans	Garage for one car	$10,000s
Small electricity-generating plant (250 kW$_e$)	200–300 houses or a small industry	1500–2000 odt wood, straw or wet wastes for biogas	5–6 medium trucks/week	Gasifier or boiler and gas engine or steam engine	Small house and garden	$100,000s
Medium electricity-generating plant (5 MW$_e$)	4000–6000 houses or a small industrial estate	20,000–30,000 odt using range of biomass fuels	40–50 large trucks/week	Gasifier or boiler and gas engine or steam turbine	Petrol service station and forecourt	$1,000,000s
Large electricity generating plant (30 MW$_e$)	25,000–35,000 houses or industrial estate	120,000–140,000 odt using dry biomass fuels	120–150 large truck/trailer units/week	Steam turbine or gas turbine or small combined cycle	Large church and graveyard	$10s of millions
Combined-cycle gas turbine or pulverized coal-fired station (500 MW$_e$)	>500,000 houses or a large industrial site	800 Mm^3 natural gas or 1 Mt coal	Pipeline or 400–500 large truck/trailer units/week	Gas turbine and/or steam turbine	Tower of London or Sydney Opera House	$100s of millions

Some materials will burn and others, such as sand and water, will not. Combustion needs oxygen to change chemically the carbon- and hydrogen-containing molecules of the fuel and hence produce heat. Therefore a fuel can be defined as a substance that interacts with oxygen, changes chemically, and during the process releases its stored chemical energy.

For example, methane (CH_4), a common fuel contained in biogas, landfill gas or natural gas, reacts with oxygen (O_2):

$$CH_4 + 2O_2 \longrightarrow CO_2 + 2H_2O + \textbf{ener gy}$$

This chemical reaction typifies the burning of any common fuel: a compound containing carbon and hydrogen interacts with oxygen (usually from the air, though there are cases when pure oxygen is used) to produce carbon dioxide and water. The amounts produced are predictable based on the atomic weights (H = 1, C = 12, O = 16).

$$[12 + (1 \times 4)] + 2(16 \times 2) \longrightarrow [12 + (16 \times 2)] + 2[(1 \times 2) + 16]$$
$$16 \quad + \quad 64 \quad = \quad 44 \quad + \quad 36$$

Thus, in this simple example, 16 t of methane produces 44 t of CO_2. Since methane has a heat value of 55 GJ/t, the combustion process releases 16×55 = 880 GJ of heat. In other words, for each 20 GJ of heat generated 1 t of CO_2 is emitted. However, various losses in the process result in lower efficiency. If the combustion system has a thermal efficiency of 60%, a fuel input of 100 GJ is needed to produce 60 GJ of useful heat output.

The greater the ratio of carbon to hydrogen in a fuel, then the more carbon dioxide is emitted per unit of energy. In terms of greenhouse gases, therefore, coal when converted to electricity typically produces around 830 g CO_2/kWh of electricity generated, oil 600 g CO_2/kWh and natural gas 400 g CO_2/kWh. But for biomass the carbon is recycled through the next crop, so the emissions are not an issue and the CO_2/kWh is in effect zero.

It is not clear how much biomass could be made available on a global basis for energy purposes. Several analyses have shown that sufficient land is available to meet all demands for food, fibre and energy. Hall and Scrase (1998),[3] for example, calculated there to be sufficient land physically available to grow crops and to supply all human needs for food, fibre and energy at current population levels, though their study did not include the social, economic and logistical constraints of such a proposal.

Sufficient labour, water and nutrients must all be available if a sustainable and economic bioenergy industry is to be developed. On marginal lands, such as the increasingly saline soils of Australia, growing crops for energy can be beneficial. Here short rotation eucalyptus plantations are grown in strips between blocks of cereal crops to help lower the water table and hence reduce the soil saline levels to bring back the natural fertility and increase the wheat yields again. The eucalyptus biomass is to be coppice harvested every 3 to 4 years and used in power-generating plants. Using energy crops to supply small, local bioenergy conversion plants in this way can also create employment in rural areas. Owing to the diverse nature of biomass and the wide variations in local conditions, any assessment of the biomass potential of a region must also consider the environmental and social factors.

Varying assumptions about future fossil fuel prices and the appropriate discount rates to use in the analysis have led to a wide range of estimates. The Third Assessment Report of the Intergovernmental Panel on Climate Change (IPCC) shows that 400 EJ/year will be feasible from energy cropping by 2050 (Table 1.2). Together with biomass estimates from forest, cereal and sugar cane residues (39 EJ/year), animal manures (25 EJ/year) and municipal solid wastes (3 EJ/year), then by 2050 about half the total annual global supply of primary energy could be met by biomass.

Sustainability issues from using biomass

Biomass in its many forms is simply a fuel or 'energy carrier' and of little value for anything else. The stored chemical energy can be converted to other more useful energy products and services such as heat, or cooling, or electricity to drive motors and provide light, or to run an engine to give motive power for transport purposes. In this regard biomass and its use are subject to the first and second laws of energy (or thermodynamics) and, in this regard, answerable to the question 'Is biomass both a truly sustainable and renewable energy source?'

The problem, as succinctly put by British environmentalist Jonathan Porritt (2000), is that politicians see environmental issues as a series of single independent obstacles in their drive to create more wealth and consumer goods. These issues are symptoms of a non-sustainable system that ignores the fundamental principles relating to the undisputable laws of conservation of energy and matter. Society has ignored the thinking behind these laws to pursue material progress, and has ended up with a system that cannot continue to function for ever. The current lifestyle of the developed world depends on cheap fossil fuels and minerals, which in turn depend very much on mining and on utilizing the limited and finite supplies of both.

Energy exists in numerous forms (including chemical, heat, light, gravitational and motion), which are continually being converted from one type to another in order to power the world economy and enable useful work to be carried out. The **first law** of energy is that it can be neither created nor destroyed during these transformations: so the total energy input must always equal the total energy output.

When biomass is combusted, the stored solar energy in it is simply transformed into thermal energy. The heat produced contains no more or less energy than that holding together the molecules of the biomass as chemical bonds. In addition the total mass of materials such as carbon, hydrogen and oxygen that were originally involved in the process will always equal the total mass that results from it. So, just like energy, matter cannot be either created or destroyed, and the amount of matter also remains constant during the transformation process.

The **second law** of energy states that the availability of energy to perform useful work becomes less as it passes through successive transformations.

Table 1.2 Projection of the technical energy potential from purpose-grown biomass by 2050

Region	Population in 2050 (billion)	Total land with crop production potential (ha × 10⁶)	Cultivated land in 1990 (ha × 10⁶)	Additional cultivated land required in 2050 (ha × 10⁶)	Available area for biomass production in 2050 (ha × 10⁶)	Maximum additional amount of energy from biomass[a] (EJ/yr)
Developed countries[b]	-	820	670	50	100	30
Latin America						
Central & Caribbean	0.286	87	37	15	35	11
South America	0.524	865	153	82	630	189
Africa						
Eastern	0.698	251	63	68	120	36
Middle	0.284	383	43	52	288	86
Northern	0.317	104	4	14	50	15
Southern	0.106	44	16	12	16	5
Western	0.639	196	90	96	10	3
China[c]	-	-	-	-	-	2
Rest of Asia						
Western	0.387	42	37	10	-5	0
South-Central	2.521	200	205	21	-26	0
Eastern	1.722	175	131	8	36	11
South-East	0.812	148	82	38	28	8
Total for regions above	8.296	2495	897	416	1280	396
Total biomass energy potential						441 EJ/yr

[a] Assumed 15 odt/ha.yr average yield and 20 GJ/odt is the typical biomass lower heat value
[b] This includes OECD countries and economies in transition
[c] Projected values and not maximum estimates
[d] Includes 45 EJ/yr of biomass currently used for traditional heating and cooking

Taking a simple example, the energy in biomass can be used to produce heat at high temperatures to produce steam, which can then be used to drive a turbine to produce electricity. The heat in the residual steam from this process could be used in, for example, a wood pulp factory to cook the wood fibre. Any heat left over from that process might be sufficient to produce hot water to wash the extracted fibres, and any heat left over from that could possibly be used – if at sufficient temperature – to heat the buildings to make the office workers more comfortable. But eventually all the chemical energy from the biomass will end up in the atmosphere as heat. The good news is that, using input energy from the sun, nature has developed the techniques to concentrate energy and matter and put them together again as renewable biomass.

Since the Industrial Revolution mankind has assumed that the laws of energy do not apply to us. As a result several environmental concerns are rapidly growing, climate change being but one. Engineering – in terms of designing systems, constructing pipes and analysing mathematically – can no longer be separated from the broader issues of sustainability and environment, especially when dealing with a natural and variable product such as biomass.

So what is bioenergy?

A number of conversion routes exist to change biomass into useful forms of energy, as shown in simplified form in Figure 1.1. Many of these will be covered in detail in later chapters. The main point to note now is that the owner of a biomass resource can work in partnership with a project developer to convert that resource into useful bioenergy projects in order to maximize the return on the investment. If the resource is a waste product, avoiding any treatment or disposal costs can lead to additional benefits, or a 'win/win' opportunity.

Terms and units used in biomass production

Biomass fuels vary with the plant species, the nature of the resource material (such as straw, wood, bark, leaves, sludge, municipal wastes, algae or manure), and the moisture content, which ranges from 95% moisture content (wet basis) for dairy farm wastes to 10% m.c.(w.b.) for cereal straw.

Physical structure

Physical structure, such as cell size and piece size, is not an important parameter when determining the energy value of biomass materials, but the **chemical composition** is. The basic **ener gy value** is measured as joules of energy in 1 g of fuel (J/g). For a given plant species or type of biomass it varies with the cellulose, hemicellulose and lignin contents, and with the content of resins and tannins present. For convenience, biomass energy values are normally quoted as MJ/kg or GJ/t.

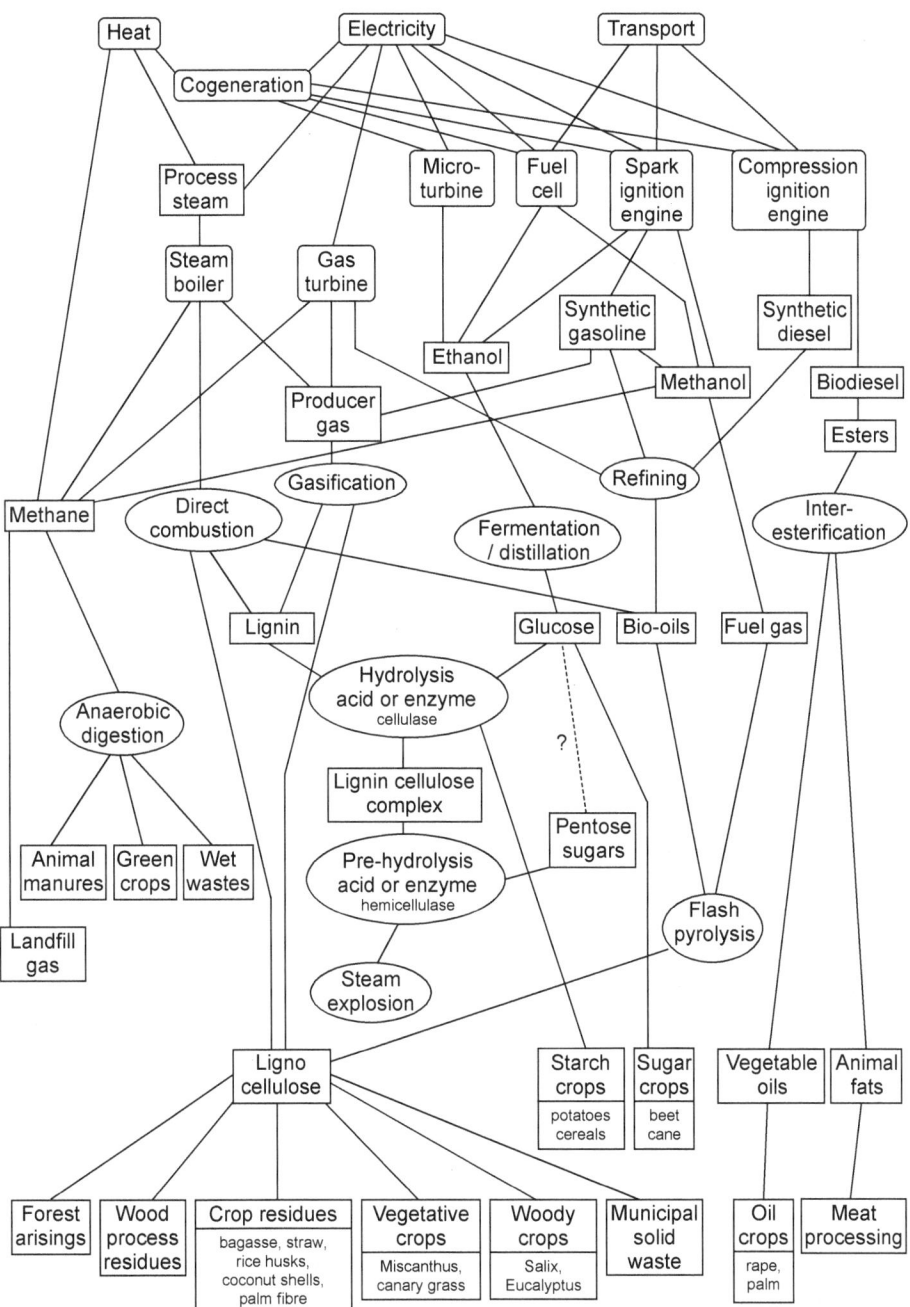

Figure 1.1 Some routes for converting a number of different biomass materials into useful energy products

Weight

This is usually measured in **kilograms** (kg) for small amounts of biomass and **tonnes** (t) for larger, commercial-scale amounts. Since biomass contains varying levels of water, it is also important to specify the **moistur e content** when quoting the weight of fuel. For easy comparison between fuels, they are usually presented as the dry weight of the biomass material: that is, at 0% m.c. This is termed **tonnes dry matter** (tdm), or **oven-dry tonnes** (odt), or sometimes in forestry jargon **bone dry weight** (bdw) or **tonnes dry weight** (tdw). **Gr een weight** is a fairly vague term, normally used to express the weight of freshly harvested biomass product, which can be up to 60% m.c. for forest biomass or around 10–20% m.c. for cereal straw. If not specifically stipulated, it can be taken to imply that the biomass has around 40–50% m.c. (wet basis). As a rule of thumb, a green tonne of woody biomass will contain approximately one third of the energy contained in a tonne of bituminous coal.

Volume

The usual metric unit used for biomass is **cubic metr es** (m^3). The old unit of a 'cord' is occasionally still used in relation to domestic firewood (128 cubic feet based on a cube being 4 ft by 4 ft by 8 ft). When individual pieces of biomass are collected together there is always a considerable voidage (volume of air in the spaces between the separate pieces of wood), which is associated with the total observed volume. This makes the simple unit of volume of limited practical use other than to calculate storage requirements for the biomass.

Density

This is normally defined for any material as its weight/volume. However, for biomass its density is sensitive to moisture content, and since biomass varies widely in moisture content as well as in material composition it is more difficult to define. The fundamental measure for biomass is **basic density**. Taking wood as an example, this is the weight of oven-dry wood contained in a unit volume of green wood, in kg/m^3:

$$\text{basic density} = \frac{\text{oven-dry mass}}{\text{volume at 50\% moisture content}}$$

The term **nominal density**, which is oven-dry mass/volume at 12% moisture content, is less often used.

Basic density accounts for any shrinkage in volume, which often occurs when wood is dried to below 10–12% m.c. Imagine a piece of freshly harvested *Pinus radiata* at 50% m.c., with a volume of 0.1 m^3, and weighing 80 kg. If it is oven dried to 0% m.c. it now has a slightly smaller volume of 0.095 m^3 and weighs 40 kg. The basic density is 40/0.1 = 400 kg/m^3 (and not 40/0.095). Basic densities vary considerably with biomass type and species and even within species depending on the growing conditions, being typically for fuelwood between 300 and 600 kg/m^3. For example, a survey of sawmills in New Zealand in 1977 gave an average basic density of wood process residues as 415.5 kg/m³, with the range varying between 400.8 kg/m³ in the Rotorua region (pine forest area) and 482.5 kg/m³ in Westland (mainly beech forests).

The range of **moistur e contents** (m.c.) normally found in biomass when used for fuel does not significantly affect the volume. The increase in actual density with increasing moisture content relates only to the additional water present. In the example above the actual density of the piece of *P. radiata* at 50% m.c. is 80/0.1 = 800 kg/m³, but the measure has little practical use as it changes with the moisture content.

Bulk density

The type and form of the biomass material mean that there can be a considerable difference in the mass of material contained in a given volume such as a shed or a trailer.

Consider a small trailer with floor and sides forming a cube 1 m × 1 m × 1 m.

- If it is filled with a large, freshly cut, 1 m³ block of **solid** air-dry wood at say 50% m.c. from a relatively dense tree species such as mahogany or jarrah, the weight of that cube might be 1 t, which would give it a density of 1000 kg/m³.

- Now imagine that 1 m long firewood logs (called **r oundwood**) are neatly stacked inside the trailer. Assuming the logs are of the same species and at the same m.c. as the block of wood above, owing to the inevitable air gaps even when stacked neatly the full load would weigh only around 700 kg. Bulk density allows for these voids between the wood pieces, and is 700 kg/m³. Thus the wood contained in a stacked cubic metre is approximately 70% that of a cubic metre of solid wood.

- If these logs were then cut into **billets** of around 300 mm length and randomly thrown in rather than stacked, the 1 m³ trailer would contain perhaps 600 kg of biomass since the remaining 14% of billets would not fit. Therefore, the bulk density of the thrown billets would be 600 kg/m³.

- If the billets filling the bin were then chipped and the small, 25–40 mm diameter **wood chips** were then loaded into the trailer until it was full, only around 400 kg of biomass would be contained, giving a bulk density of 400 kg/m³.

Bulk density varies with species, piece shape, piece size and moisture content. It is a useful measure as it affects the amount of biomass fuel that can be carried by a truck or that can be stored in a shed or on a given area of land alongside a combustion plant.

For biomass there is a relationship between bulk density and the moisture content of the material. For wood chips, for example,

$$\text{bulk density (kg/m}^3\text{)} = \frac{13{,}600}{(100 - \%\text{m.c.w.b.})}$$

Using this relationship shows that if wood (say *P. radiata*) is chipped soon after harvesting at 50% m.c. and then placed in the 1 m³ trailer, the chips would weigh 272 kg and therefore have a bulk density of 272 kg/m³. If the chips were left there for a few days to air dry down to 20% m.c., their weight would reduce since much of the water would evaporate away, resulting in a bulk density of 170 kg/m³. (The energy contained in the chips would also have changed, but not by as much as might first be thought: see Chapter 4.)

Now imagine again the solid block of green wood with bulk density of 1000 kg/m³, as mentioned above. If it was now to be broken into small pieces, or small branches were sourced from the same tree and cut into roundwood logs of around 100–150 mm diameter, then the resulting relative bulk densities within the 1 m³ trailer would be:

- roundwood logs: 700 kg/m³ (70% of the original)

- billets: 600 kg/m³ (60% of the original)

- chips (small pulpwood type): 400 kg/m³ (40% of the original)

- chunks (larger, fist-size particles): 500 kg/m³ (50% of the original)

- sawdust: 600 kg/m³ (60% of the original)

- wood shavings: 200 kg/m³ (20% of the original).

Note that the sawdust particles pack down to eliminate much of the air voids, so a higher bulk density results.

Densification

Biomass in the form of small particles such as straw, sawdust or shredded municipal green waste can be densified to increase the density to allow easier handling and storage. If the same 1 m³ trailer above were now filled with loose cereal straw it would contain around 100 kg of material. If this and more straw were then compacted into a 1 m³ bale using a high-density square baler it might weigh 400 kg. Thus the bulk density changes with the degree of compaction of the biomass, in this example by a factor of four.

Smaller **briquettes** or **pellets** can be made from sawdust and chips, and can vary from 600 kg/m³ to 1500 kg/m³ basic density depending on the equipment used for densifying the material and the biomass type. Moisture content of the biomass to be compacted usually needs to be between 7% and 14% wet basis. If higher it will not compact easily, as water does not compress, and if lower the compacted product will not bind as well.

Table 1.3 gives a range of typical bulk densities for a variety of biomass fuels. For wood fuels the lower end of the range typifies **softwoods** and the upper end **hardwoods**.[4]

Physical characteristics of pieces of wood vary depending on

- species

- growing environment

- the site within the tree (heartwood, sapwood, branchwood, etc.)

- whether cut with or across the grain (since the plant cells can be thought of as small tubes lying next to each other, their diameter being much less than their length).

Moisture content

The moisture content/dry matter ratio of biomass material varies widely (Table 1.3), and this has a significant effect on many of the energy conversion processes. For example, when biogas is obtained from an anaerobic digestion process the percentage of **solids** present in the digestate affects the gas yields. For dry biomass fuels such as straw or wood the amount of water present has a considerable effect upon the proportion of the total heat content of the wood that it is possible to recover as a result of combustion.

Moisture contents vary widely, and can range from 25% to 65% (wet basis) in wood and up to over 90% for sewage sludge and effluents from food processing. For wood, the m.c. depends on a combination of harvesting method, climatic conditions, time of year when harvesting takes place, and the length and method of storage.

$$\text{m.c. (wet basis)} = \frac{\text{total weight of wet wood} - \text{oven-dry weight}}{\text{total weight of wet wood}} \times 100$$

$$\text{m.c. (wet basis)} = \frac{\text{total weight of wet wood} - \text{oven-dry weight}}{\text{oven-dry weight}} \times 100$$

Table 1.3 Typical bulk densities and moisture contents for biomass in various forms

Type of biomass	Moisture content (% w.b.)	Bulk density (kg/m³)
Green roundwood	40–50	510–720
Green wood chips	40–50	280–410
Green wood chunks	40–50	350–530
Green sawdust	40–50	420–640
Air dry roundwood	20–25	350–530
Air dry wood chips	20–25	190–290
Air dry wood chunks	20–25	240–370
Kiln dry wood chips	10–15	160–250
Kiln dry wood chunks	10–15	200–310
Kiln dry sawdust	10–15	240–370
Wood briquettes	7–14	900–1100
Wood pellets	7–14	500–700
Straw bales	10–15	200–500
Coal for comparison	6–10	700–800

Further discussion on moisture content occurs in Chapter 4 as it is inextricably linked with storage and transport issues.

Heat value

When energy is derived from biomass fuels by combustion in air, the energy content equates to the **heat ener gy** released, also termed the **calorific value.** This measure has an upper **gross** value and lower **net** value, the difference being the energy necessary to evaporate both the water that is present in the fuel and that formed when the hydrogen in the fuel combines with oxygen during combustion. The higher heat value (HHV) (or gross calorific value, GCV) of a fuel is the amount of heat released when combusted at standard atmospheric conditions (15°C, 100 kla (or 1 bar) and 60% relative humidity). It is assumed that the products of combustion have returned to this standard temperature and pressure, and that the final water content of the exhaust conditions is in liquid form. The lower heat value (LHV) (or net calorific value) indicates the amount of **useful heat** made available on combustion. It assumes that the water vapour produced is not recondensed within the bioenergy conversion plant, and that therefore the **latent heat of vaporization** that was used to evaporate the water is still contained in the vapour and is not available to the process – which is invariably the case. The difference between HHV and LHV depends on the fuel moisture content and its hydrogen content. The LHV values are more useful when comparing one fuel with another. On an 'as-fired' basis, the energy value of biomass is generally low compared with bituminous or sub-bituminous coal.

Energy density

It is useful to know the available energy contained in a given volume of biomass when planning a bioenergy project as it affects the size of plant, the storage area needed and the transportation arrangements. The lower energy density of most biomass fuels compared with that of oil and coal is a limitation that increases their delivered energy cost. Methods of increasing the biomass energy density (for example by producing pyrolytic oils or densified pellets) have potential but tend to be costly. A comparison of the energy density of biomass and fossil fuels is given in Table 1.4.

Coal has an obvious advantage of having around five to six times more energy stored in a given volume. This also affects the burner or boiler output capacity, as a larger combustion chamber would be needed for wood to accommodate the larger volume of fuel necessary in order to give the same heat output as coal.

Terms and units relating to conversion of biomass to bioenergy

For heat and transport fuel applications, **MJ**, **GJ** and **TJ** units are most commonly used. When the biomass is converted into electricity, **kWh**, **MWh**

Table 1.4 Energy density of a range of fuels derived from their heat values and bulk densities at a range of moisture contents

Property	Fuelwood 20% m.c.	Fuelwood 50% m.c.	Fuelwood 60% m.c	Coal 5% m.c
Lower heat value (MJ/kg)	16	10	8	25
Bulk density (kg/m³)	170	270	335	600
Energy density (MJ/m³)	2720	2700	2680	15,000

and **GWh** are generally preferred, being equivalent to 3.6 MJ, 3.6 GJ and 3.6 TJ, respectively.

Capacity

The capacity of a conversion plant is expressed in terms of the maximum heat or power output as **kW** or **MW**. When this is in the form of heat the subscript th for 'thermal' is added (kW$_{th}$); when as electricity it is presented as kW$_e$. Thus a heat plant might have an installed capacity of 25 MW$_{th}$, which is used to drive a power generating plant of say 8 MW$_e$.

A conversion plant cannot always be operated at its maximum output. Because of maintenance, breakdowns and demand it also rarely operates for more than approximately 80% of the total time possible (7000 hours out of 8760 hours in a year, termed 80% **availability**). So the annual actual or estimated energy output, in terms of say GJ/year of heat or MWh/year of electricity, is often a more useful figure than quoting the plant capacity as shown on the nameplate.

Cogeneration

Cogeneration occurs when a plant produces both heat and shaft power, usually for driving a generator, when it is then also termed a **combined heat and power** plant.

Efficiency

The efficiency of energy conversion at a plant is generally quoted as overall thermal efficiency, which equals

$$\frac{\text{energy contained in the biomass fuel}}{\text{energy output of the conversion plant}}$$

Thermodynamically this is the ratio of the useful energy output to the energy input into a process. Many definitions of 'efficiency' are used, so care needs to be taken when reading a report that the meaning is clear. **Heat rate** expresses the efficiency of an engine or turbine as the fuel heating value consumed per unit of heat or electrical output. For a power-generating plant it is expressed as MJ/kWh, which is divided by 3.6 and inverted to ascertain the efficiency. Thus a gas engine with a heat rate of 36 MJ/kWh is 10% efficient whereas another design with heat rate of 18 MJ/kWh is 20% efficient. Other terms commonly used are **enthalpy (** the measure of the heat content of a substance) and **entr opy** (the measure of unavailable energy in a thermodynamic system).

Efficiencies for boilers and steam power plants are normally quoted by the manufacturers on a **higher heat value**   (HHV) basis using the HHV of the fuel, rather than a **lower heat value**   (LHV). This relates to the measurement of the heat value of the biomass fuel and its hydrogen content, as discussed above. The difference between HHV and LHV can be significant: for a plant fired with biomass at 40% m.c., LHV = HHV −15%; for a natural gas plant it is HHV − 11%; but for a 5% m.c. bituminous-coal-fired plant it is only HHV − 4%.

Gas turbines and reciprocating engines are usually quoted on an LHV basis. The difference between LHV and HHV flows through to efficiency calculations and the calculation of fuel costs, so care should be taken to determine which heating value is meant when quoted.

Biomass fuel characterization

Biomass comes in a wide variety of forms, and being of a biological nature, their characteristics vary considerably. Even woody biomass arising from the same monoculture plantation forest of just one tree species can vary significantly depending on such factors as the age of the tree, the location within the tree, whether bark is included and if so the type of bark, the moisture content, and the time since harvest. Some energy conversion systems require very specific fuelwood parameters, particularly regarding size of particle and moisture content. Others are less discerning.

In practice, when biomass fuel is delivered by truck or rail to a bioenergy plant it probably has come from a number of sources (more so than when crude oil is delivered to an oil refinery, though oil properties from different oil fields vary markedly too). Hence each load has to be sampled and measured for fuel 'quality'. Payment is made to the grower based on these measurements, as determined in a fuel supply contract between the biomass resource owner and the operator of the power plant. Just as for cereal growers, a penalty could be imposed for loads that are outside the limits of moisture content, and in extreme cases a load could be rejected and returned at the supplier's expense. So it is important to be able to sample and measure the load accurately (Figure 1.2).

The physical, chemical and combustion characteristics of biomass fuels can be determined by laboratory test procedures. This information does not necessarily need to be carried out for each truck load, but is often useful to determine the value of a resource as feedstock for a power plant. When a power plant is first planned one critical element is to ensure that there is sufficient biomass fuel available in the vicinity, and that it will last for the life of the plant (at least 20 years normally). This means assessing not only the available volumes but also the fuel characteristics. Only then can fuel supply contracts be properly negotiated.

Figure 1.2 Random sampling of each truck load delivered to the power plant for assessment of moisture content and fuel quality, on which payment is based

Proximate analysis

Proximate analysis of a fuel provides a simple breakdown of the components fixed carbon, volatile matter, moisture and ash, thereby giving an indication of the combustion characteristics of the fuel. Other than moisture content, woody biomass sources do not vary too much in these properties, regardless of species, but they differ from other types such as rice husks, straw and bagasse. Dry biomass tends to have a high volatile content and low fuel ratio (fixed carbon to volatile matter) of around 0.2 (that is, 10% of the biomass is fixed carbon and 50% is volatiles). Bituminous coal by comparison has a fuel ratio of around 1.5–2.0.

The ash content of biomass is generally low at 0.4–2.0% by weight (though cereal straw can be over 10% owing to its relatively high silica content). Higher ash contents of biomass result when poor harvesting and handling methods cause soil contamination of the fuelwood. Ash content is a key element of some burner designs, and using fuels with a low ash content can be a problem, not just during combustion as such, but to protect the grate of the burner from the heat by the ash covering the surface. If the ash content is low, the protection is less and the grate will need to be made of more expensive, heat-resistant material.

Ultimate analysis

Ultimate analysis provides the chemical composition of the biomass in elemental terms, and is used to determine combustion air requirements, flue losses and likely emission levels. On a dry ash-free basis all wood types are similar, with a relatively high oxygen content compared with fossil fuels. This is typical of lower-ranked fuels in terms of energy content.

The sulphur content of biomass is usually very low (<0.1%) compared with that of coal. It should not therefore lead to emission problems of SO_x (thought to contribute to acid rain), or to corrosion of the conversion plant components outside the combustion zone owing to high acid dew points, when sulphuric acid can be formed as the gases cool and then eventually condense.

Energy (or heat) values

These vary only to a small degree with the biomass resource type. For example, when measured on a dry, ash-free basis, hardwood whole-tree biomass is 19.6 MJ/kg and softwood whole-tree biomass is 19.4 MJ/kg and, if the bark is removed, 20.9 MJ/kg. Owing to the variation in moisture contents, comparisons with other fuels are best made in terms of net energy values. Typical examples are:

	Forest residues (40% m.c.)	Bituminous coal (5% m.c.)
HHV	11.2 MJ/kg	28.2 MJ/kg
LHV	9.5 MJ/kg	27.1 MJ/kg

Variations in heat value are greater with varying moisture content than with biomass type. However, using wetter fuel is not as grim as it looks! Consider a pile of green wood chips at 60% m.c. and weighing 1 t: then 600 kg is water and 400 kg is dry biomass. The lower heat value (or LHV) is around 6.3 MJ/kg (Figure 1.3), giving a total energy value in the pile of 6.3 GJ. If the pile is left to dry to 50% moisture content it loses some weight as the water evaporates off, resulting in a pile that looks the same, and is of similar dimensions and size, but now has only 400 kg water and the 400 kg dry biomass. It has a heating value of around 8.5 MJ/kg giving 8.5 × 800 = 6.8 GJ of total energy available for use. Now after some weeks assume it has dried to 20% m.c. and may have shrunk slightly in volume. It now still has 400 kg dry matter (though in reality this would have reduced slightly owing to respiration losses and some decomposition) but contains only 100 kg water, so weighs 500 kg total. At 14.9 MJ/kg it has in total 14.9 × 500 = 7.45 GJ of available energy. So the key point to note is that although the moisture content has dropped from 60% to 20% the energy value per kg has more than doubled, from 6.3 MJ/kg to 14.9 MJ/kg. The available energy in the pile has only increased from 6.3 GJ to 7.45 GJ since the pile has simultaneously become lighter as it dries out. This relationship between moisture content and energy value is also shown in Table 1.5.

Bulk density and volume

Although the energy content of a 1 kg piece of wood at a given moisture content is fairly constant regardless of species (that is, around 19–20 MJ at 0% and 10–11 MJ at 40% m.c.), this is not true on a volume basis. A 1 kg piece of

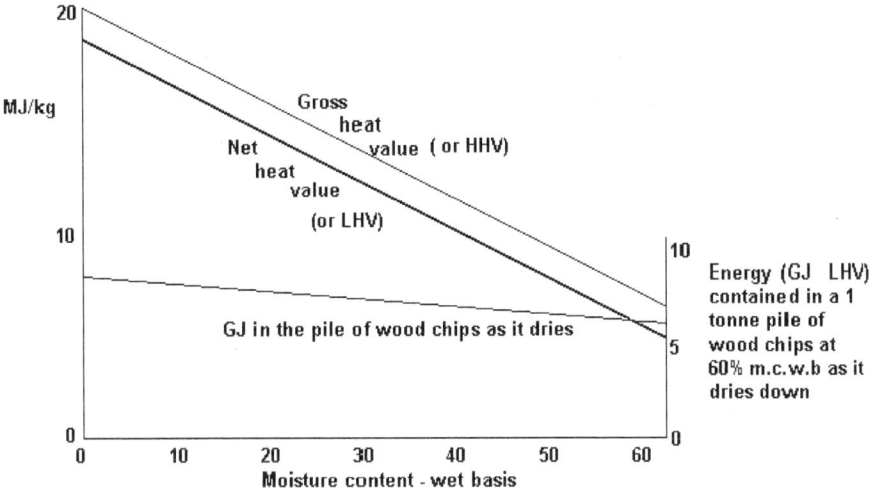

Figure 1.3 For any plant species the net heating value is less than the gross, but both are affected most by the moisture content, as is the energy content of a pile of biomass as it dries

relatively dense wood species such as matai or ebony would fit easily into one hand, but a 1 kg piece of balsa wood (as used for making model aeroplanes) would be hard to hold as it would fill a carrier bag. Yet both contain similar amounts of energy. This makes a difference both for the volume of wood to be cut, transported and stored to supply a given amount of heat and for the design of burner, which has to be able to cope with the larger volumes. Coal is much more energy dense than wood, so converting a coal-fired boiler to run on wood normally results in a large drop in output simply because not enough fuel can be fed into the combustion chamber.

For transport it is important to maximize payloads if costs are to be minimized, but this may not be easy with dry wood of lower bulk density. Consider a 15 t tare weight truck with a container bin 2 m high × 2 m wide × 10 m long. If this 40 m³ volume was filled with coal the truck might then weigh a total of 48 t (and probably be illegally over the weight limit). If filled with dense, green (that is, around 50% m.c.) eucalyptus chips it might weigh around 35 t and be about at maximum legal payload. If filled with dry wood chips of a less dense pine species it might weigh only 25 t and be underloaded, which means higher transport costs in terms of $/t per km. So knowing the bulk density of the fuel, whether it be in the form of chips, chunks, logs, branches or shredded whole trees, is useful in developing an optimum handling, transport and storage system (see Chapter 4).

Ash

Ash from biomass comes from the minerals inherent in the structure of the wood and soil contamination. Assessing how much ash there will be to dispose of is only a small reason for determining the ash content of a biomass fuel.

Table 1.5 The effect of moisture content on the net heat value of a 2000 cm³ piece of woody biomass with basic density of 500 kg/m³ and a gross heat value of 19.9 MJ/kg dry weight

Form of woody biomass used for fuelwood	Moisture content and weight (% w.b.)	Total weight of piece (kg)[a]	Weight of water present (kg)	Heat to evaporate moisture present (MJ)[b]	Heat to evaporate water from combustion of hydrogen (MJ)[b]	LHV per 2000 cm³ piece of wood (MJ)[b]	LHV/kg wet wood (MJ/kg)
Freshly harvested green wood	60	2.5	1.5	3.73	1.39	14.78	5.9
Recently harvested wood	50	2.0	1.0	2.57	1.39	15.94	8.0
Sawmill residues	40	1.67	0.67	1.72	1.39	16.79	10.0
Pallets/demolition timber	30	1.43	0.43	1.10	1.39	17.41	12.2
Air-dried biomass	20	1.25	0.25	0.64	1.39	17.87	14.3
Wood process and furniture residues	10	1.11	0.11	0.28	1.39	18.23	16.4
Oven-dry wood for comparison	0	1	0	0	1.39	18.51	18.51

[a] The total weight of the piece of wood consists of 1 kg dry matter plus the weight of water present

[b] The energy used to evaporate the moisture present, plus 1.39 MJ/kg dry matter used to evaporate the water formed from the hydrogen present in the biomass, plus the lower heat value energy content of the biomass gives the high (or gross) heat value of 19.9 MJ/kg

Characterization of ash by elemental analysis and ash fusion temperatures enables the problems of slagging, fouling and clinker formation in the burner and boiler to be predicted. When softwood is combusted on its own ash fusion may not be a problem, as the combustion temperatures are likely to be low. But when it is combusted in conjunction with coal in a 'co-fired' plant fuel bed temperatures may reach a level where slagging could occur.

Ash chemical analyses show that biomass, and straw in particular, contains high levels of alkali metals (particularly Na_2O, CaO, MgO, K_2O and P_2O_5) compared with coal. The Na_2O and K_2O in particular are present in a form that is easily volatilized during combustion and then deposited on the heat transfer surfaces of the boiler as it cools, leading to problems such as pipe blockages. It is also possible that whereas burning specific types of coal or biomass on their own proves satisfactory, when co-fired a chemical imbalance can occur, resulting in slagging.

Carbon sinks and greenhouse gas mitigation from the use of biomass

Scientists are now confident that the greenhouse effect is already resulting in climate change, and that a substantial part of the observed changes is due to human activities. Fossil fuels are abundant, and projections from the United Nations Intergovernmental Panel on Climate Change (IPCC) suggest that oil, coal and gas should all be available throughout most of this century. In spite of the threat from climate change, they will probably remain the dominant energy source for the foreseeable future.

The IPCC Third Assessment Report (2001) provided a comprehensive scientific basis for further national and international policy development and negotiation in the context of the United Nations Framework Convention on Climate Change. The message was simple:

- Working Group I confirmed that the Earth's climate is changing as a result of human activities, particularly from energy use, and that further change is inevitable. If we continue with 'business as usual', utilizing the known reserves of fossil fuels, the IPCC models (from the IPCC Special Report on Emission Scenarios) show that the atmospheric level of CO_2 will rise to a potentially devastating 540–970 ppmv[5] by 2100. The current level of 367 ppmv has already risen from 280 ppmv since 1860, mainly as a result of fossil fuel use.

- Working Group II showed that natural ecosystems are already adapting to change, but that some are under threat, and that human health and habitats will be affected worldwide. Predicted increases in global temperatures could increase agricultural production in some areas, but reduced rainfall and more extreme weather events may offset any benefits.

- Working Group III showed that the known global oil, coal and gas reserves contain over five times more carbon than the quantity already released into the atmosphere by burning all fossil fuels since 1860 until today. Many mitigation technologies already exist to encourage the increased implementation of renewable energy projects using the latest designs of conversion technologies, and the uptake of energy efficiency measures, which often have direct economic benefits because the value of the energy saved exceeds the costs of implementation. In addition, fuel switching from coal to gas, the increased uptake of nuclear energy and the possible physical capture and storage of carbon dioxide will also help to achieve reductions.

Given the abundance of fossil fuels consumed in an increasingly energy-intensive world, the rapidly growing carbon dioxide emissions from their use are of major concern. Although fossil fuels are not the only source of greenhouse gases, they remain dominant over others, including methane from ruminants and paddy fields, nitrous oxides from nitrogenous fertilizer use and animal wastes, and carbon dioxide from cement production and changing land use practices including deforestation.

The effects of greenhouse gas (GHG) emissions are becoming increasingly evident as the behaviour of the global atmosphere and oceans is becoming better understood as a result of improved predictive computer models. This has allowed improved detection techniques to be developed, and a global warming trend is now clearly evident in the observed climate.

While there are many sources of uncertainty that preclude confident prediction of future patterns of climate change at the regional level, all scenarios of future global socio-economic and technological development imply greater emissions, leading to further global mean warming. The effect of this is predicted to be a 1.4–5.8°C increase in temperatures, with average sea level rising by somewhere between 90 mm and 880 mm over the next century. (The models are not sufficiently accurate to predict a smaller range with confidence.)

No one knows what level of stabilization is necessary to avoid dangerous interference with the climate and minimize adaptation, as computer models are insufficiently developed to predict whether carbon dioxide, for example, needs to be as low as 450 ppmv or as high as 1000 ppmv. The maintenance of current levels is virtually impossible, even if we were to stop using fossil fuels today. So some climate change and adaptation to it are inevitable. Significant reductions in emissions are needed relative to 'business as usual' circumstances to avoid catastrophic results. The Kyoto agreement, difficult to ratify as it has proved to be, provided only the first instalment necessary, because the achievement of acceptable levels implies *inter alia* major changes to the world's energy system. This is often termed **decarbonization**, but from the bioenergy viewpoint this can be misleading since carbon is still emitted but then recycled.

Role of bioenergy systems for greenhouse gas reduction

Biomass has the dual advantage of acting both as an energy substitute for fossil fuels (a carbon offset) and as a means of sequestering carbon (a carbon sink). Where agricultural land is transferred to energy crop production, a net uptake of CO_2 also often results from the increased 'carbon density' of the land use and possibly in the soil too. Other forms of biomass utilization, such as landfill gas or the collection of forest residues otherwise left to decompose on the forest floor, also avoid the release of methane (a more potent GHG) into the atmosphere.

Although the establishment of new plantation forests as sinks has the potential to sequester carbon, it can only be a short-term measure for reducing the ever-increasing atmospheric carbon levels, as there is only a finite amount of suitable land available. So the concept of forest sinks makes sense as a long-term solution only if the accumulated carbon stored in the biomass is utilized, for example as heat. In this way the carbon can be recycled within the overall biomass production and bioenergy utilization system.

Forest carbon sinks

When a forest plantation is first planted into pasture or arable land, some carbon is locked up in the biomass when the trees grow as a result of photosynthesis. Carbon levels in the soil may possibly also increase over time as a result of organic matter build-up from root growth and litter decomposition.

When the forest is harvested it can be assumed that all the carbon locked up in the above-ground portion of the trees will be returned to the atmosphere over time.

- If the carbon is used for paper this return could occur in a matter of months.

- If used for furniture or buildings it may not be for many years.

- The tree components that remain in the forest after harvest will slowly decompose.

So eventually all the carbon will be released again.

If the tree, or part of it, is used for energy purposes, then this simply results in speeding up the natural process of all the biomass material sooner or later being reconverted to carbon dioxide and released. This carbon cycle goes on continuously in a natural forest as trees and plants grow and die, so that it remains in a carbon balance.

The plantation forest acts as a carbon sink only when first planted, and it must be replanted after each harvest if it is to continue this role. *The more carbon dioxide is released into the atmosphere from fossil fuel use, then the more forests will need to be planted into more land without forest cover to absorb it.* Thus carbon sinks can only ever be a temporary solution for mitigating carbon dioxide emissions. If, however, the biomass grown is used for energy purposes and thereby displaces fossil fuels, the biomass also has the value of a carbon offset since the fossil fuel carbon will remain underground. A short rotation forest (SRF) crop planted into pasture would have a lower carbon sink value than a denser, taller, plantation forest, but it could have a greater carbon offset value depending on the end uses of the forest products (Figure 1.4).

Cutting down a native forest and changing the land use to pasture or crops lead to much of the stored carbon being released into the atmosphere. If the land is subsequently reforested, some carbon is reabsorbed. In this example, when the forest reaches maturity it contains nearly the original carbon stock of 300 tC/ha. When the forest is cut down it can be assumed that the carbon stock returns to zero soon after felling, even though some wood is used – as artefacts, for example. The average carbon sink under the plantation forest over time is therefore approximately 100 tC/ha as the forest is continually harvested and then replanted, in this example every 30 years or so. Hence on average the 1 ha of land now has 100 tC/ha stored in the biomass, which is significantly greater than when it was in pasture. Similarly, if planted in SRF it would have an average carbon sink of around 20 tC/ha being harvested every 5 years or so.

The benefits of using the biomass produced to displace fossil fuel in terms of carbon emissions avoided are significantly more in the longer term if 1 ha of pasture is converted into a short rotation energy crop than if it is planted in forest purely to act as a sink. There is increasing understanding of the value of biomass in this role. Once there is an international value for carbon (when carbon trading begins in earnest) then the added revenue from carbon offsets for any biomass used for energy will make the economics far more competitive than is the case at present.

A detailed comparison of the use of biomass to generate electricity, in terms of gC/kWh and $/tC avoided, with a range of other energy resources including fossil fuels, nuclear power and other renewables is given in the IPCC Third

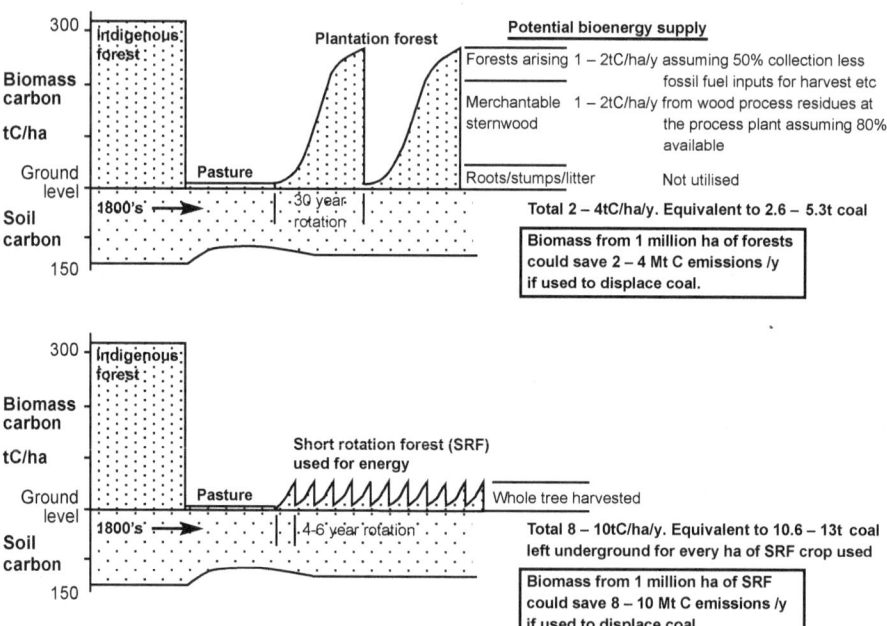

Figure 1.4 The concept of deforestation and reforestation with land-use change leading to carbon sinks and offsets by using forest arisings from plantation forests or whole trees from short rotation forests as sources of biomass

Assessment Report (www.ipcc.ch). In summary, compared with coal and natural gas-fired power plants, biomass can significantly reduce the gC/kWh emitted, and in many specific situations this would also result in cheaper power being produced.

The Summary for Policy Makers of the IPCC report states:

Low-carbon energy supply systems can make an important contribution through biomass from forestry and agricultural by-products, municipal and industrial waste to energy, dedicated biomass plantations where suitable land and water are available, landfill methane, wind energy and hydropower.[6]

So there seems little doubt that biomass has a role to play.

Notes

[1] The measure of energy 1 EJ (exajoule) is 10^{18} joules or 1,000,000,000,000 MJ (megajoules). As a guide, one litre of diesel fuel contains approximately 40 MJ of energy; 1 kWh or 'unit' of electricity is equivalent to 3.6 MJ; and 1000 MJ = 1 GJ (gigajoule).

[2] Note: the term *biofuel* is becoming more commonly used to imply either liquid or gaseous fuels used for transport rather than to describe biomass fuels in general.

[3] Specific references together with further readings are listed at the end of the book.

[4] For convenience tree species are often divided into softwoods (mainly coniferous) and hardwoods (mainly deciduous), though the classification is far from distinct and many properties overlap. Typically softwoods have higher moisture contents at harvesting, lower basic densities, lower energy content and less strength.

[5] ppmv = parts per million by volume.

[6] Solar energy was not specifically listed as, although the solar industry was growing, it was not thought it would make 'an important contribution' within the 10–20 year time frame being considered.

Chapter 2

The woody biomass resource

Biomass resources that can be used for energy production cover a wide range of materials. **Traditional biomass** is generally confined to small-scale uses in developing countries when the biomass fuel used is non-commercial (that is, not purchased but collected by the users at no charge). It includes firewood and charcoal for domestic use, rice husks, other plant residues, and animal dung. **Moder n biomass** relates to large-scale uses of commercial biomass, and usually substitutes for conventional fossil fuel energy sources where it is a cheaper option. The economics of use may well include saving on the costs of having to dispose of the biomass material in some way where it is a 'waste' product of a process. This 'negative cost' can make a bioenergy project economically feasible, whereas when the biomass has to be especially grown then purchased, it is usually cheaper to use coal, oil or natural gas instead.

Biomass includes forest arisings, wood process residues, cropping residues, animal wastes, urban wastes, gas collected from landfill sites and energy crops of various tree species, as well as grasses or cereals. The most common biomass fuel is wood, and people have been burning it for thousands of years. A quarter of the Earth's land area remains covered in forests, but of course most of that is indigenous and should be conserved and protected. Woody biomass used for energy should be sourced only from sustainably managed forest systems, forest residues or waste products. This chapter deals with woody sources of biomass, which have specific properties and form a significant proportion of the total global volume of biomass used for energy. Other forms of biomass as listed above will be discussed in Chapter 3.

In developed countries woody biomass resources are mostly used at the commercial scale in heat or power plants between 1 MW and 100 MW, but the use of firewood at the domestic heating scale is also a significant market, and common practice in both developed and developing countries. Taking South Australia as an example, a 1993/94 analysis of primary energy use showed that 7.71 PJ came from wood, and nearly 80% of this was used in homes. A previous survey in 1990 had found that nearly a quarter of households were using wood for space heating, hot water or cooking. The remaining 20% of woody biomass was used by the timber industry to produce steam for wood kiln drying and electricity generation. As a means of home heating, dry wood currently has one of the lowest costs in many regions, so long as the wood is available free or purchased cheaply in bulk, and burnt efficiently in a slow combustion wood stove rather than on an open fire.

Woody biomass consists of plant materials comprising mainly cellulose, hemicellulose and lignin, and it therefore differs from other biomass materials such as sludges, municipal waste and some agricultural/horticultural crops. Trees contain 20–30% lignin, which has almost twice the heat value of cellulose. When felled, they also contain over 50% water by mass, which can be reduced to about 10% by natural air drying over periods of up to one or two years. Woody biofuels are typically sourced from:

- forest residues from production forestry regimes such as **arisings** left over after harvesting and **residues** from wood processing activities

- purpose-grown energy forest plantations

- municipal green waste (including urban tree prunings).

Forest residues

Forest residues include the several forms of woody biomass unwanted and left over as a result of timber production and processing by the forest industry.

- **Forest arisings** are what remains after the logs have been extracted, either on the ground of the 'cutover' (the area of land being harvested; Figure 2.1) or at the 'landing' (a small site in the forest where the harvested trees from the immediate locality are taken for processing and loading onto trucks; Figure 2.2).

- **Thinnings** are generated from clearing land for forest access roads or for silvicultural operations, when a forest block is tended during its early growth stages to produce quality logs (Figure 2.3). Thinning is normally undertaken well before harvesting, and entails cutting down the smaller tree specimens

Figure 2.1 Forest arisings on the cutover after extraction of the stemwood in a managed plantation forest

Figure 2.2 Forest arisings from a *Pinus radiata* plantation collected at a landing

so that those selected to remain grow better without as much competition for light and nutrients. Extracting the thinned trees without damaging those remaining is difficult and can be expensive, especially on steeper terrain. If the small trees have a commercial value (for example as fence posts, or pulp logs or for firewood), collection is carried out, sometimes using horses to minimize soil and tree damage. Use of thinnings is common practice in Europe, Scandinavia and other parts of the world in order to maximize the returns from well-managed woodlots or forests, which are usually only a few hectares and privately owned. However, it is a labour-intensive activity and likely to be costly on a larger scale: it is therefore usually uneconomic to use the thinnings as a commercial energy source.

- **Prunings** in forests result from removal of the lower branches once or twice during the growth of the tree in order to produce a more valuable butt sawlog with fewer knots in the sawn timber and hence of greater structural strength when milled.

- **Wood pr ocess r esidues** are produced at sawmills, pulp mills, fibreboard plants or furniture manufacturers, for example, as a result of the processing of the logs and timber (Figure 2.4). They include bark, sawdust, shavings, wood chip rejects and offcuts. At the Mount Gambier sawmill, for example, a 10 MW$_{th}$ wood-burning power station has been operated by Forwood Products since 1956. It uses wood waste collected from several nearby mills in the two boilers, which can produce 25,000 kg/h of steam. Some steam is used for timber drying in the mill, and the balance drives steam turbines to generate electricity. About 80% of the power produced is used on-site, and the surplus is sold to the local power company. This plant is a good example of the use of waste biomass to cogenerate both electric power and process heat (Chapter 8).

Figure 2.3 Thinnings being extracted from a 6-year-old *P. radiata* plantation

- **Wind-thr own** trees from natural attrition and storm damage, together with stands remaining after forest fires from lightning or human stupidity, can produce woody material with little value other than as an energy source. However, because storms, fires and tree mortality are not predictable and therefore cannot be guaranteed as a regular and consistent fuelwood supply, such biomass resources are rarely able to be utilized in commercial bioenergy plants. Fallen trees can be manually cut into firewood and sold on the domestic market by firewood merchants.

- **Black liquor** is a waste product generated by the pulp and paper making industry which can be combusted, pyrolysed or gasified for use as a biomass

Figure 2.4 Wood process residues accumulate in the yard of a factory

energy source. It is mainly the lignin fraction of the wood that cannot be utilized in the pulping operation, and disposal of it onto land or into waterways can cause environmental problems. The usual solution is to convert this waste to useful energy by combusting it. As an alternative a fluidized bed, fast pyrolysis process to convert the black liquor into a 'bio-oil' has been developed at the University of Melbourne. The bio-oil can then be processed into transport fuel substitutes such as biodiesel.

Many of these forest residue materials are often not utilized, and are therefore considered to be 'wastes'. Forest arisings are often left to decompose on the forest floor, and wood process residues have to be carted away from the processing site to landfills or burnt in the open to dispose of them. Such biomass material can be collected and used cleanly and efficiently as a fuel source where it is economic to do so, thereby both reducing the methane greenhouse gas emissions resulting from landfilling or during natural decomposition, and avoiding air pollution from burning in an uncontrolled state.

It is important to note that indigenous forests, and any by-products from any trees that are harvested from them (either by clear felling or using the more sustainable selective logging system), should not be used for bioenergy purposes. This is a sensitive issue in many parts of the world, and it must be made clear to all stakeholders concerned, and the public in general, that *woody biomass as a source of renewable energy does not involve damaging native forests*. In fact it can have the opposite effect by helping to conserve them if the provision of fuelwood from sustainably managed forest residues or specially grown energy plantation forests avoids the need to cut down native forests. Rough judgement could be made as to the acceptability of a potential resource by assessing whether or not Greenpeace would approve of its use for energy purposes.

Short-rotation energy forest plantations

Short-rotation energy forest plantations (SRF) offer the potential for woody biomass to become a more significant source of renewable energy in many parts of the developing world including Asia, Africa and South America. Purpose-grown energy forests have also been closely evaluated and recently grown commercially in many developed countries. The best species of trees to suit these woody biomass energy crops are those that are fast growing and therefore produce high energy levels in terms of GJ/ha.yr. They can be harvested as single-stem trees in the usual way and then replanted; alternatively some species, mainly hardwoods, are suitable for growing under a coppicing regime.

The practice of coppicing (from the French word *couper*, meaning 'to cut') has been undertaken for centuries in Europe to provide such things as posts, firewood and fence palings, and many ancient coppice stands still exist. It involves harvesting the trees a few years after planting them and then allowing them to sprout again from the stumps, followed by subsequent harvesting of the regrowth some years later. This can be on a 2–10 year period or 'rotation', and several rotations can be conducted before replanting becomes necessary due to sufficiently high plant mortality or new and improved genetic stock becoming available (Figure 2.5).

A large proportion of nutrient uptake in a forest is recycled through the leaves or needles via litter fall, and at harvest relatively few nutrients are removed from the site, since only the C, H and O are consumed during the combustion process of the wood and these are easily replenished from H_2O and CO_2. Nitrogen, however, may need replacing as this can be lost as NO_x during the combustion process. It is theoretically possible to return other nutrients and trace elements through the ash and develop a sustainable production system. Hence some species, if carefully managed and depending on the soil characteristics, will grow under a short-rotation regime with minimal fertilizer inputs and without a significant decline in soil nutrient depletion or biomass yields. The optimum length of rotation is determined by the species, soil type,

Figure 2.5 Manual harvesting of the fifth rotation of 3-year-old regrowth of coppiced *Eucalyptus brookerana* at the age of 15 years

planting density, wildlife management, harvesting machinery available and, in some regions, the need to limit the height of above-ground biomass regrowth to reduce its susceptibility to wind damage.

When coppicing SRF, limited information is available regarding optimal stump height, stump diameter and slope of cut required. To minimize stump damage and disease infestation, and hence reduce tree mortality, cutting at a height of 150–200 mm above the ground at an angle of up to 45° appears to be suitable for many species, but this cannot easily be achieved by mechanical harvesting. Thinning of the multitude of regrowth stems to three or four dominant ones in order to give a larger mean stem diameter is possible, but it tends to be labour intensive and is therefore not usually recommended if the total biomass is to be chipped and used for boiler fuel. It could be beneficial if the stem wood is to be separated out and sold for other higher-value markets such as timber or pulp.

Optimal coppice regrowth of deciduous crops such as *Salix* is most likely when the trees are harvested in the winter. However, for eucalyptus the spring or early summer period is usual, though late frosts should be avoided as they may damage the young regrowth. To create a year-round supply of fuelwood, harvesting the trees throughout the year seems a more logical approach than harvesting all the area at one time and then having to stockpile the material for use throughout the year. Year-round harvesting has been shown to be feasible, at least experimentally, for certain species under specific conditions, but no known commercial operations have been harvested in this way.

The wood sourced from energy forests can be burnt to generate steam for electricity, and heat for cooking, water and space heating, or can be used in charcoal manufacture. Growing SRF crops on a large scale has received renewed interest in Australia, the UK, New Zealand, the USA, Scandinavia and other parts of the developed world. With careful husbandry and sustainable management techniques employed, SRF have the potential to improve agricultural productivity, reduce soil erosion, and improve degraded land (such as dryland saline soils). Energy forests can diversify farm revenue streams, but the financial viability for growers over the long term is currently uncertain in most regions.

Municipal green waste

Municipal green waste (MGW), sometimes termed **in-city for ests,** has been identified as a potential fuelwood supply. It comprises the green and woody portion of municipal solid waste (MSW), and includes tree trimmings and gardening wastes (grass clippings, pruning and leaves), both domestic and municipal, as well as the waste wood and paper component of domestic rubbish. Rather than dump it into landfill sites or burn it in the open air to dispose of it, it can provide a useful source of energy. The MSW should first be sorted and its components recycled wherever possible if it is economic to do so. Any remaining MGW can then be separated out and used for compost, for mulch, or as a waste-to-energy resource as a last resort. Waste-to-energy processes using a number of technologies are common in Europe in particular, and are covered in detail in Chapter 7.

Prunings are also produced in reasonably large volumes from orchards (Figure 2.6). The cost of collection can be high, but in places such as California, where many wood-fired power plants are already operational, these annual prunings are added to the biomass fuel resource.

MGW does not include material from industrial or agricultural sources, or treated wood waste from the MSW stream. It therefore excludes potentially contaminated materials. Traditionally MGW has been disposed of in landfills or by open-air burning, which is now discouraged. The use of MGW for energy production has the advantage of reducing waste loads to municipal landfills and hence reducing the greenhouse gas methane that arises from its decomposition. Elimination of this high-volume and bulky material has been highlighted as a significant factor in saving landfill space and meeting the recycling targets of many regions.

Environmental implications of the use of woody biomass

Several developing countries in Africa, such as Kenya, and in Asia, such as Nepal, derive over 90% of their primary energy supply from traditional biomass. In India it currently provides approximately 45% and in China 30%. Most is collected by rural dwellers, but a portion is marketed in urban areas (Figure 2.7). At the domestic scale in developing countries the use of firewood in cooking stoves is often inefficient, and can lead to health problems. The use of

Figure 2.6 Orchard prunings, if left to decompose on the ground, are a nuisance for subsequent operations, and are better either mulched or removed

appropriate wood stove technology to avoid emissions and improve human health, as well as to reduce the demand for firewood, which is often scarce, is a no-regrets solution. In addition there is opportunity to develop the concept of growing SRF to supply the firewood, although this is not always culturally acceptable, and competition for land for food and fibre, together with complex land ownership issues, may prevent this approach. In addition there are cultural problems in countries where the men, who are traditionally responsible for managing the land, have little interest in growing firewood crops since obtaining firewood, however arduous, is the responsibility of the women.

Modern small-scale bioenergy applications at the village community scale are gradually being implemented, leading to better and more efficient utilization

Figure 2.7 The saleyard of a firewood and charcoal merchant in the city of Accra, Ghana

of the available fuelwood, which in many instances complements the use of the traditional biomass fuels and provides rural development. A US Department of Energy study (Interlaboratory Working Group, 1997) identified the use of bioenergy technologies in all countries as critical for minimizing the costs of reducing carbon emissions. The three areas specifically recognized as having most potential were the cofiring of wood and coal in coal-fired boilers, biomass-fuelled integrated gasification combined-cycle units (BIGCC) for use particularly by the forest industry, and the production of the biofuel ethanol from the hydrolysis of lignocellulosics. Estimates of annual carbon offsets by 2010 from the uptake of these three technologies in the USA alone were 16–24 MtC, 4–8 MtC and 13–17 MtC respectively. It was also claimed that fossil fuel energy savings from the use of each of these technologies should cover the associated costs in the near term, with co-firing giving the lowest cost and technical risk at present.

Co-firing, in which woody biomass is blended with pulverized coal at up to 10–15% of the fuel mix, is already being implemented to some degree in, for example, Denmark and the USA. However, in many situations it remains uneconomic because coal is cheaper than the biomass. In addition the additional costs needed for combustion plant conversion and the resulting reduction of output capacity are deterrents. However, major environmental benefits can also result, including the reduction of atmospheric emissions of oxides of sulphur (SO_x) and nitrogen (NO_x).

High-yielding, short rotation forest crops can produce stored energy equivalents of over 400 GJ/ha per year at the commercial scale, leading to very positive input/output energy balances of the overall system. Liquid biofuels (Chapter 9), when substituted for fossil fuels, will also directly reduce CO_2 emissions. Therefore a combination of biomass fuel production with carbon sink options can result in maximum benefit from mitigation strategies. This can be achieved by planting traditional forest or energy crops, such as short rotation coppice, into arable or pasture land, thereby increasing the carbon density of that land (a **carbon sink**), while also using all or part of the biomass produced to be used as a source of energy (a **carbon of fset**) (as shown in Figure 1.4). The role of forests for greenhouse gas mitigation is covered in detail in a Special IPCC (2000) report on *Land Use, Land Use Change, and Forests*. The link between these low-cost carbon sinks and eventually using some of the biomass grown for energy purposes is relevant here. Once the limited area of available land is covered in forests planted as a sink to absorb carbon dioxide from the atmosphere, no more planting will be possible. So carbon sinks can only ever be an interim measure to offset fossil fuel use. Hence recycling of the carbon by using the biomass to displace fossil fuels may then become feasible. Economic mechanisms such as carbon trading to link a forest sink project with a bioenergy project have been evaluated and are likely to become popular once carbon trading begins in earnest.

Converting the accumulated carbon in the biomass for energy purposes, and hence recycling it, alleviates the critical issue of maintaining the biotic carbon stocks over time, which is essential if a forest carbon sink is to prove effective.

Increased levels of soil carbon may also result from growing perennial energy crops. An additional benefit is that the decentralization of heat and electricity production using woody biomass to supply local conversion plants creates employment in rural areas.

Certain woody crops grown to produce biomass are often quoted to have high dry matter yields, at least theoretically. However, commercial yields are often lower than expected from those produced in small plot research trials. In Sweden, for example, 16,000 ha of coppice *Salix* species were planted (Figure 2.8) with expected yields of 10 odt/ha per year. After around 2000 ha had been harvested for the first time during the winters of 1996–1998 it became apparent that an average yield of only 4.2 odt/ha per year had been achieved. It was hoped that the average yield would increase over time as growers became more familiar with the agronomic requirements of this new crop.

Methods of identifying the most appropriate *Salix* species for a specific site can be based on non-destructive yield measurements together with measurements of the fuelwood characteristics of selected species. Correct species and clone selection to meet specific soil and climatic site conditions are necessary in order to maximize energy yields in terms of GJ/ha per year. With better genetic selection, crop management, and grower experience once viable markets for the product are established, it is anticipated that commercial yields will eventually approach the 10 odt/ha per year originally predicted.

Perennial woody energy crops normally require less direct and indirect energy inputs per hectare to plant, grow and harvest them than do annual food crops. Energy balance ratios comparing each unit of energy input required to grow and harvest SRF crops with the units of energy contained in biomass fuels

Figure 2.8 Large areas of *Salix* are grown and harvested in Sweden, with the coppice regrowth evident in the foreground

produced from them are in the region of 1:15 to 1:30 depending on crop yields, which is very acceptable.

The utilization of wood process residues for energy reduces the amount of material going into landfills, and reduces degradation of the landscape. These positive attributes need to be balanced against the likelihood that a small amount of additional greenhouse gas emissions may arise from the transport of material between sites, and that community issues associated with increased trucking and potential dust emission problems could result.

Environmental issues relating to the extraction and use of forest arisings are complex, and include interactions between many different factors such as risks of soil erosion, degraded receiving water quality, nutrient removal, biodiversity, the use of chemicals, and the addition of wastes (such as ashes and sludges) as soil or nutrient amendments. Critical to the long-term use of forest arisings, however, is the need to demonstrate long-term sustainable production of the forests. Current evidence suggests that sustainable production can be achieved through prudent forest management, which considers the conservation of organic matter between rotations, minimizes site damage from vehicle access, introduces effective weed control, and avoids nutrient depletion.

Evidence from the UK suggests that, provided the local landscape values are not high (as they would be in a national park or reserve), cultivation of short rotation energy forests would not be visually intrusive assuming the plantations are sensitively sited and designed. Furthermore, such young forests can benefit biodiversity through the creation of wildlife habitats. However, the environmental benefits of establishing short rotation energy crops within the landscape would require further investigation for any specific region.

Woody biomass resource assessments

Unlike fossil fuels, woody biomass is most often found in dispersed small quantities, such as individual industrial sites or small farm woodlots. Locating, securing and collecting sufficient woody biomass to supply a central power plant can be a logistically difficult and costly process unless purpose-grown energy crops can be produced extensively in specific regions near to the centre of energy demand. Also, if a power plant is located near to a large traditional forest plantation area, then it may be possible to source forest arisings for fuel.

An assessment of the **technical potential** of the woody biomass resource from plantation forests in a region is relatively easy to undertake, given that the ages and areas of the plantation forest estate are known. Hence the area likely to be harvested each year as the trees mature can be estimated well in advance of the harvest date. Quantities of wood process residues from industrial waste sources are more difficult to estimate with any degree of accuracy, as the availability, type, quantity and quality are extremely variable, and are often specific in both location and time. The other difficulty to be overcome in assessing the future biomass resource of a particular region over time is estimating the potential future establishment of energy plantations. This will depend on the land availability, the regional distribution of this land, the share of available land use likely to be reserved for other purposes, the productivity of the available land, the genetic improvements of the plants, the water resources available in each region, and the local infrastructure of roads and electricity supplies.

Following assessment of the resource there is always the possibility of some variation occurring in the volumes of biomass becoming available for energy as a result of changing world market prices for saw logs and pulp logs, and also for local fossil fuels and electricity. Hence it is more difficult to estimate the **economic potential**   with any degree of accuracy. If the commodity price rises, some plantations will be harvested earlier to capture the higher returns; if the price drops, some plantations will be left standing for longer with the hope it will rise again within a year or two. The actual **market potential**   of the biomass resource is also variable. If the costs of coal, oil and gas increase in the future, then biomass becomes more competitive and the market potential changes. Since the biomass is normally only a by-product from the forest and is relatively low in value compared with logs, there is some degree of risk that sufficient fuelwood supplies will not always be available. For this reason a bioenergy project developer would be wise to first secure reliable biomass supplies by setting up long-term supply contracts with the resource owners.

However, it is the **socio-economic potential**   that is most likely to occur in reality. This is the amount of biomass likely to become available when all the social and environmental factors have been considered along with the economic factors. It could be for example that in future, owing to growing concerns over climate change, society may be willing to pay more for sustainable sources of energy than for fossil fuels. Some government policy initiatives and incentive mechanisms will then need to be created, such as a target for a minimum portion of electricity and heat to be generated from biomass sources, or a high feed-in tariff guaranteeing the price paid to the bioenergy plant developer. Thus from a consumer's perspective the market price for what would be attractive solely on a financial assessment is exceeded.

Sources of woody biomass are subject to the development of a logistical supply chain to enable them to be used to provide energy services (Figure 2.9).

Sources of woody biomass available from production forestry consist of material left after pruning, thinning, harvesting, log making, and the subsequent processing of logs into sawn lumber, pulp and paper or panel products. New Zealand is a good case study.[1] It has around 1.7 million ha of plantation production forests (around 7% of total land area) and harvests over 18 Mm^3 of stemwood logs a year. The annual harvest is expected to increase to about 30 Mm^3 by 2010 for use within the sawmilling, veneer log, and pulp and paper production industries, as well as for exporting unprocessed logs. This anticipated rapid increase in wood products will also give rise to a larger potential source of woody biomass.

Current estimates of SRF in New Zealand are around only 1000–2000 ha, grown mostly in small plots for domestic firewood. In addition during the last two decades over 20,000 ha of pulpwood forest, mainly *Eucalyptus nitens* for harvest around 12–15 years old, have been established in regions close to ports to provide export pulp wood chips. Increased planting is expected to continue, and the arisings after harvest of the stemwood could become a viable biomass resource. The future use of SRF plantations for energy production will probably be from the use of these arisings, as well as from SRF grown specifically for the land application and treatment of liquid wastes. There are several commercial-scale plantings linked with sewage treatment and meatworks effluent (one being over 100 ha). However, use of the biomass for energy purposes on-site has not yet been successfully achieved, as coal and gas remain too cheap for it to compete even where transport distances are short. Planting SRF as a dedicated energy crop solely to provide greater fuelwood supply security may occur in the longer term, especially if there are carbon credits to be had as additional sources of revenue, but without such incentives it will probably remain uneconomic for some time.

The MGW resource in New Zealand and indeed most other countries is not well assessed, and currently little information exists on current or future volumes. However, MGW is unlikely ever to become a significant biomass stream as volumes are always limited.

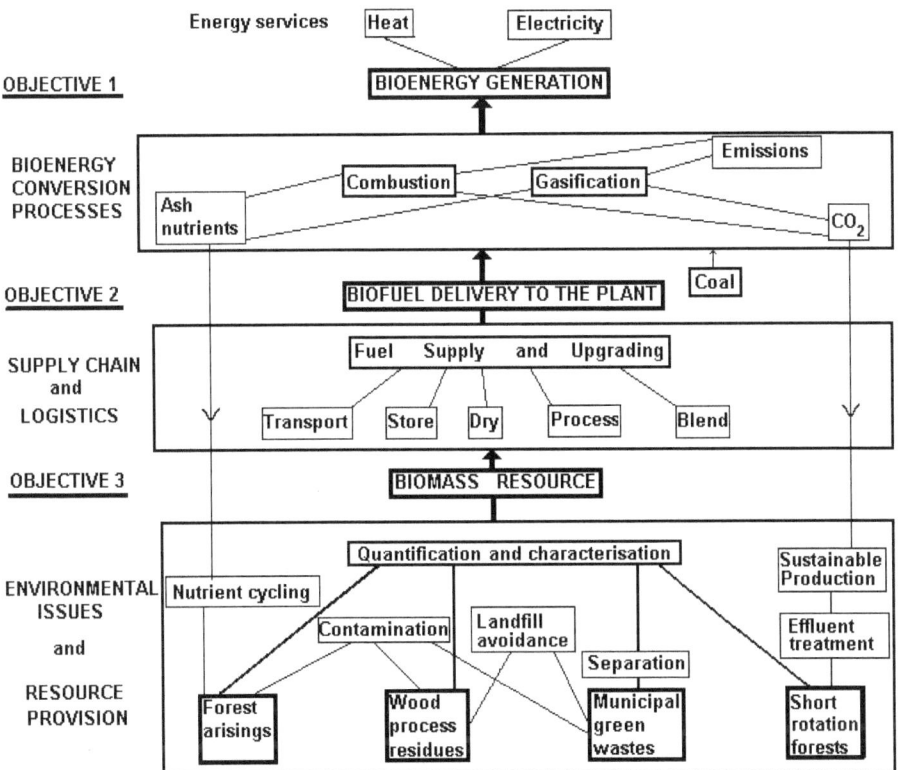

Figure 2.9 Woody biomass resources together with the environmental implications of their use require a complex logistical process before they can be converted into useful energy services[2]

Characteristics of woody biomass

These are variable in nature, and relate partly to specific site growing conditions such as soil type and climate. The tree species and the proportion of plant components present in the fuel (leaf, branch and stem) also partly determine the nature of the biomass. Particularly important characteristics are moisture content, ash content, volatile matter, bulk density, annual volume available and the chemical composition, since the specific nature of these characteristics will affect the requirements for the conversion technologies (Table 2.1).

The inorganic composition of biomass fuels and their tendency to cause slagging and fouling within furnaces and boilers have attracted considerable international research attention since the 1990s. Failure to address potential slagging and fouling issues can cause costly outages of the conversion plant. Studies undertaken on wood process and forest residues indicate that slagging and fouling can be minimized provided boilers are appropriately designed for the specific inorganic composition of the fuels to be used.

The characteristics of forest arisings are affected by the relative contribution from leaves and needles, branches, stemwood and bark, as each of these various components of the tree has different properties. The contribution of these components to the total biomass fuel mix and the degree of soil contamination present will depend on the nature of the forest operation and the residue collection system employed. Traditional forest arisings typically consist of medium-size to large branches together with some damaged stemwood and small or poor-quality rejected trees. Hence they usually require comminution (size reduction) prior to use. Arisings usually have a relatively high moisture content compared with wood process residues unless they have been stored and allowed to dry prior to collection. Artificial drying is normally uneconomic, but leaving the material for some weeks after harvest usually results in a lower moisture content.

A common feature of woody biomass feedstocks is that they tend to change over time owing to changes in the wood process technologies used on a site, competition from other markets for the available resources, and changes in logging and harvesting practices. The nature of typical woody biomass feedstocks (high moisture content, variable ash contents, high volatiles and relatively low calorific values), along with their inherent variability and unpredictable characteristics, means that careful attention must be paid to matching the fuel type to the design and operation of the conversion plant system. The more consistent and predictable the fuel is, then the more closely the conversion plant can be designed to match it, resulting in higher overall efficiencies and avoiding the need for and extra cost of over-engineering the system in order to handle a variety of fuel sources.

To ensure that woody biomass fuels can be used with minimum operational problems, quality information on the fuel availability and its characteristics is required. Research has focused on relating fuel parameters to combustion performance under consistent conditions in order to evaluate fuel performance

Table 2.1 Fuelwood characteristics of typical wood process residues and forest arisings from *P. radiata* (% of total analysis except where shown)

	Sawmill bark	Wood offcuts and sawdust	Forest residues
Proximate analysis			
Total moisture content	38.9	36.7	52.9
Ash content	3.5	0.3	3.2
Volatile matter	39.7	52.2	36.8
Fixed carbon	17.9	10.8	7.3
Heat value	13.0 MJ/kg	12.9 MJ/kg	9.0 MJ/kg
Ultimate analysis (wet basis)			
Carbon	32.9	32.29	23.27
Oxygen	21.2	26.66	17.8
Hydrogen	3.3	3.9	2.8
Nitrogen	0.2	0.13	0.05
Sulphur	0.02	0.007	0.005
Ash composition			
SiO_2	59.5	32.6	69.8
Al_2O_3	14.7	9.2	14.2
Fe_2O_3	3.3	13.6	3.1
CaO	7.3	17.0	3.3
MgO	1.9	4.9	1.1
Na_2O	3.4	4.0	3.9
K_2O	6.6	9.3	9.2
TiO_2	0.4	1.3	0.4
Softening temperature (°C)	1126	1070	1175
Spherical temperature (°C)	1200	1150	1220
Hemispherical temperature (°C)	1220	1170	1275
Fluid temperature (°C)	1380	1190	1410

Note: The above properties as measured by New Zealand Forest Research are given for fuels likely to be received at a biomass conversion plant prior to any pre-treatment

in different commercial conversion configurations, especially with regard to the design parameters for boiler grates to minimize slag formation (Chapter 5).

Security of fuelwood supply

The long-term availability of various woody biomass sources differs.

- For wood process residues the development of competing markets, improvements in wood-processing technologies to increase the yield of saleable products from each log and the increasing demand for fuel for on-site energy production for kiln drying mean that there will be less 'waste' biomass available.

- The future availability of forest arisings will vary with the growth and age characteristics of the specific forest, the silvicultural regimes employed, the harvesting practices used, the collection systems developed, the nature of the terrain and competition for the biomass resource (for example, reject logs or branches could be collected and used for wood chips).

- Growers contemplating planting energy crops will need to assess their land availability, the opportunity cost of using the land for traditional purposes and the economics of growing material specifically for the energy market. Investors will also need to assess similar issues, as insecure fuel supplies could adversely affect the future business prospects. Farmers are unlikely to enter into a long-term crop-growing agreement unless attractive prices are to be paid, since this would reduce their management flexibility in being able to change the mix and rotation of crops grown in order to improve their incomes. In Europe such decisions are based largely on the varying subsidies, grants and commodity prices over the medium term. Recent experience, however, from the British Yorkshire ARBRE project, which has contracted growers for 1500 ha of *Salix* (Chapter 6), suggests that 15-year contracts are not impossible to negotiate.

If a greater uptake of bioenergy projects and their successful development are to be achieved, then it may become critical to secure adequate supplies of biomass of an acceptable quality by using long-term contracts, and possibly passing the risk of maintaining supply on to fuelwood supply merchants. Methods for more effectively predicting the impacts of changes within the forest industry on the estimated quantities of biomass available for energy production would be useful. Important variables that impact on such estimates of potential fuelwood supplies are the predicted harvest volumes, the allocation of logs to various end uses, the amount of recycling of residues within the wood-processing sector, and the nature and characteristics of the harvesting regimes employed. Better estimates should instil greater confidence in investors that fuel will be available over the expected lifetime of the project. Any analysis that makes bioenergy projects more 'bankable' can only be of assistance to what is still a fledgling industry.

Approximately 25% or more of the valuable total log volume harvested ends up during processing as bark, shavings and sawdust. Typically well over 50% of these residue volumes is already reused within the wood processing sector (for products such as particle board and pulp chips) and so is not available for energy production. Even so, estimates of wood process residues used for energy production are significant in many countries with forest industries. Estimates are normally based on current and predicted energy requirements, as few detailed assessments of residue yields and volumes exist. Factors affecting these estimates could be the quantity of logs exported and whether or not they are debarked prior to shipping.

Estimates of the amount of forest arisings available include the residues already collected at the landings and roadsides during the harvesting operation. Material that could be collected from the cutover after stemwood harvest should also be considered (though collection from the cutover is assumed to be more costly). Then much more biomass would be available.

Cost of the biomass

The cost of supplying wood process residues for energy may range from negative, where there are other treatment or disposal costs involved, to around $6–15/m³ for wood chip that could otherwise be sold for pulping. Negative costs result where the producers of the waste residues are prepared to use them on-site or pay for their transport to an energy plant instead of incurring even greater landfill disposal charges.

A typical price for wood process residues can be assumed to be $1–2/m³ of solid green material, which is equivalent to around only $0.1–0.2/GJ delivered. This assumes minimal transport costs over short distances and that much of the material is used on-site where it is produced. It excludes any opportunity costs, although these may exist in some circumstances. Where there is a market for trading supplies of residues, this will tend to optimize supply availability for minimum costs.

The costs of delivering forest arisings to an energy plant are dependent on

- whether they are taken from the cutover, the roadside or the landings
- the total amount of material available
- the nature of the collection and transport equipment used
- whether the collection of residues is centralized
- the requirements for any pre-processing or drying
- the haulage distance
- the handling and conveying systems adopted at the energy plant (Chapter 4).

Typically delivered costs range from $1.50 to $4.50/GJ for an 80 km average haulage distance; a significant part of the total cost is in the handling and transport operations. Over recent years effort has gone into refining modelling systems to improve the prediction of delivered fuel cost for forest residues, and a range of $1–2/GJ is now considered achievable in practice. These costs can be further reduced by increasing efficiencies, maximizing truck payloads, and improving the management of fuel in the supply chain to ensure that it is delivered to a conversion plant in optimal condition and at least cost.

For forest residues to be used in preference to fossil fuels without government incentives, further cost reduction measures need to be identified. The above cost ranges should be regarded as indicative only, as many site-specific variables will influence the economic viability of extracting forest arisings in different parts of the world.

For a 3-year short rotation coppice production system, an indicative discounted cost for producing fuelwood at the farm/forest gate has been assessed to be around \$25/t (fresh weight basis) or around \$2.80/GJ. Harvesting represents a major component of the costs of the coppice feedstock, and further evaluation of cheaper extraction systems is needed. The high cost of biomass fuel supplied from SRF is currently limiting its use as a primary fuel source for energy projects in many regions. Valuing other beneficial attributes of SRF energy crops is needed to improve the overall economic feasibility.

Case study 1

Orbost Power Co-operative, forest arisings, Victoria, Australia

Orbost is situated 300 km from Melbourne in the centre of the Gippsland forest area of eastern Victoria (Figure 2.10). Several sawmills operate in the region, and around 300,000 t per year of hardwood pulp chips are exported out of the port of Eden or transported five hours by road to the Amcor pulp mill at Geelong, north of Melbourne. Currently, millable logs are being recovered for around \$13–15 per green tonne, which includes purchase from the state government (\$1.50–2/t), collecting operation (\$6–7.50/t), and transporting an average of 100–120 km (\$5–6/t). Many logs, however, are rejected and left in the forest.

Approximately 1.6 Mm^3 per year of the most valuable logs are milled at present, and the remainder are left to rot on the ground. It has been estimated that 800,000 t per year of arisings in the form of these reject logs are left in the forest with no current commercial value (Figure 2.11). This results in a volume of waste material effectively free on-site that could be utilized as an energy resource.

The Orbost Power Co-operative was established in 1993 with the aim of utilizing the reject logs for electricity generation by collecting them, passing any reasonable ones through a sawmill to first extract any useful timber product, and then using the remaining 90% or more of the total volume for fuelwood in a 10 MW$_e$ power plant.

The forest is situated towards the end of a transmission line where quality of power supply is poor in terms of fluctuating voltages and power outages. The development of a power plant in the vicinity would help to boost the quality of supply. Currently the coal-fired power station at Morwell provides power to the grid that services Melbourne to the west and a few small towns in Gippsland to the east. There is a proposal for a new pipeline passing near Orbost to take natural gas from Bensdale to Sydney. If this occurred it would probably be cheaper to develop a 10 MW$_e$ or larger integrated gas-fired combined cycle power station compared with using the wood waste in a bioenergy power plant. In addition some of the sawmills could be converted to run on gas to replace their current diesel generation systems.

The sawmills are not running at full capacity at present, and so could handle any additional reject logs. As an alternative a contractor in the region is keen to

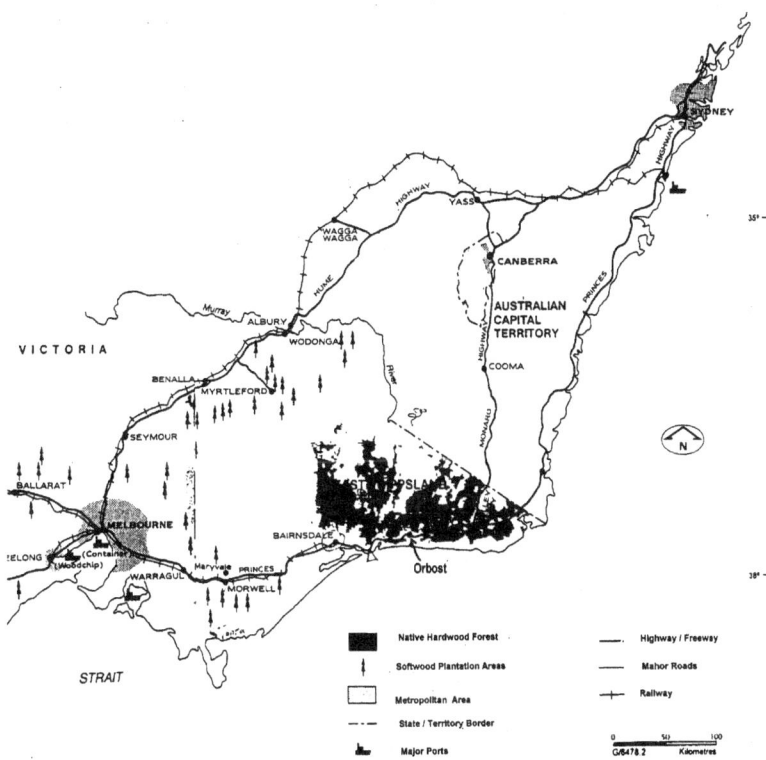

Figure 2.10   Location map of Orbost Power Co-operative

Figure 2.11   The forest arisings resource of Orbost, Victoria, including examples of logs rejected for millable timber owing to imperfections or too short a length

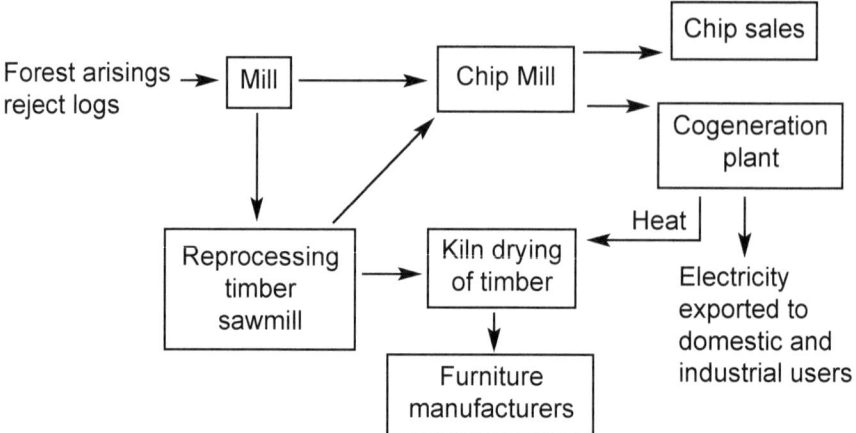

Figure 2.12 Integrated cogeneration project outline for a new mill. If this does not eventuate, the residual logs will go direct to the existing timber mills

establish a new milling facility specifically to handle the arisings material (Figure 2.12). (If this does not occur the residual logs with some value could still go direct to the existing timber mills.) Much of the residue resource is also of too poor quality for pulping, and even for those logs that might be suitable the distance from the ports is too great for it to be viable for sale as export pulp chips. So it would only be economic to collect the resource at all if a wood-fired power plant were developed in the locality.

A preliminary cost estimate showed that fuelwood could be delivered to a central site for around $15 per wet tonne, which equates to around $1.80/GJ. This could not compete with the delivery of local cheap brown coal to a new coal-fired power plant or with a combined cycle gas plant unless a premium price existed for green power generation. Another possibility is for a power company to seize the opportunity to help meet its obligations under the Australian federal government's Mandatory Renewable Energy Target (MRET), which became law in 2001. The legislation empowers electricity retailers to purchase at least 9500 GWh per year (2% of the total demand) from new renewable energy sources.

Security of fuelwood supply from the Orbost forest over the medium term appears not to be a problem, based on a 1993 CSIRO report, although long-term fuel supply contracts would obviously need to be put in place by the developer with the state-owned forest resource owners. In theory there should be no public concerns or environmental issues raised if these available arisings were to be utilized for fuelwood since the forest is managed sustainably under the Regional Forest Agreement, which was negotiated with all stakeholders and signed off by the Victorian government in 1997. The reseeding process used after harvest is to leave selected specimen trees unfelled at frequent intervals (Figure 2.13); these then provide good seedstock for natural reseeding of the area. However, some environmental groups are still not in agreement with the use of any indigenous forest resource for bioenergy, even if it is deemed to be

Figure 2.13 Selected specimen trees, left standing after clear felling to provide seeds for subsequent crops

a waste by-product as in this instance. They claim that adding value will only encourage greater felling of native forests for both timber and biomass. At the time of writing, concerns by environmental groups over using residues from 'old growth forests' for fuelwood may prevent the Orbost arisings from being utilized at all.

When the Orbost power project was first mooted, the eucalyptus forest was being harvested for structural timber framing. Since then competition from New Zealand *Pinus radiata* has caused a change in the local wood product market, so the timber is now used mainly for furniture manufacture, with some of the residues from this process being exported as hardwood chips.

If a cogeneration plant is to become a feasible proposition, the heat would need to service a central kiln drying facility, whereas the sawmills tend to have their own facilities and use their own waste wood residues. There is limited demand from other users for heat in the region at present, but kiln drying is growing. In order to offset increasing competition for sawn timber from softwood plantations, the sawmillers are now looking to provide kiln-dried speciality timber for such applications as furniture manufacture.

The economics of the power plant concept will only work if 7–10% of the current forest arising volume can be recovered as high-value timber and there is an energy market for the remainder. A relatively large local mill of 400,000 t/yr would be required to provide sufficient fuelwood as a by-product, though a further 100,000 m³ of sawdust residues could be purchased from the existing sawmills and also used in the power plant. (Currently some of this sawdust is carted 150 km to provide heat for a cheese factory, which is apparently a cheaper option than other forms of landfill disposal.)

This project exemplifies some of the difficulties in using what appears to be a free woody biomass resource. After eight years of reports, evaluations, interest by overseas investors and enthusiastic support by the local community supported by 'Business Victoria', it does not look as though the project will proceed. Meanwhile the logs continue to decay, and other coal and gas energy sources are used instead, continuing to generate greenhouse gases.

Case study 2

Auspine sawmill drying kiln, Tarpeena, South Australia

Auspine is a fully integrated plantation products company; its primary business is sawmilling, timber preservation and wood chip exports. At its sawmill plant in Tarpeena, two existing 7 MW cyclone burners have been used for many years to burn sawdust (Figure 2.14). The heat produced is used in a thermal oil heater to operate eight timber drying kilns, which each take a 35 m³ charge of three packs of *P. radiata* (Figure 2.15). Energy consumption for kiln drying varies with species and moisture content. Approximately 4.5 MJ of heat is needed per kg of moisture removed. *Eucalyptus* timber typically requires around 6.1 GJ/m³ to dry it over a three-day period, whereas *Pinus* can be dried in just 12 hours. Over 700,000 m³ of sawn timber is dried each year in this relatively large mill.

In 1998 the company purchased an 8 MW heat plant from a local sawmill company in receivership. Valued at around $3 million, it was purchased at auction for around $50,000 including fuel conveyor. It can burn a range of fuels including hogged bark, shavings and sawdust, all generated on-site. No power is generated at present, but there is the opportunity in the future to produce activated carbon for sale as a reductant as used by the gold and nickel refining

Figure 2.14   Two existing 7 MW sawdust cyclone burners behind the new 8 MW heat plant

Figure 2.15 Timber stacks being prepared for kiln drying

industry, for small-scale industrial heating, for wastewater treatment and for air filtration. The plan is to convert the sawdust to carbon while using the volatiles produced for heat supply to the kilns. However, the capital investment needed was too high for the company to proceed without a partner.

Auspine would like to develop a 1–2 MW pilot plant at its Tarpeena sawmill based on the CSIRO fluidized bed system for carbonization and cogeneration of heat through the burning of the volatiles. In this process green wood is heated to 120°C initially, then when dry it is further heated to 230°C, at which stage the volatiles are driven off. The residual fluid-bed carbon can then be used to produce either briquettes with an energy value of 30–32 MJ/kg or activated carbon. At present 100,000 m³ per year of Jarrah (*E. marginata*) residues are used in Western Australia to produce 24,000 t per year of charcoal to meet current demands, though there are rising public concerns about using native forest resources for such practices. The oil mallee crops now being grown in Western Australia to reduce dryland soil salinity levels may also be used in future for this product as well as for power generation (see Case Study 18).

Auspine also plans to construct a 60 MW bioenergy plant at Tarpeena, fuelled with wood residues from its harvesting operations as well as bought-in biomass. Investment costs are estimated to be $50 million. The plan is to establish an 'energy park' in the vicinity to take advantage of both the heat and the power to be generated.

Notes

[1] This section is based on the summary of a 2001 report prepared for the New Zealand Energy Efficiency and Conservation Authority by J Gifford, B Cox and the author. See www.eeca.govt.nz.

[2] Only combustion and gasification conversion technologies are shown, though a range of others exist, as covered in Chapter 7.

Chapter 3

The non-woody biomass resource

Other than woody biomass resources originally derived from forests, biomass resources include

- agricultural crop residues

- energy crops

- animal residues and human-derived wastes such as sewage sludge

- municipal solid wastes and landfill gas.

The challenge is to provide the sustainable management, conversion and delivery of bioenergy to the marketplace in the form of modern and competitive energy services. Since biomass in all its forms is widely distributed, it has good potential to provide both rural and urban areas with a renewable source of energy.

Agricultural crop residues, animal manures and municipal wastes usually have a disposal cost associated with them. Therefore 'waste-to-energy' conversion processes for heat and power generation (Chapters 6, 7 and 8), and even in some cases for transport fuel production (Chapter 9), often have good economic and market potential. They have value particularly in rural community applications, and are already used widely in countries such as Sweden, Denmark, Netherlands, the USA, Canada, Austria and Finland.

Crop residues such as straw, bagasse and rice husks, if they do not have to be returned to the land for nutrient replenishment and soil conditioning, could be better utilized in the future for heat and power generation. This could be either in co-fired systems with coal or gas, or in specially designed and appropriate biomass conversion equipment with low emissions. Such technology is well proven. In many countries the crop residues produced from such crops as winter cereals, rice, sugar cane and sorghum are usually ploughed back into the soil or burnt to waste in the open air, left to decompose or grazed by stock. Agricultural and biomass studies have concluded that it may be acceptable to remove and utilize a portion of these residues for energy production, hence providing large volumes of low-cost biomass material. These crop residues could be processed into liquid fuels, or combusted or gasified to produce electricity and heat. For example, over 11 million wet tonnes of bagasse are produced annually in Australia alone, with an energy content of around 120 PJ.

Animal manures and industrial organic wastes are currently used in many locations to generate biogas. For example, in Denmark there are at least 19 decentralized, community-scale biogas plants used for electricity generation. Biogas can also be used for cogeneration, to supply heat for direct heating of buildings or as a transport fuel.

Agricultural crop residues

Large quantities of crop residues are produced annually worldwide, and are often under-utilized. These include rice husks, bagasse (fibre from sugar cane), maize cobs, coconut husks (copra), groundnut and other nut shells, and cereal straw. Coconut shells and nut residues tend to be used only on a small scale, whereas larger quantities of rice husks, bagasse and straw are usually accumulated in one place. Such wastes tend to be relatively low in moisture content (10–30% m.c. wet basis) and therefore more suited to direct combustion or gasification rather than to anaerobic digestion, which better suits wet wastes such as tomato skins, meat processing wastes or reject fruit.

Rice husks

Rice husks are one of the commonest global agricultural residues. They make up 20–25% of the harvested rice grains on a weight basis, and are usually separated out at the processing centre. Indonesia alone, for example, produces around 8 Mt per year. The husks have a relatively high silica content, which can cause an ash problem and possible slagging within the boiler on combustion. However, their homogeneous nature lends this biomass resource to more efficient conversion technologies such as gasification, which requires a uniform fuel quality for best results. Several commercial examples exist.

Bagasse

Bagasse has considerable potential as a biomass fuel since it arises mainly at sugar factories, where flows of bulky volumes of biomass, typically around 300,000 t per year, in the form of sugar cane, are already well organized and often based on many decades of experience. Each fresh tonne of sugar cane brought into the factory for processing yields around 250 kg of the residual fibre, which has a useful energy content of around 10 MJ/kg. Any country that grows sugar cane therefore has a significant biomass energy resource available in the form of the crop residue remaining after sugar extraction (Figure 3.1), and one that has already been collected and delivered in effect free to the processing plant. Most sugar factories use this bagasse as a source of heat for raising steam to 'cook' the cane and extract the sugar and possibly generate enough power for use on-site. However, since there are such large volumes, they have tended to 'waste it efficiently' just to avoid accumulation of surplus material on-site and to avoid disposal costs.

Annual bagasse volumes vary, and are produced only during the cane crushing period, which normally lasts 6 to 7 months. As well as using the heat from combustion many sugar factories generate around 2–3 MW$_e$ on-site for their own use. At present only a few export significant quantities of surplus

Figure 3.1 The fibrous bagasse residue after extraction of the sugar from cane creates a major disposal problem since there are large volumes and, unlike sugar beet pulp in northern Europe, there is little demand for it as livestock fodder

power because of the operational and contractual difficulties of being able to sell the power only during the cane-crushing season. The potential for generating 20–30 MW$_e$ all year round by using wood wastes in the 5–6-month non-crushing season has created recent interest, particularly in countries where independent power producers have been established following electricity industry reforms and privatization and are willing to partner a project.

Studies conducted in Thailand, Jamaica, Brazil, Zimbabwe and elsewhere have shown that optimization of bagasse combustion for energy in a large sugar factory could provide fuel for up to 30 GW$_e$ of generating capacity. In addition, the utilization of some of the cane trash (the tops and leaves usually left in the field during harvest, or burnt off before harvest to provide easier access for the machines) could almost double this capacity. Of the world's 1670 sugar mills, more than 800 have greater than 5 MW$_e$ potential export generating capacity, and many are situated in India, Pakistan, S.E. Asia, China, South Africa, Central America, the Caribbean and South America, where electricity is in short supply. The flows of materials and energy in a sugar plant are worth highlighting, with regard to the potential bagasse supply being a co-product (Figure 3.2).

In Australia there are 31 sugar mills, and if all were upgraded to utilize the bagasse and woody biomass in the non-crushing season efficiently for cogeneration of heat and electricity, then the technical potential of total plant capacity would be 3.4 GW$_e$. The 20 TWh/yr of electricity that could be generated would reduce carbon emissions by over 16 MtCO$_2$, assuming that electricity from coal-fired plants would be displaced. A 1998 study by the Sugar Research Institute, Queensland, identified conventional high-pressure steam

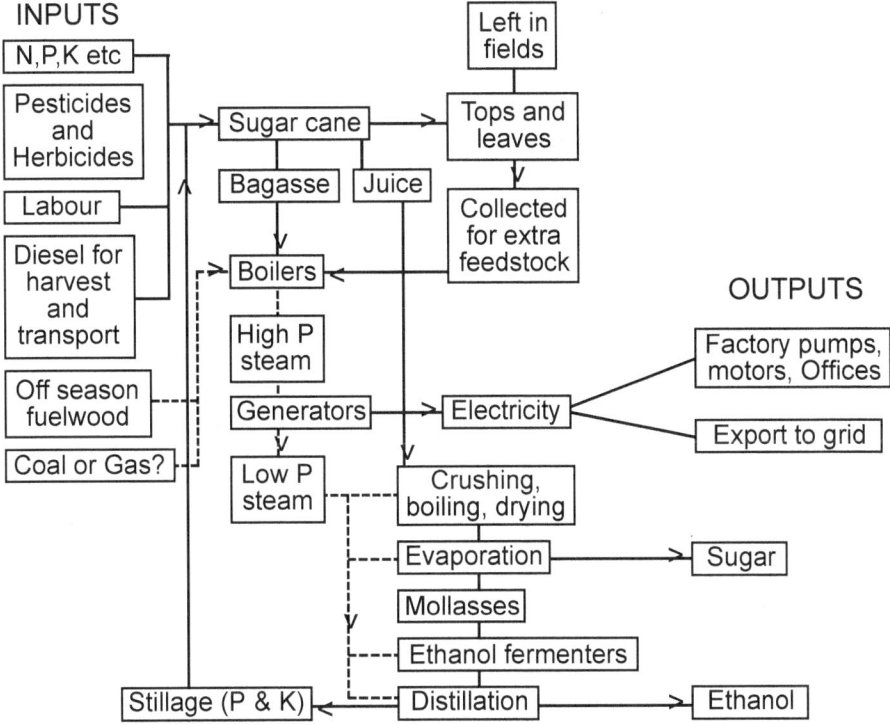

INPUTS

OUTPUTS

Figure 3.2 Energy and material flows during the sugar cane production and processing operation

turbine technology and integrated gasification combined cycle gas turbine technology as best suited for cogeneration using bagasse and fuelwood over the long term. A large capital investment is needed, and the breakeven costs range between 3 and 3.5 c/kWh, which compares with coal-fired power at around 2–3 c/kWh. However, it would be competitive with many other renewable energy technologies. For many plants the sale of the electricity would provide a useful second revenue stream, which one day might even surpass that from sugar. However, not all mills are able to benefit owing to the lack of close proximity to power lines with adequate capacity to carry the extra power. Upgrading these lines over long distances can make it a non-viable proposition. The significant investment involved is another constraint.

There is a link between using high-pressure steam to generate power and the need to upgrade the process plant in order to cope with the resulting higher steam pressures. A joint approach between a power company, a sugar company and perhaps a third party investor is therefore warranted where the sugar company would have insufficient capital expenditure capability to upgrade the plant. A power-generating company also has to consider the prospect that a sugar company it wishes to partner in a new power plant development might not survive, since the sugar industry is not buoyant. The project risks need to be shared within both the power and the sugar industries.

Such partnerships could lead to different sugar cane management practices to increase the biomass available at different times of the year, but this will be a slow process as traditional management and ownership attitudes will remain for some time. Current cane harvesting practices were developed to leave around half the biomass in the field to reduce subsequent disposal costs. If there is a demand created for these residues for energy purposes, then the volumes of bagasse that could become available as feedstock for additional power generation for export off site could increase. This assumes there is no soil nutrient deficiency risk as a result of collecting and removing this material, which also helps to retain valuable soil moisture and suppress weed growth. Research to ascertain how much can be removed without detrimental effects is under way. However, since with the traditional harvest method of pre-burning little trash remains, there should be no problem in removing it instead. The additional revenue from electricity generation could also change the agronomy of sugar cane production, since both sugar and fibre yields could become equally important. High-fibre cane varieties grown in high-density plantings, and refining of the harvesting method to integrate field trash recovery, may become commonplace to meet the demands of new cogeneration plants on-site.

Sugar factories are well experienced in the handling of large volumes of biomass material. Queensland sugar mills, for example, own, operate and maintain 4100 km of narrow gauge (610 mm) railway used to deliver the cane to the mills (Figure 3.3). Where light rail is uneconomic, as the cane fields are less concentrated, road transport prevails, and bins have been designed to be easily loaded onto trucks after filling by the harvester in the field.

Figure 3.3 Transport of sugar cane from field to factory by small rail and bins is a well-developed biomass transport operation

Sugar cane is a C4 plant, as is sorghum, which means that it has a better photosynthetic efficiency and hence ability to convert carbon dioxide using sunlight than do other more common C3 plants. It also usually requires only minimum inputs of pesticides and herbicides compared with cereal and other crops. Whether it can be considered to be grown on a truly sustainable basis is debatable, as some nutrients such as N need to be added in the form of fertilizer to replace those removed in the crop. However, if the stillage or effluent from the crushing and distillation process and the ash from the combustion of the bagasse were to be returned to the fields (particularly where the cane trash was also removed for energy purposes), then only N would be in deficit. Increasing public and scientific concerns that growing monoculture crops is not sustainable and that emissions from biomass-fired power stations need stringent controls also need to be addressed.

Cereal straw

Cereal straw from small cereal crops such as wheat is produced at around 2.5–5 t/ha depending on crop type, variety and the growing season. Maize and sorghum stover can be higher yielding. These crop residues range from 10% to 40% m.c.w.b., giving a typical heating value of 10–16 GJ/t wet weight. In terms of comparative gross energy values, 1 t of straw equates approximately to 0.5 t of coal or 0.3 t of oil. It has a higher silica content than other forms of biomass, leading to ash contents of up to 10% by weight.

After harvesting of the grain, most straw is either burnt in the field or incorporated into the soil, as there is only a limited demand for it for animal bedding, stock feed, mushroom compost or garden mulch. Uncontrolled burning in the field is a cheap method of disposal, and can help to reduce the incidence of disease carry-over to future crops, but is now banned in European countries for environmental air pollution reasons and because of the risk of road accidents caused by drifting smoke. Incorporation by chopping and additional cultivation operations is more expensive, and many long-term evaluations show that there is little benefit to the soil or its organic matter content as a result, since the straw consists mainly of cellulose (C, H, O) and has a very high C:N ratio. Hence the utilization of straw for energy purposes has increased, with Denmark leading the world with thousands of straw-burning facilities in place for district heating (3–5 MW_{th}), industrial process heat (1–2 MW_{th}) and domestic heating (10–100 kW_{th}). The straw is normally stored on the farms, only being delivered to the central heating plants as needed or used on the farm at the small scale for grain drying, animal house heating, and space and water heating for residences (Figure 3.4).

In the UK approximately 14 Mt of cereal straw are produced annually, over half being surplus to present demand. This portion has the technical potential to supply 100 PJ/yr or 1% of the UK primary energy source. If the straw is assumed to have zero economic value, and with the costs of collection at around $25/t for raking, baling, etc., then the energy in the straw, usually stored as large round or square bales, would cost around $2/GJ assuming a moisture

Figure 3.4. An example of one of many types of straw-fired burner installed in a Danish farm for heating the dwelling (see inset) and animal houses. Note the door for ash removal

content of 15–20% w.b. Cartage to a central conversion plant site might add another $3–5/GJ if within 25 km average distance, leading to a high electricity-generating cost of around 8–10 c/kWh. This may be an economic option in Denmark and elsewhere, owing to the high power prices and various forms of subsidy, but is unlikely to be competitive in many other cereal-growing countries. Direct combustion of the straw for process heat in nearby plants (such as in factories that produce malting barley for breweries) may be more economic, but in many places straw is unlikely to compete with coal or natural gas at current prices and without carbon emission externality costs being included.

Part of the reason for the relatively high cost of straw is that, unlike bagasse, it is not delivered to the plant as a normal part of the cereal harvesting process. Additional collection, transport and handling operations are required. Although straw has a relatively high energy density (GJ/t) compared with other forms of biomass, owing to its low moisture content, even when baled it has a relatively low mass density. So the energy density per truck load is 10 to 20 times less than that for the same truck carrying coal or oil, even if the maximum possible payload can be achieved with the straw bales. Conversion equipment, whether for heat or electricity, is also more expensive than for the same output capacity when using fossil fuels (in terms of $/kW installed capacity) since the bulk of the straw necessitates relatively larger plant and more complex conveying and feeding equipment.

A range of straw pellets and wafers (Figure 3.5) with a greater mass density than bales has been developed in an attempt to reduce transport costs and also

Figure 3.5 A range of commercially produced straw pellet types and sizes, the longest shown being around 300 mm long and 75 mm in diameter

to enable automatic stoker feeding to occur, particularly at the smaller domestic scale (10–30 kW heat output). Many specialist pellet burners are on the market, and are suitable for wood or straw pellets, but the cost of the total system is relatively high. The big advantage is that the pellets can be delivered in bulk by small truck to the dwelling or small business as required and then fed automatically to the boiler, just as conveniently as can heating oil or gas.

Energy crops

Various annual and perennial crop species have been identified as having high efficiency properties when converting solar energy into stored biomass, which can then be converted into heat, electricity or transport fuels with zero or very low net carbon emissions. Many traditional food crops can also be grown specifically as energy sources, including sugar cane, corn (maize), wheat, sorghum, and vegetable oil-bearing crops such as sunflowers, rapeseed (canola) and soya beans. Most of these crops are grown to produce liquid fuel sources: that is, they are harvested, processed, and the product converted into fuels such as ethanol (a petrol substitute) or biodiesel. The by-products can then be used for heat and power generation. The most widely grown energy crops are sugar cane (there is even a special high-fibre species known as 'energy cane') and maize (or corn). High-yielding crops, especially from C4 plants, can give stored energy equivalents of over 400 GJ/ha.yr at the commercial scale, leading to very positive input/output energy balances for the overall system. However, the relatively low energy yields per hectare for many oil crops (around

60–80 GJ/ha.yr for the oil), compared with crops grown for cellulose or starch/sugar (200–300 GJ/ha.yr), have led the US National Research Council to advise against any further research investment in this area. Ethanol from maize and other cereals in the USA and from sugar cane in Brazil, and biodiesel from oilseed rape in Europe, are all being commercially produced but are subject to commodity price fluctuations, and depend heavily upon government support.

Crops grown specifically for energy supply purposes tend to have higher delivered costs in terms of $/GJ of available energy than if utilizing residues from food crop. In addition land used exclusively for biomass production will need to have an additional opportunity cost factored in to the economic analysis to account for the lost income from producing food or fibre crops from it.

Planting energy crops into arable or pasture land in some circumstances can also increase the carbon density of that land, while also yielding a source of biomass. Utilizing the accumulated carbon in the biofuels for energy purposes, and hence recycling it, alleviates the critical issue of maintaining the biotic carbon stocks over time, as is the case for a forest sink. Increased levels of soil carbon may also result from growing perennial energy crops such as miscanthus or reed canary grass, but to assess this a detailed life cycle assessment is warranted for specific crops grown in different regions.

The future for growing energy crops is good. There is enough land available to provide the world's population with all its needs for food, fibre and energy (though equitable distribution of these basic necessities is another issue yet to be resolved). The challenge is to integrate crop production to give all three products. Oilseed rape, for example, produces oil that can be used for cooking or energy, an edible high-protein meal, and straw, which can be used as a paper pulp or combusted for heat.

The future role for 'designer biomass' by developing suitable genetically modified crops cannot be ignored. Certainly the concerns over genetically modified organisms entering the food chain and the environment without full and proper evaluation are of considerable concern. However, biotechnology is here to stay, and new crops do indeed have great potential (if introduced only after thorough investigation). Imagine several attractive C4 plant species becoming available that have nitrogen-fixing ability, consume relatively little water, are high yielding and easy to harvest, and can be grown extensively to produce protein, carbohydrates, fibres and lignin. These products could all be processed through a 'biorefinery' to give a range of industrial, edible and energy products. The major issues of sustainable production, lack of biodiversity, species contamination and monoculture need to be carefully considered, but with some innovative thinking and careful research crops could be produced a lot better than they are now in traditional agriculture.

A summary of the potential for energy crops, from both short rotation forests and agricultural crops, is illustrated in Figure 3.6.

There has also been research on growing green crops such as kale and other brassicas as feedstock for biogas plants, some being stored as silage to provide feedstock all year round. The cost of production, however, is prohibitive at this

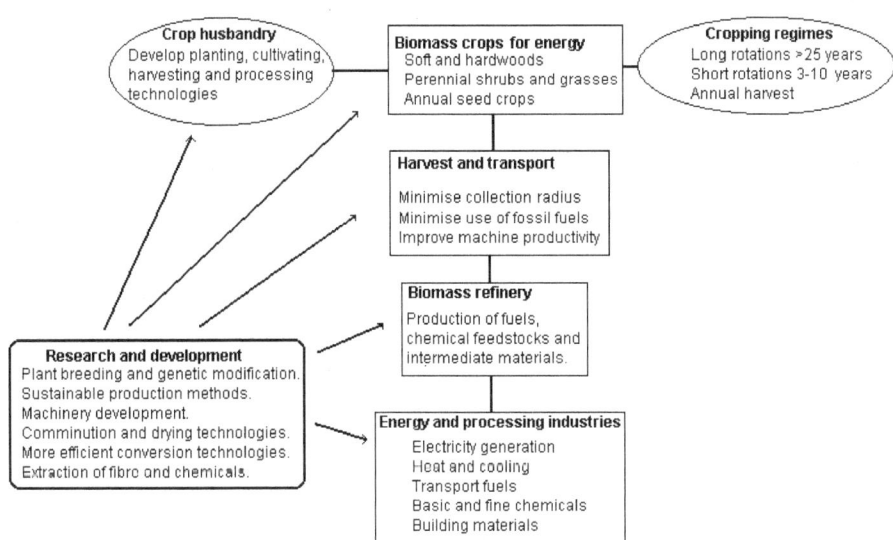

Figure 3.6 An overview of the technical potential for energy cropping and the resulting products

stage compared with the use of waste organic materials that are virtually free on-site.

Animal wastes

There are a wide range of animal wastes that can be used as sources of biomass for energy. The most common sources are manures from pigs, chickens and cattle (in feed lots) because these animals are reared in confined areas, generating a considerable concentration of waste.

- A 500 kg dairy cow produces around 35 kg/day of manure, of which around 4.5 kg (or 13%) is total solids and 87% water depending on diet.

- A smaller beef steer produces 25 kg/day, also at around 87% moisture content, giving 3.2 kg/day solids.

- A fattening pig gives 3.3 kg/day of manure but at only 9% solids, which is equivalent to 0.3 kg/day solids.

- A laying hen gives 0.12 kg/day at 25% solids or 0.03 kg/day.

In the past much of this waste has been recovered and sold as a fertilizer or simply spread on agricultural land as an organic fertilizer, but the introduction of tighter environmental controls on odour and water pollution means that better forms of waste management are now required in many countries. This provides further incentives for waste-to-energy conversion opportunities, but before designing a conversion plant the volume and characteristics of the animal wastes available must be carefully assessed.

A common method of converting these animal waste materials is via anaerobic digestion (Chapter 7). The product from anaerobic digestion is 'biogas' that can be used as a fuel for internal combustion engines to generate electricity, or burnt directly for cooking, or for space and water heating. Food processing and abattoir wastes are potential anaerobic digestion feedstocks, their energy value depending on the components present, proteins having around 20 MJ/kg, carbohydrates 11 MJ/kg and fats over 30 MJ/kg.

Sewage sludge

Domestic and municipal sewage from mainly human waste is a source of biomass energy that is very similar to the other animal wastes previously mentioned, the only difference being that, at least in developed countries for health reasons, it has normally been subjected to primary treatment before use. Energy can be extracted from raw sewage using anaerobic digestion to produce biogas, but more usually the higher solid content sewage sludge that remains after primary treatment is digested, or it can be dried and incinerated, or pyrolysed to produce 'bio-oil' (see Case Study 10).

Municipal wastes

Millions of tonnes of household waste are collected each year, with the vast majority being disposed of in landfill dumps. In practice, municipal waste is waste collected by or under the authority of a local authority, whereas urban waste encompasses MSW as well as industrial and commercial wastes collected and disposed of privately into the same waste stream. The term 'MSW' is usually used in a loose way to designate both types in order to simplify the terminology, which is the case here.

MSW is first and foremost, as indicated by its name, a waste issue:

It is the result of industrial societies organized as open systems operating in a predatory mode with respect to their environment. The word 'waste', through Old French g(u)ast, comes from the Latin vastus, which originally meant an uninhabited and uncultivated expanse of land (as in the English vast), that nowadays we call wilderness or, more ideologically, the environment. By extension the word meant land that had been rendered uncultivated and uninhabited or sparsely populated as a result of war, or predatory use, which we now call environmental degradation. By further extension, in the eighteenth century, the word waste took as well the meaning of a superfluous and lavish abundance of something and, a century later, the meaning of worthless surplus material (such as the extra sheets of good but now worthless paper left after a printing run). It is only in comparatively recent times, essentially towards the beginning of the industrial revolution, that the word came to mean the useless products of any industrial process from which no further economic value can be generated and that therefore must be rejected somewhere in a waste land, a tip, a dump, or nowadays a sanitary landfill.[1]

Considering MSW as a renewable energy resource is perhaps debatable. It is renewable, in that more MSW is being produced each year, but the industrial societies producing the waste are not sustainable in the true sense of the word since they do not function as closed systems integrated with their environment. As necessary changes to a more sustainable society occur, the meaning of waste will disappear. Nature does not produce waste, and a properly integrated society should not produce waste either.

MSW is thus only at best a transient resource, and a poor one at that, since by definition waste products are negative-value products that cost more and more to return safely to the environment or to reuse in some way. The key issues are the high entropy of wastes (being widely dispersed in both the energy and order senses of the word) and the high cost for the inputs incurred to reorder the waste material to make something valuable out of it.

MSW is not an energy resource *per se* but the end stage of many very complex and ever-changing production and consumption processes. It 'contains nothing in particular and a bit of everything in general!'[1] As part of the transition towards more sustainable forms of society, the processing of MSW is a fractionation and refining process generating a range of commercial co-products, of which energy, in the form of heat, gas, oil or power, is only a small component.

The 'waste management policy' of any nation should be:

- to ensure that, as far as practicable, waste generators should meet the costs of the waste they produce

- to encourage the implementation of the internationally recognized hierarchy of r eduction, r euse, r ecycling, r ecovery and r esidual management   of waste.

In some countries waste 'incineration' or mass-burn is the preferred form of disposal, primarily an alternative to usually preferred landfills where land is scarce or they are difficult to implement. Whatever energy is recovered during the incineration process remains a minor by-product in both thermal and economic terms. The chief benefit of incineration, as commonly practised in over-populated Europe, is to shrink the amount of solid waste to be landfilled down to ashes and slag, and hence transfer most of the mass that would otherwise have been landfilled to the atmosphere and to waterways.

The contribution of MSW treatments to a transition towards renewable forms of energy supply will never be significant since energy recovery is but a small component of waste management. Owing to the complex nature of MSW treatment, there is no single set of processes that can be easily regrouped and reviewed under a simple 'energy technology' label. The general field of MSW technologies includes the broad areas of incineration and refuse-derived fuels (RDF) and other emerging more complex processes for materials, fuels and energy recovery.

The composition of MSW varies according to the location, degree of recycling and type of the collection service. Organic wastes from catering establishments such as restaurants and hotels have around 70% moisture content, 5% non-combustibles and a heat value of 5 MJ/kg. Domestic, commercial and industrial rubbish, with contents that include packaging, paper, wood scraps and prunings, has around 25% moisture content, 10% non-combustibles and over 13 MJ/kg. Typically, refuse is an even mix of both, and is a crude, unrefined source of biomass fuel. Hence the biomass resource in MSW comprises mainly putrescibles, with garden rubbish and paper averaging over 70% of the total MSW collected. This fraction is sometimes termed **municipal gr een waste** (MGW).

Thermochemical conversion of MGW for heat and power generation, or biochemical anaerobic digestion to a gaseous fuel, uses the same processes as for other forms of biomass, the main difference being the degree of prior separation required. MGW can be converted into energy by direct combustion, or by natural anaerobic digestion in a landfill site. At these landfill sites the gas produced by the natural decomposition of organic wastes is approximately 50% methane and 50% carbon dioxide. On a well-designed and managed landfill site the gas can be collected from the stored material and scrubbed and cleaned before feeding into gas pipelines for reticulation as a heating source or for use by internal combustion engines or gas turbines on-site to generate heat and power.

Industrial wastes

Industrial wastes from the food and fibre industry include a large number of residues and by-products that can be used as biomass energy sources. These waste materials include those stemming from meat processing and rendering, wool scouring, pulp and paper making, fish processing, fruit canning, confectionery and vegetable processing. Only a small proportion of these wastes is currently converted for energy use owing to the high capital cost of plant (such as biogas digesters), the lack of financial incentives and the poor attitudes of many companies towards their environmental obligations. These wastes are usually disposed of in landfill dumps, with the business paying for their disposal. As these costs rise there is increasing interest in 'waste-to-energy' solutions as the conversion technologies improve, emissions are controlled and the value for the green energy increases.

Liquid waste streams are generated by washing meat, fruit and vegetables, blanching fruit and vegetables, pre-cooking meats, poultry and fish, wool scouring, dairy whey, grease trap wastes, other cleaning and processing operations and spent brewery and wine-making wastes. These effluents contain sugars, starches and other dissolved and solid organic matter, but in a fairly dilute form. The potential exists for these industrial wastes to be anaerobically digested to produce biogas, or fermented to produce ethanol, and several commercial examples of these waste-to-energy conversion routes already exist. Often there remains a final stillage (or dilute effluent) disposal problem to be overcome. Tertiary treatments can improve the water quality sufficiently for it to be granted consents for discharging into nearby waterways. Alternatively there is growing interest in land treatment by irrigating the effluent on to growing crops. For sewage, farm and some industrial effluents there are health concerns about applying them to edible crops or even to pastures ultimately consumed by meat or milking livestock. The solution is to link in the land treatment of effluent with the growing of energy crops, thus removing the material from the food chain.

Landfill gas

Landfill gas is an adventitious fuel that is a by-product of current landfilling practices and hence occurs only after MSW has been disposed of in a totally non-renewable way. It is also an extremely inefficient way of recovering energy from MSW. In the long run, as the use of landfills declines, landfill gas will quickly disappear as a resource. It is thus of an inherently transient nature, but will be covered here as many examples of landfill gas plants already exist and more are planned.

Landfill gas is generated when the organic material of refuse is decomposed anaerobically by a mixed population of the bacteria that are naturally present in the anaerobic conditions of a compacted landfill from which any residual oxygen has been used up. Conditions in a landfill are far from ideal for the bacteria compared with a carefully engineered anaerobic digester, where they are encouraged to multiply under optimum conditions. Therefore the landfill gas process is relatively slow, with a retention time of years rather than days. Methane gas may still be produced up to 50 years after a landfill is sealed.

Costs of landfill disposal vary with the nature of the waste, its volume, the legislation governing disposal procedures, the availability of landfill tip sites nearby, and the distance the product must be carried for disposal.

Approximately 50–200 m^3 of landfill gas is produced per tonne of MSW put into the landfill. The wide variation is due to the proportion of biological matter contained in the refuse which is typically around 50% but varies from country to country and even from village to village. The gas comprises 50–60% methane and 40–45% CO_2, plus non-methane volatile organics and traces of halogenated organics. On well-designed and managed landfill sites of sufficient capacity to warrant it, the gas can be collected and used. Usually, however, the gas migrates through the heap and dissipates into the atmosphere, which is unacceptable since each molecule has a global warming potential 21 times that of a CO_2 molecule over a 100 year period. It is a growing source of global methane at around 240 MtCO$_2$ equivalent per year. Using site-specific models that take into account local conditions such as soil type, climate and methane oxidation rates to calculate overall methane emissions, it has been estimated that, globally, landfill methane contributes approximately 4% of total global warming potential and is growing rapidly but with substantial variations from country to country. So if it can be captured and used for energy use, and hence the methane converted to CO_2 in the process, then landfill gas provides a win–win situation.

Methane emissions from landfills vary considerably depending on the

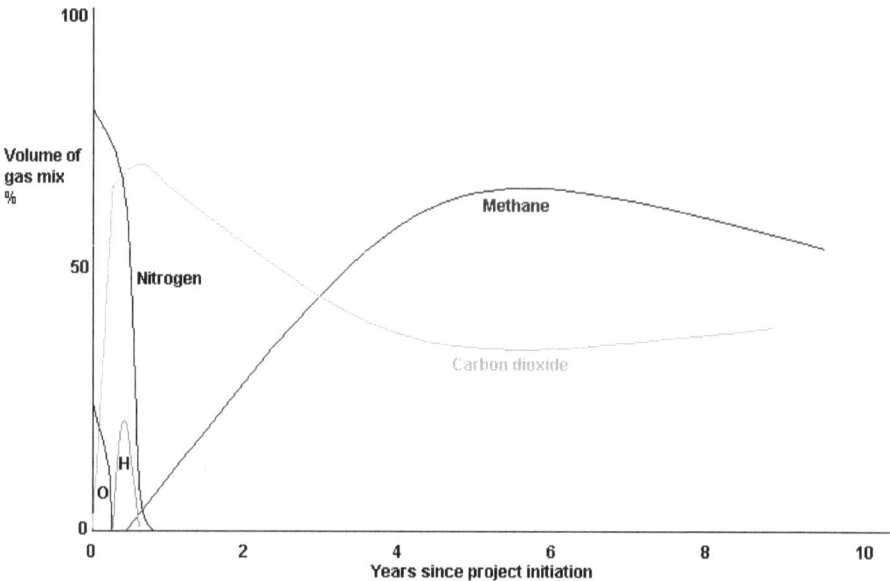

Figur e 3.7 Changes in landfill gas composition over time

characteristics of the waste (composition, density, particle size), moisture content, nutrients, microbes, temperature and pH. Data from field studies conducted worldwide indicate that landfill methane production ranges widely between 0.003 and 3000 g/m^2 per day. Not all landfill methane is emitted into the air; some is stored in the landfill, and part is oxidized to CO_2. The composition of the landfill gas changes with time, which must be taken into account when designing a collection and utilization system (Figure 3.7). As the oxygen levels deplete, the anaerobic bacteria begin their work and methane is produced. After some years the volumes of methane produced decline, as does its proportion in the gas mixture.

Minimizing the volumes of materials going into landfills is the goal of many countries, and the use of garden refuse for mulch and compost, recycling of glass and metals, and utilizing any combustibles for 'waste-to-energy' should be encouraged. Nevertheless, the majority of municipal and industrial wastes end up in a landfill. The aim therefore should be to avoid methane emissions for both environmental and safety reasons, since the gas is flammable and has caused explosions in nearby buildings after seeping through the ground.

Landfill gas can be collected and used for a wide range of energy purposes. Many successful examples exist, as it is often cost-effective to collect and combust. By using the gas to run gas engines driving generators, electricity can be produced for around 2 c/kWh, which is one of the cheapest sources of bioenergy. The availability of this resource, however, is substantially constrained owing to the limited number of suitable sites and the need to try to minimize waste production by recycling it in the future.

Case study 3

Straw-fired heating boiler, Woburn Abbey, UK

The Duke of Bedford's residence and outbuildings at Woburn Abbey in the UK were converted from oil-fired heating to straw-fired heating in 1987, giving a payback period of less than 3 years (Figure 3.8). Up to 20 round bales are consumed daily in the winter to provide heat not only for the large residence

Figure 3.8 Woburn Abbey and its outbuildings are all heated by straw produced from the local estate

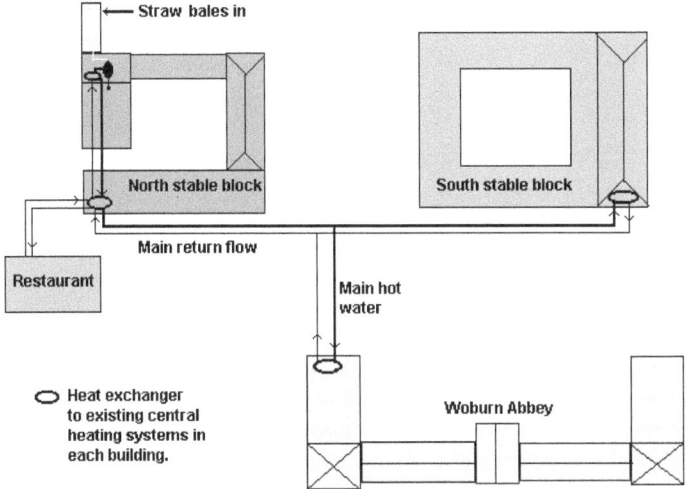

Figure 3.9 The small district heating scheme, showing several buildings, including the main residence, linked by underground hot water pipes conveying heat from the boiler to heat exchangers connected to the internal heating system of each building

Figure 3.10 Round straw bales from the Woburn Estate are loaded onto the conveyor – a twice-daily task in winter – then shredded automatically and fed to the boiler

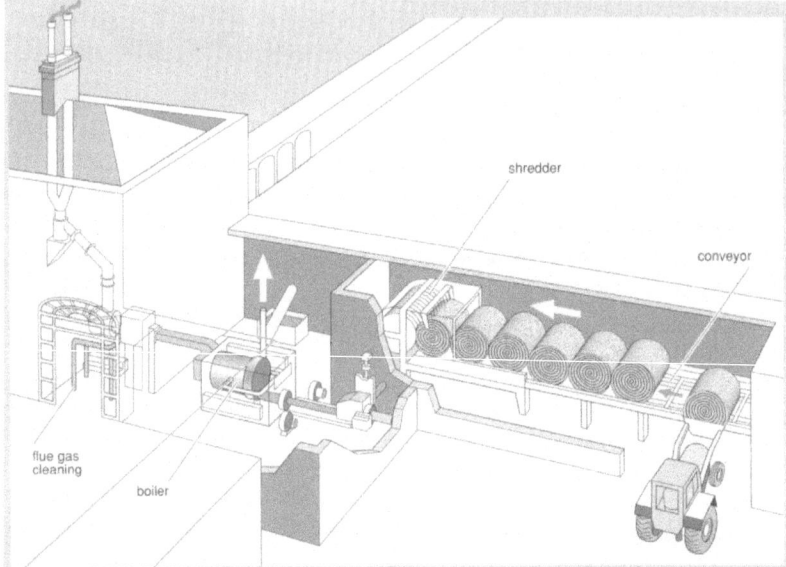

Figure 3.11 Chopped straw passes through the fire wall in the covered auger conveyor that feeds the 250 kW straw-fired boiler
Source: Open University, UK

Figure 3.12 Covered screw auger feeding straw into the boiler. Note the safety device to prevent any blow back passing through the fire wall.

but also for outbuildings such as the restaurant and stables. This required the design and installation of a small district heating scheme with a network of underground, lagged hot water pipes carrying the heat from the straw-fired boiler (Figure 3.9). The bales are placed on an automatic conveyor (Figure 3.10), which shreds them before feeding the chopped straw to the boiler by auger conveyor through a safety fire wall (Figure 3.11). Other straw burner designs are fed using whole bales, but since these tend not to burn uniformly, it is more difficult to maintain a constant heat output.

Case study 4

Bagasse and MGW cogeneration plant, Rocky Point Sugar Mill, Australia

Construction of a 30 MW_e biomass-fired power station began in mid 2000 next to the Rocky Point Sugar Mill in south-east Queensland, Australia. The present 110-year-old sugar mill facility has evolved over time as technology improved. By 1999 a very expensive capital upgrade was necessary if the mill was to remain competitive. This included replacing the very old steam drives of the sugar crushing equipment, other outdated processing equipment, and the boilers, which were old and inefficient. So a partnership was formed with Stanwell Corporation Ltd so that the mill owners could afford all the necessary modern boiler and turbine plant as well replacement of the steam drives with the latest Haglunds electro-mechanical drives. This increased the efficiency of the sugar extraction process and reduced the energy demand.

Over 120,000 t/yr of bagasse is available at around 40–50% moisture content and with a net heating value of around 9–10 GJ/t. Since the cane-crushing season occurs for only part of the year, local wood process residues and MGW will be used to ensure an all-year-round fuel supply. When cane is being crushed and steam is taken off for use in the sugar mill, the cogeneration plant will generate 22 MW_e, of which 2.5–3 MW_e will be needed for the cogeneration plant auxiliaries, 1 MW_e for the biomass handling and comminuting process, and 4 MW_e for the sugar mill.

The stakeholders

The cane growers

In order for this new co-production system of sugar and green energy to prove successful, a series of changes to the overall cane production and processing system will need to be addressed by the growers:

- New high-fibre cane varieties should be planted, and at higher densities than before, in order to maximize the biomass grown.

- Burning of the standing cane to ease harvester access should be discontinued, both from the environmental perspective and also since this reduces the available biomass yield.

- Field trash recovery should be undertaken to provide greater volumes of biomass.

- Year-round cropping should be further investigated to provide a constant supply of feedstock rather than just for 24 or so weeks a year during the crushing season.

The sugar mill owner

The Heck Group is a visionary host company. Management realized that the whole plant needed to be upgraded, both to improve the productivity and to reduce the energy demand of the process so that more steam would be available to generate power. By undertaking a joint venture with Stanwell, a very costly upgrade was partly avoided.

The plant needed to be redesigned so that the bagasse could be conveyed directly from the mill to the boiler and not to an intermediate storage area for later use. Steam is generated for use by both the mill and the power-generating turbine, so agreements on priority use had to be reached. For example, if for some reason steam becomes limited for a period, would it be more profitable to produce sugar or electricity? Collection of the woody biomass resource began as the plant was being constructed (Figure 3.13), since over 180,000 t/yr will be required, assuming it is around 30–40% m.c. and 11–13 GJ/t. It will be passed through a Savalda hammer mill to comminute it to <40 mm piece size suitable for delivering to the boiler. Tertiary-treated water will be used in the boiler, and negotiations and consents were needed from neighbouring landowners.

Figure 3.13 Stockpile of municipal wood waste retrieved from the Brisbane region by the fuel supply contractor

Most of the power produced is exported and sold in blocks as 'green power' to provide a second revenue stream for the sugar mill owners and the co-developers. Meters have been installed to the electricity network, to the mill and to the fuel processing and combustion plant, but green credits for all uses were approved, not just for the power exported.

The power generator

Stanwell Corporation Ltd is the owner and joint developer of the cogeneration plant with the mill owner. Approximately 30 GWh/yr of electricity will be used on-site to power the cogeneration and sugar process and a further 150 GWh/yr exported, which will be adequate to power over 12,000 homes (if consuming 12,000 kWh/yr each on average). This investment will help Stanwell to meet the government's mandatory renewable energy target obligation. The project will displace coal-fired power generation and hence avoid almost 130,000 tCO₂ emissions per year.

No old growth forest residues will be used as fuel for this project, Stanwell having draughted a biomass policy outlining stringent standards and auditing procedures relating to the use of green and wood residues. The Brisbane-based company Wood Mulching Industries has been contracted to supply the fuelwood using a large part of the wood waste in south east Queensland that would otherwise be burnt or consigned to landfill. The delivered fuelwood is purchased for $15/t by the plant owner from the supplier, who also charges the waste producers $25/t to take it away.

Plant designer and manufacturer

The contract for plant construction was awarded to the world's largest supplier of power plants, Alstom Power Limited, to design, supply, manufacture, deliver, erect and commission the 130 t/h bagasse wood-fired boiler plant, the wood-handling system and the 30 MW$_e$ turbine generator set (Figure 3.14). The balance of plant was also to be provided, including fuel storage and handling, water treatment and deaeration, ash and dust storage and removal systems, low- and high-voltage electrics, instrumentation and controls, burner management systems, structural steelwork, low- and high-pressure pipework, emission controls, civil engineering and buildings and cooling water systems. Integration of the new technology with the existing plant was critical to minimize disruption to the plant operation (and to the neighbouring Beenleigh Rum distillery, which is provided with steam and electricity from the plant as well as sugar). Notice to proceed was given to Alstom in February 2000 and completion took around 21 months. Approximately 40 people were employed in the construction, with several permanent positions established including fuel supply personnel.

Figure 3.14 Artist's impression of completed facility, showing bagasse storage loft in foreground, old boiler and stacks, new boiler, stack and turbine house and stack and ethanol production plant on extreme right
Source: Alstom Ltd

Financial investors

Capital investment cost was around $25 million, or around $850/kW of installed power. This is more expensive than a new coal-fired plant, estimated to be around $600/kW, and higher than the cost of other similar biomass projects, with some estimates as low as $550/kW. The developers originally thought

$600/kW or so was achievable, but the additional costs incurred by its being a pioneering project drove up the costs. With more experience and greater acceptance of the concept, $600/kW should be achievable. An approximate breakdown of costs was:

- civil engineering/building works: $2.5 million

- boiler: $6 million

- turbine generator: $8.5 million

- fuel and ash handling: $4 million

- balance of plant: $4 million.

Contracts

Contracts turned out to be more complex than previously experienced by Stanwell for a power station project, as they had to include:

- a site lease

- fuel supply agreement for the bagasse

- fuel supply contract for the wood wastes

- electricity and steam sales to the host

- electricity sales to the grid

- network connection and access to the grid – which were particularly complex as it is a monopoly and has no customer focus, the wholesale electricity market rules are new and everyone is still learning, and it was not clear how to pass on the benefits of embedded generation

- water supply consents and discharges – which were thought to impact on the saline ponds of neighbouring prawn farmers

- operation and maintenance arrangements for the host to undertake, but Stanwell to provide technical oversight and review.

Wood wastes are comminuted on-site and then fed into the boiler. An integrated material-handling system was designed by Alstom to minimize handling difficulties (Figure 3.15). Fuel is comminuted and delivered to a receiving pit and screen, from where it is conveyed to either the storage bin or, if full, to the fuel stockpile in the yard. From here it is conveyed back again to the boiler via a surge bin to even out the flow rate. Ash from the boiler is mixed with the mill mud from washing the cane and returned to the land.

From experience with the Rocky Point project, Stanwell Corporation consider that developing a biomass project is hard work. The capital investment cost turned out to be considerably higher than first calculated, mainly because of all the unforeseen approvals and contractual agreements that had to be negotiated.

Conveyor to fuel storage bin

Conveyor from storage bin

Surge bin

Fuel receiving (with screening)

Fuel stockyard

Bypass for fuel circulation

Figure 3.15 Fuelwood-handling system as integrated into the boiler plant
Source: Alstom Ltd

The experience gained from developing a new technology energy project has been shown to benefit future projects and bring down the costs over time. Based on other technologies, for every doubling of total installed capacity the generating costs are reduced by 20%. Hence although the generating costs of Rocky Point at around 3 c/kWh cannot compete with local coal-fired plant at present, in future similar plants could well do so.

Notes

[1] *New and Emerging Renewable Energy Technologies in New Zealand.* Published by EECA (Energy Efficiency and Conservation Authority) and the Centre for Advanced Engineering in 1996.

Chapter 4

The supply chain: harvesting, transport and processing

Growing or procuring the biomass resource is generally well understood (Chapters 2 and 3). Converting the resource into useful forms of bioenergy using a wide variety of processes is also reasonably well advanced, and there are many examples of mature technologies (Chapters 5, 6 and 7). Often the major challenge for a project developer is to deliver the biomass fuel to the conversion plant gate in a form that can be utilized easily, meets a set of fuel quality standards, and is at the lowest cost in terms of $/GJ delivered. This chapter covers the closely interrelated aspects of fuel quality, harvesting, transport, processing and delivery.

Harvesting operations, transportation methods and distances to the conversion plant have a significant effect on the energy ratio balance and cost of the overall biomass system. When the biomass is grown as an energy crop, or is a by-product of a crop grown primarily for other purposes, harvesting it and collecting it from the field are a key operation. Ideally conventional crop-harvesting machines should be used where feasible, as they are normally reliable and well proven. For example, oilseed rape used for biodiesel is harvested using conventional cereal combine harvesters, and vegetative grasses used for combustion can be cut with conventional crop mowers or windrowers, then baled using hay balers. Using such machines can help spread their fixed ownership costs over more hectares and longer harvesting seasons. For example, the owner of a baler could seek contracts for baling silage in the spring, hay in the summer, cereal straw in the autumn and *Miscanthus* in the winter. This would give all-year-round work and spread the fixed costs over a greater number of bales per year, thus minimizing the costs per bale.

Forest harvesting systems

Use of the manual chainsaw remains the most common method of harvesting trees. However, for safety reasons as well as economic ones, there is a trend towards more mechanized systems. Once cut, the material has to be removed from the site, and many designs of extraction machine are available for use on a wide range of terrains and with various sizes and species of tree. What suits one forest may not suit another. Some common designs and methods could be easily adapted to harvest energy forests as well.

Figure 4.1 Feller buncher with accumulating head modified for harvesting of SRF poplars in the USA

Harvesting

Harvesting is by manual chainsaw or by feller buncher (Figure 4.1).

Feller buncher   :[1] a purpose-built forest vehicle with a front-mounted felling and accumulation head. It uses either a chainsaw-type cutting head or a hydraulically actuated guillotine to cut through the stem, and then holds the trees using hydraulic arms until dropping them onto the ground ready for collection.

Extraction

Extraction of the cut material from the forest to a road or to a 'landing' for further processing can be accomplished in numerous ways. Forwarders are capable of moving through a plantation, depending on the tree spacing, and removing selected material. It is also possible to fit a felling mechanism to their grapple arm to allow both felling and extraction of the thinnings to be integrated into a single process. Forwarders are becoming common in forestry operations mainly for collecting residue material and small wood pieces, but they can also be used for extracting timber logs. Logging trailers pulled by agricultural tractors can perform similar tasks to a forwarder but at a lower productivity rate.

Forwar der tracto r: a four-wheel-drive agricultural tractor with linked logging trailer and grapple used to extract shortwood, logs, cut stems and small trees and carry them clear of the ground.

Forwar der : a purpose-built, frame-steered forestry vehicle with integral timber platform and grapple to load the logs or trees (Figure 4.2).

Cable haulers

Cable haulers are used to extract forest material on steeper terrain where wheeled vehicles cannot safely go. Costs of harvesting are higher and, unless the trees are pulled out as whole trees, extraction of residues is not usually economic.

Tractor-mounted cable crane   : uses a tower and tractor power take-off (pto) drum winch, and is capable of extracting loads of stemwood or whole trees on steep country, the biomass being totally or partially clear of the ground.

Figur e 4.2 Forwarder in Sweden extracting forest arisings and small SRF stems to the landing ready for chipping and then loading into trucks

Figure 4.3 Cable hauler tower in *P. radiata* plantation extracting whole trees down to landing for processing (Note bulldozer used for anchor only, as no suitable trees were available on this site for this purpose)

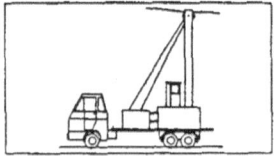

Truck-mounted cable crane   : uses a tower with drum winch, and is also capable of extracting loads totally or partially clear of the ground. It normally has a greater load capacity than the tractor-mounted systems (Figure 4.3).

Skidders

Skidders are simple machines designed to pull harvested stems or whole trees to the landing for further processing or loading onto transport vehicles. They are usually capable of extracting whole trees by lifting only one end of the load clear of the ground during extraction. As a result, soil and stone contamination is often a problem.

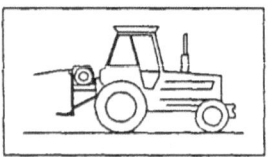

Agricultural tractor winch skidder   : a four-wheel-drive forest tractor fitted with a rear-mounted winch, which is powered by the tractor pto.

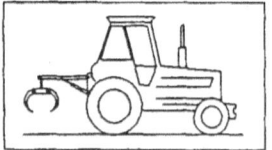

Agricultural tractor grapple skidder   : also a four-wheel-drive forest tractor but fitted with a rear-mounted grapple used for skidding out the trees.

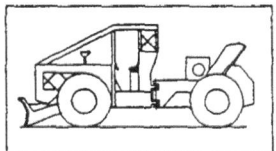

Articulated winch skidder   : a purpose-built, four-wheel-drive, frame-steered forestry vehicle, with a drum winch as an integral part of the design.

Clam bunk skidder   : a purpose-built, frame-steered forestry vehicle with a platform-mounted hydraulic clam and integral grapple.

Processors

Processors are used, once the whole trees have been delivered to the landing, to strip the limbs from the stemwood, cut off the tops, and cross-cut or 'section' the stemwood by cutting it to the desired lengths as specified. The logs are then ready for transporting to the sawmill or timber-processing plant. The residues remaining at the landing are then returned to the forest by bulldozer, dumped, burnt or used as a biomass resource. In this last event they are normally chipped after leaving them for a period on the landing to dry, and then transported to the heat or power plant.

Bed pr ocessor : a two-grip unit mounted on a forwarder chassis and capable of delimbing and cross-cutting whole trees to specific lengths.

Grapple pr ocessor : a single-grip unit mounted on a forwarder chassis or on a semi-portable fixed platform.

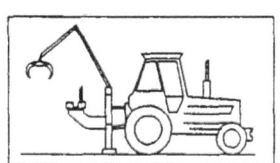

Tractor-mounted pr ocessor : a two-grip bed-processing unit mounted on an agricultural tractor.

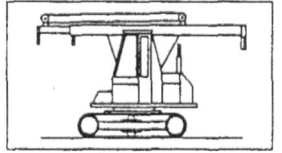

Sliding boom pr ocessor : a purpose-built processor capable of delimbing and cross-cutting but not always to specific lengths.

Short rotation forest harvesting

Coppice harvester : a purpose-built unit powered and drawn by a four-wheel-drive agricultural tractor, or self-propelled, and capable of felling, bunching and processing single stems or coppice regrowth.

One of the key aspects of a successful biomass production system is matching the right harvesting machinery and method to the type of plantation. For SRF regimes the method of harvesting can have a large effect on the life of the plantation, the total biomass produced, and the overall feasibility of growing SRF.

There are few purpose-built harvesting machines commercially available for use in SRF plantations but, because of increasing interest in SRF crops as a source of fuelwood and fibre, new machines are being developed. Machinery adapted from harvesting agricultural crops is being used to harvest predominantly small-stemmed crops such as *Salix* (willow) grown in Europe, Scandinavia, the UK and northern America. Other machines originally designed for forestry operations are being used to harvest larger-diameter SRF trees mainly for fibre production in places such as the southern USA, Central America and South Africa.

Specialized equipment prototypes have been built for felling and bunching small-diameter SRF trees, including the Canadian FB7 and the Irish Loughry coppice harvesters. Direct harvest/chip machines such as the Claas forage harvester with specialist cutting head (Figure 4.4) are being commercially operated in Europe for the production of willow fuel chips, but are yet to be evaluated with other species and under different conditions.

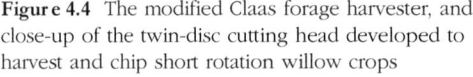
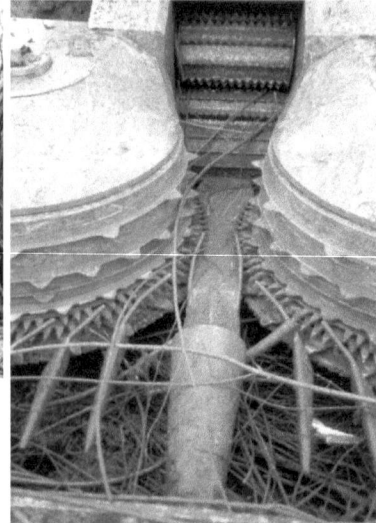

Figure 4.4 The modified Claas forage harvester, and close-up of the twin-disc cutting head developed to harvest and chip short rotation willow crops

The decision as to which harvesting method to use will often be a compromise between maximizing sustainable yields and minimizing costs. Points to consider during harvesting are the immediate and long-term effects from stump damage, soil and root compaction, and the damage caused to trees left unharvested. Consideration must also be given to the form in which material is required for further processing.

short rotation forests tend to have many smaller trees, up to 5000 per hectare, so that harvesting them individually is very time consuming. Traditional forest-thinning machines can be used, and also feller bunchers, but the work rate (or productivity) in terms of tonnes per hour or hectares per hour is generally slow and hence expensive in terms of $/GJ harvested. Harvesting multi-stemmed coppice regrowth as opposed to single-stem trees can be even more problematic.

Harvesting is a considerable cost in the overall growing of SRF: typically production accounts for 25% of the total costs, transport 25% and harvesting 50%. Hence harvesting has to be carried out efficiently and with the right equipment to minimize costs. High capital costs are associated with specialized machines giving high productivities and efficiencies, so they must be kept operational as long as possible, which is often impractical. In some circumstances less specialized equipment such as a chainsaw can therefore be cost competitive, even though it is more labour intensive. Productivity and efficiency may be compromised, but in developing countries, where labour is relatively cheap, this is a practical option.

The method of harvesting and the final market for the biomass material influence the extraction method. It is undesirable to have fuelwood material that is contaminated with soil because of difficulties encountered during and after combustion. Smaller whole trees offer advantages during extraction by forwarders by minimizing the risk of soil contact and contamination because they can be lifted from their felled site, rather than dragged.

Chipping and chunking

Comminution is usually carried out by chipping or chunking (when sharp cutting edges are used to cleave or shear the biomass into engineered particles), or by employing a blunt impacting tool to crush or shred the material, producing particles of indistinct geometry and often termed hogging or shredding:

- cutting: chipping

  chunking

- impact: shredding/hogging

  hammermilling (Figure 4.5)

Various forms of comminution system are commercially available for use by the forest industry and can easily be adapted for biomass use.

Figure 4.5 High-speed swinging hammermill blades on rotor (with top cover removed) used to produce minimum particle sizes, which can then pass through the holes of the surrounding screen to the discharge spout

Trailer mounted chipper   with auxiliary power source rather then being connected to a tractor pto.

Tractor-mounted chipper   powered from the tractor pto.

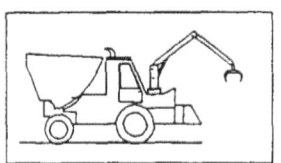

Self-propelled chipper   is a purpose-built forest vehicle with front-mounted grapple to feed the chipper unit mounted on the front or the side of the vehicle, and with a rear-mounted chip bin.

Forwarder-mounted chipper   is a large, self-propelled comminution unit with integral power source and chip bin, mounted on a forwarder chassis.

Heavy-duty trailer-mounted chipper   with integral power source mounted on a heavy-duty chassis and trailed by truck or forest vehicle.

Truck-mounted chipper   with auxiliary power source and grapple, mounted on a truck chassis. Chunkers can be obtained in similar format to the chippers above but there are fewer commercially available machines manufactured, the following being the most common.

Truck-mounted chunker   with integral power source and grapple, mounted on a truck chassis, and used in conjunction with an elevator for loading the chunks that are too heavy to blow.

The strength of wood affects the power required to reduce it from large solid wood pieces to chips or chunks. A wide range of comminution machines exist, such as chippers, hammermills and shredders. Each type consumes different amounts of energy per tonne of comminuted biomass processed. Those with sharpened edges to the cutting blades will need regular maintenance to minimize energy inputs.

Energy inputs of chippers tend to increase with shorter chip length and lower moisture content, whereas the reverse is the case with hammermills. Hardwoods, with shorter fibres, tend to require more energy than softwoods to produce the same size of chip. There are considerable increases in energy costs when comminution is required to create small particles with dimensions less than 70–150 mm. Greater size reduction requires higher energy inputs because the number of cuts into a single piece of wood increases significantly. Over a range of machines, moisture contents and materials, the energy input needed to comminute one oven-dry tonne (odt) of roundwood logs to 25 mm nominal chips varies between 5 and 250 MJ. Since 1 odt contains approximately 20 GJ of available energy, comminution consumes only a very small proportion of the total. Minimizing the energy input is an important factor, but it has to be balanced against the time taken. The chipper machine productivity in terms of tonnes processed per hour is an important selection criterion, as are the maintenance costs and the capital costs.

Chipping equipment ranges from large stationary machines developed for the pulp industry down to small tractor-operated designs suitable for on-farm use for woodlots (Figure 4.6). Selection of comminution equipment should fit into the overall handling and delivery system, and relates very much to end product specifications. The end use of the product and the energy inputs

Figure 4.6 Tractor-mounted chipper for small-scale applications at 5 odt/h and grapple-fed or manually fed. Note that the energy inputs into the total system (shown here as diesel exhaust emissions) should always exceed the energy outputs (shown here as wood chips) to give a positive overall energy balance

involved should be considered when selecting the comminution method or technique most appropriate for an individual situation.

The comminution method and moisture content together can affect the drying and storage characteristics of the material, which inevitably affects the fuel quality, though there are also other factors used to evaluate the quality of fuel. There is a need for a quantitative standard for classifying fuel quality with respect to its handling and burning properties. However, a classification system would be difficult to establish because of the variety of fuel particle types required by different energy conversion systems. The ability of a furnace to burn various types of biomass depends on its design. Some combustion units are exclusively wood chip burners, whereas others are suitable for burning combinations of wood chips, bark, sander dust, straw, coal and so on. Bark and foliage along with other irregularly shaped material such as over-sized chips and twigs can create problems for handling equipment, and possibly cause blocking of conveyers and storage silos.

Apart from the material with the required particle size range there will also be some smaller and larger material. Comminuted woody material can therefore broadly be divided into three major size categories:

- **Acceptable material** is suitable for the final end use, its dimensions being within an acceptable range. Requirements for chips for pulp may be different from those for combustion, and will vary according to the design of conversion unit being used.

- **Fines** comprise small components including bits of bark, foliage and inorganic impurities. Chips produced from SRF whole trees, forest arisings and stumps can have a high portion of fines, and their effect on the end use of the material is variable. For fuel use, chip bulk density is the most critical characteristic. Foliage and soil contents can have an effect, whereas bark contamination is not normally a problem, though it is for pulp chips. By reducing the portion of fines more air can circulate through a pile of chips, aiding moisture loss, minimizing pile temperature and micro-organism increases, and consequently minimizing biomass losses.

- **Oversized material** is typically too large for its desired end use and is excluded by a screening process on size (length or diameter) and sometimes weight. A high proportion of oversize chips can be produced when small-diameter, often dry and stringy material, such as branches and forest arisings, is comminuted. This can create problems for conveying equipment and cause bridging in silos and hoppers. Oversize material generally presents a greater problem for smaller-scale installations where material flows are not large and openings are narrower since the overall equipment size is smaller.

There are times when it is beneficial to compact small biomass particles such as sawdust into pellets or larger briquettes for ease of storage and handling. This is the opposite process to that of comminution. A system that involves comminution followed by briquetting/pelleting is basically breaking up solid biomass into small pieces in order to put it all back together again! This is costly in terms of both equipment and energy inputs, and should be avoided unless there are significant benefits to be obtained.

System options

Numerous systems are feasible for harvesting, collecting and processing biomass with the intent to use all or part of the material for energy purposes. Several sample systems are shown in Figures 4.7–4.10, based on forest arisings

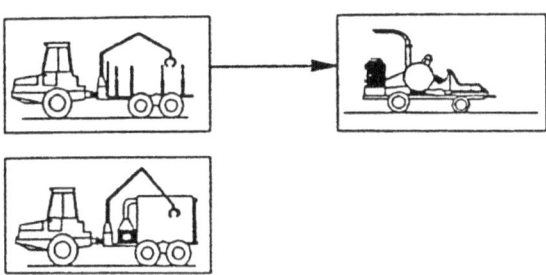

Figure 4.7 Two systems for harvesting arisings from the forest cutover using either a forwarder to extract the arisings to a heavy-duty trailer chipper or a forwarder/chipper

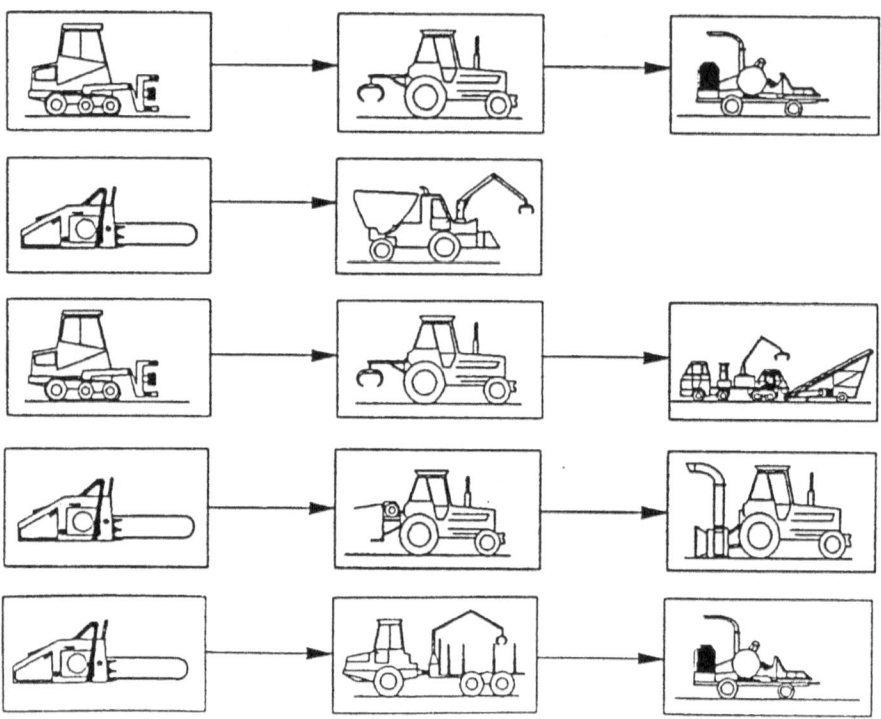

Figure 4.8 Five whole-tree harvest and extraction options with comminution either at the landing or in the forest using a chipper forwarder, mobile chunker or mobile chipper

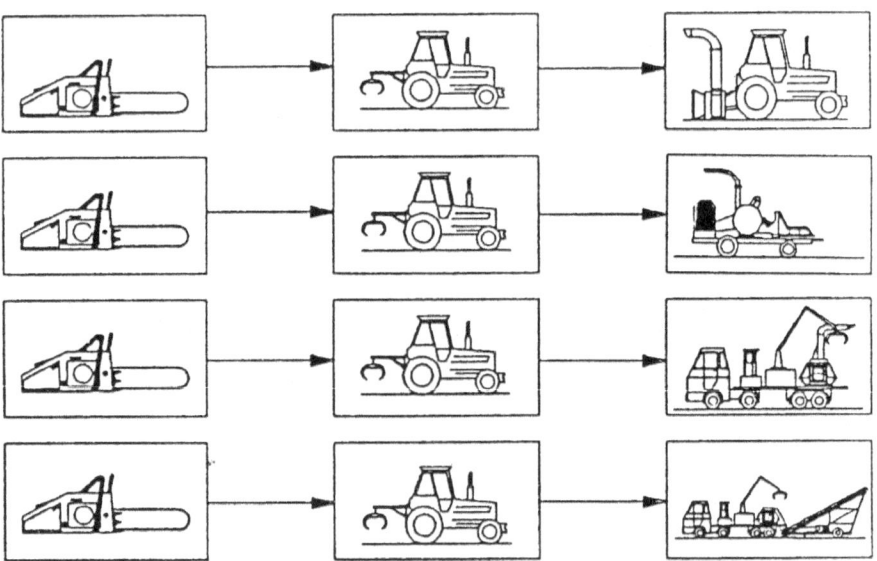

Figure 4.9 Four whole-tree harvest, extraction and comminution options based on chainsaws and small tractor skidders so perhaps more suited for use with small-scale thinning operations or selective harvesting methods

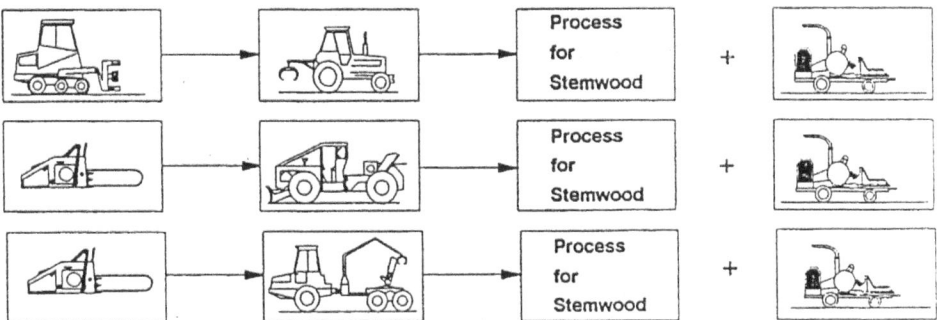

Figure 4.10 Three integrated harvesting options of stemwood for timber along with the residues being collected and chipped for use as biomass

as an example and using the same symbols as above to depict the machines involved. To be most effective in terms of fuelwood delivered to the power plant at the lowest cost, the productivities of the various·machines (in terms of t/h) need to match each other to avoid expensive delays and downtime.

Drying the biomass fuel

There is an interaction between moisture content, transport, storage and conversion of biomass, as referred to in Chapter 1. When using dry forms of biomass such as wood, the heat required to raise the temperature and evaporate off the moisture has to be generated by the wood itself. Since a typical heat plant (furnace or boiler) is designed to maintain sufficiently high exhaust gas temperatures to avoid condensation in the stack, this heat is usually not recoverable (though there are exceptions, as when using steam condensing turbines). Hence the thermal efficiency of the overall system will be reduced when using fuels with higher moisture contents. The heat lost in the exhaust gas is also directly attributable to the moisture content of the fuel and also directly affects the efficiency of the system. The loss will vary from less than 2% of the total heat input when the fuel is at around 8% m.c. to nearly 15% when the fuel is at 50% m.c. So the drier the biomass fuel the better. In addition, losses in thermal efficiency occur because unburnt fuel is carried into the ash (possibly 0.5% loss) and surface heat is lost from the actual plant at around 3% in modern plants (but this varies with the plant design, and the surface area and temperature of the external radiating surfaces).

In addition to the fuel moisture content, biomass contains a proportion of hydrogen atoms, which react to form water during the combustion process. The heat required to raise the temperature and evaporate this extra water also becomes unavailable to the system and has a similar effect on the overall heat recovery efficiency. The effect on the combustion system efficiency of both the free moisture and that formed by combustion of the hydrogen can be calculated (Table 4.1).

Table 4.1 Typical heat losses during combustion due to the moisture content of the biomass fuel and combustion of the hydrogen contained in the fuel

Moisture content of the fuel (% wet basis)	Heat loss as a percentage of higher heat value			Total losses (%)
	Moisture content (%)	Flue gas, ash and radiation losses (%)	Combustion of hydrogen (%)	
0	0.0	11.5	8.1	19.6
10	1.3	11.5	8.1	20.9
25	5.2	11.5	8.1	24.8
40	10.6	11.5	8.1	30.2
50	14.9	11.5	8.1	34.5

The environmental ambient conditions (mainly temperature, air movement and humidity), together with the structure of the tissues of the specific biomass, determine the release of moisture from the biomass by transpiration and evaporation and the rate of release.

Transpirational drying   is the loss of water via the foliage of the plant. It occurs continually while the plant is growing. For example, 4–5 m tall, 3-year-old *Eucalyptus saligna* trees grown as a short rotation forest crop can each transpire over 30 litres of water during a sunny summer day and down to perhaps 2–3 l/day when the weather is overcast and cooler. Transpiration can continue for some time after harvesting if the cut trees are left entire, a process that can be utilized to lower the biomass moisture content with minimum cost inputs.

Evaporative drying   is the loss of water from the cavities of plant cells, which occurs mainly as a result of evaporation of the moisture present. Energy, normally from the sun, is required to evaporate the water during the drying process. Evaporational drying can be achieved by the natural ventilation of air passing through the biomass when stored outside in piles. A large open or covered storage area for the woody biomass is therefore necessary as drying usually take several weeks or even months unless forced air and heating are used. The **equilibrium moistur   e content**   is the point at which the moisture content of the biomass is in balance with the relative humidity of the surrounding air, so it will normally vary by up to 10–15% m.c. day by day. When only the moisture in the cell walls remains, this is known as the fibre saturation point, and is normally between 20 and 26% m.c.w.b. The rate at which drying occurs   to **fibr e saturation point**   varies with plant species, ambient temperature and humidity and, for woody biomass, whether or not there is bark cover left on the logs and branches. Below fibre saturation point can only be reached if additional energy is added artificially to release the water molecules that are held hygroscopically in the cell walls. This energy input amounts to 2.43 MJ per kg of water evaporated.

The smaller the piece size the greater the ratio of surface area to volume, which favours greater initial moisture loss. In woody biomass the cells are longer 'with the grain', so cutting across the grain gives faster moisture loss. In addition the geometry of the pieces (that is, the chip shape) affects the way in which the pieces arrange themselves when placed into piles. This determines the ventilation circulation and hence the rate of moisture loss from the pile. Shrinkage can occur at lower moisture contents in wood when cut mainly tangentially to the growth rings rather than along the grain. Subsequent moisture uptake can result in re-swelling.

If the dry woody biomass is left outside, the rain will wet the outside of the material. If stored in large piece sizes as logs or chunks with good air movement between, this will soon dry again by evaporation. Conversely, a smaller piece size (for example when stored as sawdust, shredded bagasse or rice husks) will result in an increased exposure of cell cavities.

The overall system of delivery to the conversion plant

Wet biomass such as animal slurries and biofuels for transport fuels can be stored in tanks and pumped from place to place through pipelines as for traditional oil products. Road, rail and sea transport is also possible using tankers mounted on truck decks, on trailers pulled by tractors or in oil tanker type vessels. Some slurries with higher solid contents of up to around 10–15% may cause problems when pumped through pipelines as they can cause blockages if the system is poorly designed. Also storage of organic biofuels such as biodiesel over periods of several months may lead to biological degradation and some energy content deterioration. But overall the handling and storage of liquid biomass materials is relatively straightforward.

Most forms of dry biomass can be handled in a number of ways. For example, cereal straw can be handled loose, chopped, baled or briquetted. The chosen method depends very much on the material, its moisture content, the transport distances involved, the storage method, the storage period and the scale and type of conversion plant. Selecting the components of the overall system to minimize the total delivered fuel costs is a difficult process. Poor selection can lead to expensive fuel.

Case study 5

Harvesting and processing systems for short rotation willow coppice in the UK

Short rotation coppice (SRC) willow (*Salix*) has been identified as a potential sustainable energy source, but there is no proven, cost-effective system in place for delivering the fuel in the most acceptable form to either small- or large-scale users. A study was therefore undertaken to develop an optimum system design configuration:

- for a large-scale operation to supply a 10 MW$_e$ power station

- for an on-farm scale of operation to supply a 400 kW heating plant.

In this case study willow was the crop evaluated, but similar principles apply for a range of biomass crops. It is fairly well known how to grow trees and burn the wood. Harvesting the trees and processing the fuel is the difficult part. It was therefore intended that this study, undertaken in 1997, should serve to assist members of the British woody biomass industry by endeavouring to obtain consensus on the best way forward for SRC, based on the existing level of knowledge. This in turn would enable investors and other organizations funding harvesting system development to direct the very limited funds available towards providing the best solutions.

To lead the way forward, an efficient and reliable coppice harvesting and handling system needed to be identified. A number of prototype machines had been evaluated under UK conditions over recent years. In addition a series of scientific studies had been conducted on SRC yield production, drying, storage and transport options. The aim of this study was to bring together all the relevant information, synthesize it, and then present a series of concept options to the industry to determine the optimum harvesting and processing system.

Harvesting SRC crops for energy purposes cannot be considered in isolation from the subsequent activities necessary to get the material delivered in the most suitable form at the conversion plant and at minimum cost. In addition the establishment, layout and agronomic management practices of the crop should be determined in part by the harvesting method anticipated. Willow is deciduous and normally harvested during winter after leaf fall. In addition, under British conditions, poor weather at this time can restrict machine access for harvesting operations, which therefore were assumed to operate for only approximately 50 days each year. The assumption was made that the fuel would be required over a 12-month period for electricity production. The Swedish bioenergy system, by comparison, uses SRC willow to supply district heating plants for only a few months of the year, with a bottoming cycle steam turbine (Chapter 5) used to generate some power when excess heat is available. This avoids the major problem of biomass storage since the winter harvest period for willow closely matches the period of peak heat demand.

In order to produce a high-quality biomass fuel on a sustainable basis and to minimize the costs in terms of £/GJ of fuel delivered,[2] the whole process therefore needs to be viewed as a system. The harvested crop can be quantified in terms of GJ/ha.yr, but losses during the harvesting, storage and transport operations will reduce the energy available for conversion to heat and power, and so they should be minimized.

Concepts and scale of operation

A number of broad harvesting categories for the SRC willow, based partly on the scale of operation, were identified:

- Large-scale cut and **chip** , self-propelled machine (Figure 4.4). Chips stored on farm or at conversion plant.

- Large scale cut and **billet**,[3] self-propelled machine. Billets transferred to simple on-farm storage initially (Figure 4.11), then later comminuted to the final form required at the conversion plant.

Figure 4.11 Billets of willow stockpiled on farm and covered ready for later comminution or transport to power plant (Note camera case to show scale)

- Large-scale **stick** harvesters, self-propelled machine (Figure 4.12). Where no indoor storage facilities exist for large volumes of material, as is the case on many farms, sticks can be stored outside in piles, in tied bundles or as

Figure 4.12 Segerslatt Empire 2000 self-propelled Swedish willow stick harvester on large tyres to avoid soil compaction during harvest
Source: Dr Jorg Gigler

Figure 4.13 Bales of willow sticks formed by the Swedish Balapress designed for small forest arisings but as might also be expected from using round hay balers

compressed large square or round **bales** (Figure 4.13). Comminution to produce the biomass in a form suited to feeding into the conversion plant could then be undertaken later, either on the farm if for local use or to maximize transport payloads, or at the power station.

- Small-scale, trailed, stick harvester for grower/contractor use on smaller areas using conventional agricultural tractors as power units (Figure 4.14).

Depending on the scale, comminution can be accomplished

- on the farm if the fuelwood is destined for local use

- on the farm in order to maximize transport payloads

- at the power station if it is cheaper overall to do so – this would also provide a consistent fuel quality to suit the conversion plant design in terms of piece size and moisture content.

This case study was based on a commercial power plant operator buying in the SRC willow fuel from local growers under contract using the following assumptions:

- combined-cycle gas turbine plant, capacity 10 MW$_e$ gross

- fuel demand: 5 odt/h at 8000 h/yr (assuming over 90% availability) = 40,000 odt/yr

Figure 4.14 Fröblesta tractor-powered stick harvester suitable for small-scale harvesting

- fuel storage: need 34,000 odt stored on a hard standing area of >100,000 m², the other 6000 odt being fed into the plant directly on delivery during the harvest period of 50 days

    Storage area **filled** over 90 days at 50 odt/h, and trucks operating 8 h/day

    Storage area **emptied** over 275 days at 5 odt/h for 24 h/day

- Drying: fuel moisture content of <20% (w.b.) required by the gasifier.

The final desired form of woody biomass feedstock material is determined by the fuel specifications of the conversion plant in terms of acceptable moisture content and particle size ranges. Delivery of fuelwood to the plant site can be in a wide range of forms since the final processing operation can be conducted on-site immediately prior to feeding the feedstock into the conversion plant. A wider range of fuel moisture contents may also be acceptable if the desired level can be reached by blending wet biomass with dry, or if the plant has some form of artificial drying facility. However, a cost penalty to the grower for supplying wet fuelwood is likely.

Comparison of alternative SRC biomass fuel supply systems

The standing crop to be harvested and processed was assumed to be 2 year old willow coppice yielding 40 wet t/ha in a 12 ha field with an average distance of 1.5 km to the roadside. Several harvesting and supply chain systems were modelled from the standing crop through to delivery at the power station gate an average distance of 39.5 km away from the farm road gate. Additional costs for loading and chipping or shredding at the power station were included as a separate cost. In the analysis, the costs presented are for oven-dry tonnes (odt) delivered to the power station. In the base case, the amount the growers were paid for the standing crop was taken as £20/odt.

The different systems result in the coppice being delivered as bales, billets or chips, and as green or partly air-dried material. Allowances for dry matter losses in store were made at 4% per month for chips, 2% per month for bales and sticks and 1.5% per month for billets.

The eight scenarios were chosen to represent practical options. The cut and chip harvesters and the Empire 2000 stick harvester had been evaluated in British field trials, and the performance data are based on these results. A one-pass cut-and-bale machine for felling and baling coppice material in the field does not yet exist so, in order to model this option, three scenarios were presented (A for round bales; B and C for large square bales) based on experience from baling straw. They illustrate what might be achievable if such a machine were to be built. The concept would be a cutting head attached to the front of a commercial baler such as the Claas Rollant or the Hesston square baler, with the cut sticks being possibly crimped or billeted before being fed into the bale chamber directly. Dropping the cut sticks onto the ground for subsequent collection, possibly after a period to allow some drying to occur, would have some advantages but was not considered as it would be difficult to achieve without high field losses and damage to the stools by the baler pick-up.

The assumptions for capital cost, work rate, etc., for each machine used in the eight scenarios in the model are listed in Table 4.2.

In all system scenarios (except B and E, which involved transporting the biomass 41 km directly to the power plant), the transport distance assumptions used were:

- travel in field: 5 km (at 8 km/h average)

- travel on farm track to farm store: 0.5 km (at 10 km/h average)

- travel from farm store on farm track: 0.5 km (at 10 km/h average)

- travel on unclassified roads: 9.5 km (at 40 km/h average)

- travel on single-carriageway class A/B roads: 30 km (at 55 km/h average)

- total travel distance one-way: 41 km.

System descriptions of the eight scenarios compared in the model

A: Large round bales to farm store

Since no actual machine has been tested on SRC, assumptions were based on the Claas Rollant baler with its option for using net wrapping instead of twine. The scenario was based on the known performance of the baler when used to

Table 4.2 Key assumptions of machine costs and performance rates used in the comparative harvesting and transport systems

System	A Mower/ round baler (farm store)	B Mower/ square baler (direct deliver)	C Mower/ square baler (farm store)	D Claas harvester (chip and farm store)	E Claas harvester (direct and store at plant)	F Empire 2000 stick (farm store and then chip)	G Austoft harvester (chip and farm store)	H Austoft harvester (billet and farm store)
Machine cost (£k)	70?	200?	200?	182	182	91	190	190
Hourly harvester cost (£/h)	36.51	78.38	78.38	65.12	65.12	44.75	82.81	80.00
Work rate (odt/h)	10	16	16	10	10	7	10	10
Road transport payloads (odt)	10.2	12.0	12.0	9.0	7.7	4.3 sticks 8.6 chips	9.0	9.6
Total loss during storage (%odt)	12	12	12	24	24	9	24	9

bale straw. Some account was also taken of British trials undertaken on the Swedish Balapress using SRC willow and forest arisings. The steps involved in the supply chain were assumed to be similar to those outlined for well-established straw-baling and handling systems. Mowing/baling was assumed to cost £36/ha with the mower attached to the baler and a simple crimping or billeting system installed before the baling section. From the 12 ha field used in the analysis, 960 bales of 500 kg each could be produced. The bales are dropped in the field for later collection by loader onto a flatbed trailer for delivery to the headland or farm store. A 6 month average storage period was assumed before the bales, now at 30% m.c.w.b., are collected and transported by HGV (heavy goods vehicle), a 35 t low-bodied flatbed trailer with an 8.4 m long deck and able to carry 40 bales at three high giving a 14.6 t payload. (Alternative options, not examined, could be for the baler to carry the bales directly to the headland for stacking and intermediate storage, or for a loader to transfer the bales one bale at a time from the field and store them along the headland for later collection, assuming the HGV could gain access to this point.) The HGV is unloaded by a front loader at the power plant, and the bales are then fed into the boiler directly using an automatic conveyor. In total 24 round journeys are required.

B: Large square bales direct to power plant

This scenario is based on a Hesston high-density square baler, which has become the industry standard for handling straw in Danish power plants and also for industrial straw energy applications in the UK. It was assumed that these bales were bound with twine and that 600 bales were produced from the 12 ha field, each weighing 800 kg. The steps involved in the supply chain were those broadly used for conventional straw baling systems, except that it was assumed that a self-loading bale carrier would not work satisfactorily amongst coppice stools. So the assumed modified system was:

- mowing, by using a Claas header or similar attached to the baler

- baling, directly after mowing without dropping the cut material onto the ground

- Fastrac tractor and loader to take single bales 0.5 km across field to load 35 t gross articulated low-bodied flatbed trailer HGV with 30 square bales per load, giving a payload of 24 t with the material at 50% m.c.w.b.

- HGV travels 1 km on farm tracks at 10 km/h average, then continues to the power plant on the road types as outlined above

- unloading of bales at the power plant by front loader. In total 20 return trips will be required.

C: Large square bales to farm store

As for system B above but with a front loader and tractor/trailer to transport the bales 1 km on average from the field and unload at the intermediate farm store. After an average storage period of 6 months, the bales, now at 30% m.c., were loaded by the front-end loader onto a 35 t gross HGV and transported 40 km to the power plant, where they were unloaded by a front loader. Each load carries 30 bales, giving a payload of 17.1 t, and 20 return trips are required.

D: Claas forage harvester (cut and chip)

The short rotation coppice material is cut and then chipped directly involving the following steps:

- direct cut and chip at a harvesting rate of 10 odt/h (0.5 ha/h) then blow into one of two 15 m³ trailers pulled alongside by two 85 kW tractors

- in-field trailers transported 1 km then tipped at intermediate farm store

- chips pushed into a heap on hard standing by loader and stored for 6 months on average

- tractor front loader used to fill 38 t articulated bulk tipping HGV with 60m³ volume to give a 15 t payload of air-dry chips around 40% m.c.w.b assuming a bulk density of 0.25 t/m³ fresh weight

- HGV takes chipped biomass to power plant, where it is tipped. A total of 27 round trips are required.

E: Claas forage harvester with no storage

This uses the same harvesting machine as system D, but the material goes straight from the field to the power plant using three 45 m³ trailers to collect the material direct from the harvester in the field. This illustrates a harvesting operation suitable for dry soils but only if within a reasonable transport distance (40 km) of the plant. Wet chips at around 50% m.c.w.b. are delivered.

The steps involved are :

- direct cut and chip then blow into trailers alongside

- in-field transport and road transport in three trailers with payload 13.5 t pulled by three Fastrac tractors, then tipped at the plant. In total 36 round trips are required.

The 480 t of wet chips delivered would require a storage area of 900 m², for which land rent is included in the analysis. The stockpile was assumed to be 3 m high.

F: Empire 2000 stick harvester

The costs per hour and work rates for this system were taken from British Forestry Commission calculations. The steps are:

- sticks cut and carted to the headland by the harvester at 7 odt/h (0.35 ha/h)

- loader loads trailer at 8.7 t payload, then transported by tractor to intermediate farm storage

- stacked by loader then on-farm storage for 6 months on average, reaching 30% m.c.w.b

- delivered as and when required at power plant, chipped direct into 38 t HGV tipping trailer, giving payload of 12.6 t

- driver and articulated truck arrive to collect the trailer when full for delivery to power plant. 26 return journeys required.

G: Austoft harvester (cut and chip)

As for the Claas chipper system D above but using the Austoft modified sugar cane harvester at the assumed higher cost of £82.81/h and a productivity rate of 10 odt/h based on Forestry Commission trial data.

H: Austoft harvester (cut and billet)

This is the same system as for systems D and G above but with the Austoft harvester modified to produce billets rather than chips. The hourly cost is slightly lower, as less power is required for billeting than chipping and less knife maintenance is required. Billets result in lower storage costs, owing to reduced losses, but the other input data remain similar (Table 4.2). The work rate and machine costs were assumed to be the same for the Austoft harvester in both cut-and-chip and cut-and-billet modes of operation, but the product bulk density was slightly lower for the chips.

As noted above, the dry matter losses during storage vary with time. For a given period (e.g. 6 months) the losses would be greater for chips, than for bales, than for billets or sticks. Where the biomass is taken directly to the power plant for immediate use the dry matter losses are negligible but would be similar to losses in farm storage if stored there for any time.

A summary of the analysis (Figure 4.15) shows that systems E and B had the lowest delivered fuel cost. This is not surprising as there was no intermediate storage for these options, the fuelwood being delivered directly to the plant for immediate use. This option is appropriate in some circumstances, but it is more likely that some storage will be needed to provide the plant with fuel supplies all year round.

Where storage of fuelwood is necessary to provide an all-year-round energy supply to a plant, the costs will be increased significantly owing to the additional handling operations and storage losses that result. This confirms the benefits of having a mix of fuels from a range of sources so that the amount of fuel going straight from source to plant can be maximized. For a wood-fired plant such sources might include wood process residues, forest arisings and SRC. In some circumstances it may be appropriate to mix fossil fuels with the biomass to ensure a reliable, low-cost supply. An example of such a flexible

Figure 4.15 A comparison of delivered costs per tonne of oven-dry fuelwood to the power plant gate (£/odt) for the eight alternative system options evaluated

fuel plant is the 39 MW$_e$ cogeneration plant commissioned in 1998 by Carter Holt Harvey at their Kinleith pulp and paper plant in New Zealand using both natural gas and wood and bark residues (see Case Study 7). This reduces the risk associated with obtaining secure supplies of fuelwood for the long term, and gives the plant operators greater confidence in the fuel supply chain.

This study has shown that, for SRC systems where fuel storage is required, innovative cut-and-bale or billet systems could be competitive with chip systems and are worthy of further investigation. The big advantage of billets and bales is that the fuel delivered to the plant in these forms has a lower moisture content than do chips from the same source, and hence suffers less degradation in store over any given period. However, chipping and screening at the plant to produce a consistent quality fuel will be necessary. The disadvantage of wood chips is in maximizing truck payloads owing to their low bulk density, especially when the chipped material is dry.

Recommendations resulting from the study analysis

The following recommendations were based on using state-of-the-art technology, and assumptions that a fuelwood supply is required all year round for power generation but that harvesting SRC willow is feasible in winter only.

Recommendation 1

Bale the cut SRC sticks or chunks into large round or square bales using existing agricultural baling machines mounted on a tractor/mower to give a one-pass operation with either:

- bales dropped on to the ground for subsequent collection by tractor/front loader and then transported to trailer or truck, or

- bales carried to headland on baler for later collection by truck.

It can be assumed that large-scale wood-fired power stations will probably use a mix of existing forest arisings and wood process residues as their main fuels, together with fuelwood from SRC. Therefore delivering the range of biomass material in a single standard form is essential for cheaper and easier handling at the plant. If one truck delivers biomass in the form of bales and the next as chips, it will be more difficult to handle and process than if all trucks deliver bales or all deliver chips. It must be borne in mind that for larger plants there may be one truck arriving every 5 to 10 minutes, so there is little time to handle the material: hence it is essential to have a standard form at delivery.

Bales could be an ideal method of densifying the material to provide transport economies in terms of maximizing truck payloads. The costs of baling are, however, likely to remain uneconomic unless higher throughputs can be achieved than at present and the balers can be used for other crops or purposes during the year in order to spread the fixed costs of ownership.

Baling of SRC would also be suitable for the small grower supplying the power plant if a contractor in the locality could be hired or if the baler could be supplied by the owners of the plant for hire by the growers. This bale system could possibly be used by a small farmer growing SRC to supply heat demand on-farm or nearby if the bales could possibly be fed into a whole bale burner as developed for straw bales in Denmark.

Recommendation 2

Develop a billet system around the Austoft sugar cane harvester and conveyor with support tractor/trailer running alongside, or a mounted bin to carry billets to the end of the row for unloading hydraulically.

The advantages of this system are:

- It uses existing equipment, the base machine being available for chipping as well as for billeting (with some modifications).

- Harvesting costs are around £10/odt, which is cheaper than using whole-stick harvesters.

- The work rate of around 0.25 ha/h will require sixteen 10 hour harvest days to complete 15 ha of 10 odt/ha crops, so the harvester could be shared with four or five other growers on a syndicate basis if fifty harvesting days were available during the season.

- For a larger machine with work rate of 0.4 ha/h and assuming fifty 20 hour working days, the large-scale plant would need three similar harvesters to harvest 1000 ha/yr.

- Harvesting costs for billets delivered to the power plant would probably be similar to those for chips at approximately £70/odt, but storage of billets rather than chips over long periods provides an overall advantage as a result of lower dry matter losses while in store, lower energy inputs and cheaper drying.

- Acceptable soil compaction can be achieved even on wet sites thanks to the use of half-tracks.

Case study 6

Collection and transport of *P. radiata* forest arisings in New Zealand

A New Zealand study was conducted by Massey University and Forest Research to provide indicative costs, and the relativity of the various cost components, when delivering biomass from forest arisings to an energy plant with an average cartage distance of 80 km. A number of supply scenarios were considered, and a computer transport model was used to compare the delivered costs in NZ $/GJ.[4] It was assumed that the arisings were sourced from forests within the Nelson region with an average transport distance of 80 km to the energy conversion plant located on the outskirts of the small city of Nelson. The price paid to the forest owner for the residues was $20/t.

Systems of transport and handling compared

- A: Landing residues: load arisings onto an on-highway truck; transport to energy plant; unload and chip.

- B: Cutover residues: forward to landing; load onto an on-highway truck; transport to energy plant; unload and chip.

- C: Landing residues: load onto off-highway truck; transport to central processing yard in forest (5 km or less); unload; chip with mobile chipper direct into on-highway tipping truck; transport to energy plant; unload chips.

- D: Cutover residues: chipper forwarder to collect and transport chips to intermediate stockpile (assume 10% dry matter loss); front-end loader to load on-highway truck; transport to energy plant; unload chips.

- E: Cutover residues: chipper forwarder to collect and transport chips to landing; transfer into bins; collect bins with automatic loading hook truck; transport to energy plant; unload bins and tip.

- F: Cutover residues: forwarder to roadside; stockpile; chip using mobile chipper; stockpile chips (10% dry matter loss); front-end loader to load on-highway truck; transport to energy plant; unload chips.

- G: Cutover residues: forwarder to roadside; stockpile; chip using mobile chipper into on-highway truck; transport to energy plant; unload.

The model used was similar to that described in Case study 5, allowing for travel distances to be included over a series of road types (including tracks, B roads and motorways) as occurs in practice. The main assumptions used to provide detailed data for the transport model in the New Zealand study were as follows.

The harvest and process machines used in one or more of these systems were:

- 15 t forwarder with modified load space

- 20 t excavator-based grapple loader

- mid-sized rubber-tyred front-end loader with hi-lift bucket for top-loading chip trucks

- electric/hydraulic knuckle-boom unloader based at the energy plant

- Morbark EZ 30 mobile drum chipper, trailer mounted

- Bruks electric-powered drum chipper (70 cm), fixed installation at the energy plant

- Kockums/Bruks chipper forwarder.

The costs of each machine were calculated using the following common assumptions:

- discount rate 9%

- all the in-forest equipment working just one shift per day

- 235 working days per annum

- all equipment at the energy plant working two shifts per day.

The on-highway truck had a maximum legal payload of 23 t, and the off-highway truck – used only on private forest roads – had a 40 t payload.

There were significant cost variations between the systems (Figure 4.16), with delivered costs ranging between $NZ5.7/GJ and $NZ2.3/GJ. A major part of the delivered cost is in handling and transport, which could be reduced by increasing operating efficiencies and ensuring that truck payloads are maximized. Options A and B did not include intermediate storage with its need for double handling, and these produced the cheapest delivered biomass. This implies that the additional costs involved with attempting to lower fuel moisture content by having intermediate storage periods were not offset by lower payloads to give cheaper transport. However, possible improved combustion was not taken into account. Those systems using the imported chipper forwarder (E and F) proved to be the most expensive.

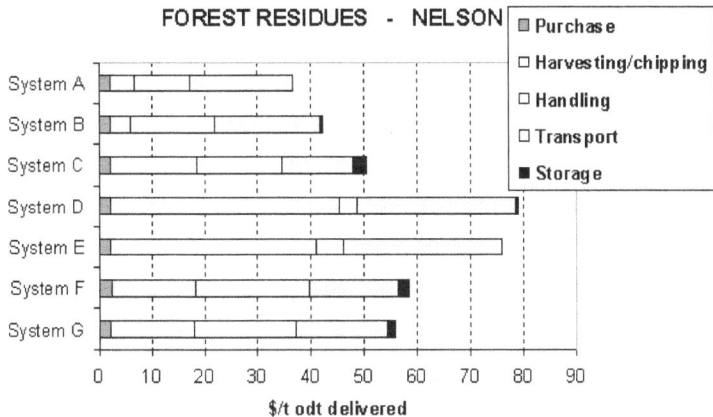

Figure 4.16 Summary of costs of delivered fuel from forest arisings from eight different systems

Notes

[1] The diagrams and definitions were based on ETSU (1990) *Wood Fuel Supply Strategies.*
[2] Costs for this case study are in pounds sterling. £1 = $US1.3 approximately
[3] A billet is a piece size that is bigger than a chip but small enough to allow the bulk material to be handled by bucket loaders and on belt conveyors. It is large enough to allow air to ventilate naturally when stored in a stack, and therefore prevents spontaneous heating.
[4] The values used in this case study are in $NZ. $NZ1 = $US0.50 approximately.

Chapter 5

Thermochemical conversion by combustion and the steam cycle

The process of converting biomass into useful products such as electricity and heating can involve both primary and secondary conversion technologies. In primary technologies, the raw biomass fuel is converted either directly into heat or into more convenient fuels or **ener gy carriers** in the form of gases (such as CH_4, H), liquid fuels (such as CH_3OH, C_2H_5OH) or char (C). Secondary technologies convert these energy carriers into the final desired form of energy product such as the transport of goods and people or electric light. Primary and secondary technologies may be inseparable in certain specific applications.

Primary conversion technologies include combustion (Figure 5.1), gasification, pyrolysis (Chapter 6) and hydrolysis. Secondary conversion technologies include the following:

- **Boilers** take the heat of combustion and use it to raise the temperature of water or convert it into steam. This hot water/steam can be used to supply process heat in a factory, or to drive a steam turbine to generate electricity (Figure 5.2).

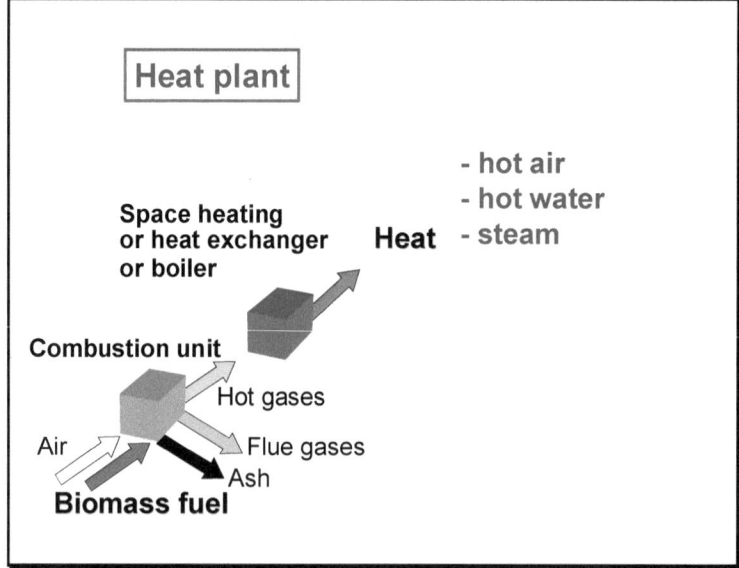

Figure 5.1 Combustion for heat generation

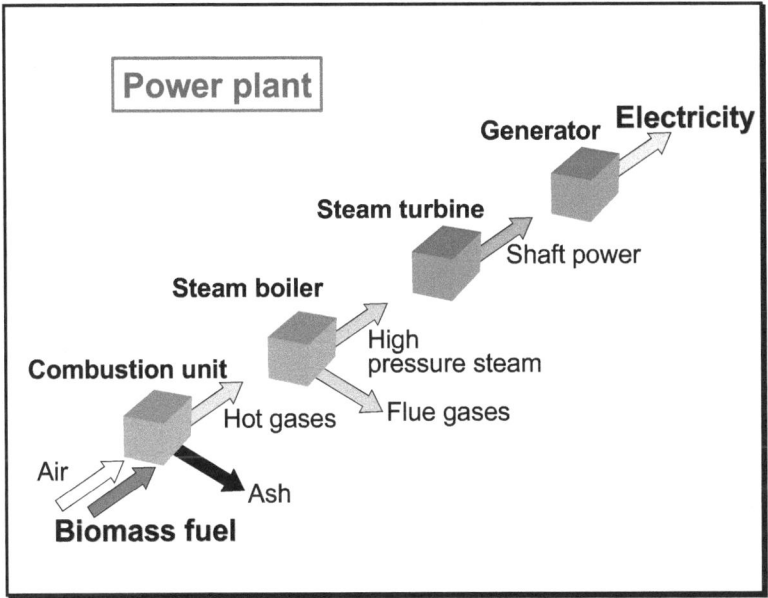

Figure 5.2 Combustion for power generation

- **Gas turbines/gas engines** . Primary gasification processes produce 'producer' gas (CO, H and perhaps some CH₄) with low to medium heat value. It can be combusted directly in a gas turbine or internal combustion engine, coupled either directly to a generator for electricity production or to a gearbox to provide a mechanical power drive. In recent years there has been emphasis on improving the efficiency of thermal electricity generation, which led to the development of *combined-cycle* plant. This recovers the heat from the gas turbine exhaust to produce steam for use for the generation of additional electricity through a steam turbine.

- **Internal combustion engines** used for transport purposes can be fuelled by alcohol fuels, biodiesel, producer gas or pyrolytic oil (Chapter 7).

Thermal conversion by combustion and secondary conversion using the steam cycle are covered in this chapter. Other forms of primary and secondary conversion are covered in Chapter 6. All of the secondary conversion technologies discussed are mature and well proven using fossil fuels, but their interaction with biomass fuels, and the degree of emissions produced, require further investigation to seek improvements. In practice, both primary and secondary technologies normally require some modification so that they are suitable for biomass fuel applications. Modification of processes for application to woody biomass is relatively straightforward, especially for combustion and gasification, although the main differences of higher moisture content, lower energy density and greater percentage of volatiles need to be accounted for when designing the conversion plant. Application to other biomass feed stocks such as straw or municipal solid waste (MSW) can be more difficult, as their form and characteristics move further from that of coal, for which the technologies were originally designed and are well proven.

The combustion process

Combustion involves the burning of a fuel in excess air (to provide the oxygen necessary for the chemical reaction to occur) to produce heat (Figure 5.1). The process includes the volatilization of combustible vapours from the biomass, which then burn as flames, while the residual material in the form of a carbon char is burnt when more air is added.

When 'dry' biomass is combusted several basic phases occur:

- heating and drying

- distillation of volatile gases

- combustion of these volatiles

- combustion of the residual fixed carbon.

This is the same process whether small domestic stoves or large industrial furnaces are the appliances used. An open fire at home may be romantic on a winter's evening, but it is very inefficient, with only around 10–15% of the energy contained in the firewood ending up as useful space heating in the room. The volatile gases produced simply go up the chimney, along with much of the heat. This same problem also occurs with traditional open wood fires used by millions of people in developing countries for cooking and heating. They consume the firewood so inefficiently that firewood sources are severely depleted, and the smoke causes major health problems. Many solar cooker designs have been proposed and developed as alternatives, but these entail cooking in the heat of the day, which is not well accepted.

The use of more efficient wood stove designs is the answer for both developing and developed countries, resulting in greatly improved thermal efficiencies of up to 60%. The principle is that the heat breaks down the plant cells and drives off the volatiles as it also does in an open fire, but then instead of the volatiles being released into the atmosphere, they pass through a high-temperature zone (above 630°C) where secondary air is also present. Here the gases are combusted and release more heat. Not only is this more efficient, it also results in less release of polluting chemicals to the atmosphere. Coal combustion is a similar process to that of wood, but coal has a higher level of fixed carbon and a much lower content of volatiles. Typical volatile matter as a proportion of the total (volatiles + fixed carbon) is:

- softwoods: 76.6%

- hardwoods: 80.2%

- bituminous black coal: 37.4%.

This is an important difference, which has an impact on commercial large-scale wood burner designs since most of the combustion process for biomass relates to combustion of the volatile fraction, which takes place in the heating chamber space above the bed. Coal combustion occurs mainly in the bed since most of its energy is stored in the char[1] fraction.

An interesting evaluation of softwood chips compared with coal as fuel in a 3 MW$_{th}$ fixed-bed boiler was undertaken by FEC Ltd in the UK in the 1980s (Table 5.1). As might be expected, a smaller volume of coal was consumed to give the same amount of heat, but atmospheric emissions were higher than for wood. Dry wood performed better than wet wood.

Table 5.1 Comparison of combustion properties of wet and dry forest arisings with bituminous coal as fuel for a 3 MW$_{th}$ boiler with exit temperature of 220°C

	Dry wood 25% m.c.	Wet wood 50% m.c.	Coal 10% m.c.
Plant efficiency (gross; %)	75	65	84
Gross heat value (as fired; MJ/kg)	14.7	9.7	28.6
Heat input (GJ/h)	14.4	16.6	12.8
Fuel input (kg/h)	962	1709	449
Volume of fuel (m³/h)	1.3	1.5	0.56
Ash produced (kg/h)	9.6	11.4	44.9
Volatiles released (kg/h)	572	677	138
Air required (theoretical; kg/h)	4570	5410	4210
Waste gas (theoretical; Nm³/h)	4006	4720	3430
SO$_2$ emitted (kg/h)	1.9	2.3	13.5

To obtain the same fixed boiler output of 3 MW$_{th}$ significantly greater volumes of biomass need to be handled than coal owing to the lower bulk density and heat value of the biomass, and the effect of moisture content on flue gas heat losses. Increased fuel volume is likely to be the limiting factor when converting existing coal-fired combustion equipment for use with biomass, and a degree of boiler derating would be expected. Coal-handling conveyors can handle woody biomass if it is large enough, but bagasse and straw may necessitate the use of specialist machinery.

'Bridging' of the material in converging sections by binding together of the particles, particularly where there are long twigs and sticks, is often a problem. This can be overcome (at extra cost) by the use of vibrators or hopper discharge aids, and by screening oversize material from the fuel before entry.

For dry agricultural residues such as straw, bagasse, rice husks and palm oil kernels the factor that most limits their greater use is the high cost of transporting, storing, handling and processing. Relatively low-cost and low energy-consuming compaction technologies have been developed recently, though the pellets, briquettes and wafers cannot compete with coal or gas in regions where these fuels are also available.

Biomass fuels vary load by load, especially regarding moisture content, and this affects the combustion rate, making it difficult for the plant operator to obtain efficient plant output all the time. The moisture content of the delivered fuel varies as a result of harvesting method, climatic conditions, time of year and storage method, and this can have a significant rate on the burning of the fuel. For higher moisture contents, special furnaces can be designed with partially water-cooled grates and preheating of combustion air. Suitable safety margins will be necessary for the fuel residence time to ensure that combustion is completed. The soil contamination levels also vary, depending on the skills of the biomass harvester operators. Higher soil means higher ash, so the combustion equipment needs to be able to operate within the range of ash contents envisaged. Fuel screening before combustion will help to remove some of the soil as fines.

To obtain similar boiler output values, between 10% and 30% more air is required for wood fuels than for coal, depending on moisture content. Most air supply systems on existing coal-fired boilers will normally be adequate if the boiler is used to combust biomass, especially if derating of heat output occurs as a result of other factors such as grate size. The degree of refractory lining of the furnace and level of preheating necessary increase with the fuel moisture. Alternatively, co-firing of wet fuels with coal can increase the burning rate and hence reduce the need for refractory lining and preheating. The wetter the fuel, the more coal is required.

Designs of combustion systems

Combustion has been the traditional mode of biomass utilization, and numerous examples of such systems exist around the world, ranging in size from 2 kW_{th} domestic stove burners to 50 MW_{th} power stations (Figure 5.3).

During the combustion process, slagging and fouling of the furnace and boiler plant can be particularly troublesome for fuels with high metal alkali contents such as straw and forest arisings. It is thought that the alkalis volatilize during combustion and condense on the relatively cool transfer heat surfaces as metal alkali salts. These elements can react with other compounds to form fusible or sticky products, and the sticky surface then captures more gas-borne particles and the layer builds up, reducing the efficiency and output of the boiler. The solid deposition of combustion ash on boiler heat transfer surfaces is to be avoided as they act as insulators, and their surface temperature can increase until they become semi-molten. Regular cleaning to overcome the problem demands frequent boiler shut-downs, which is unacceptable. The ratio of alkali oxides (Na_2O and K_2O) to silica (SiO_2) in the ash can indicate whether fouling is likely. The fouling potential of a range of biomass fuels, based on the ratio of alkali oxide to silica ratio in their ash, shows that slagging and fouling are likely to be relatively minor problems.

Most biomass combustion systems are used for process heat applications, and some for cogeneration of both heat and power (Chapter 8). Relatively few are used exclusively for electricity generation. Up to 70 MW_e installations can

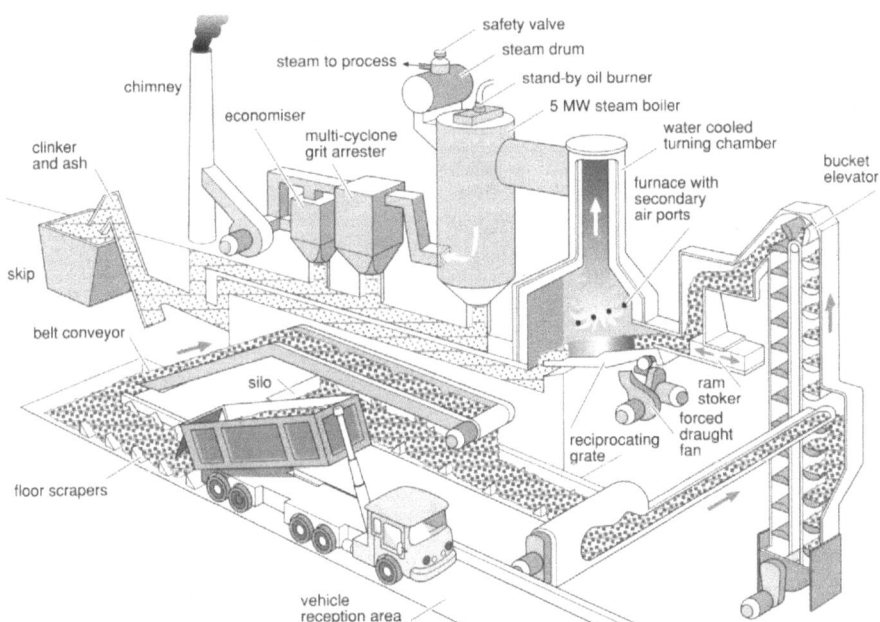

Figure 5.3 Wood-fired combustion plant showing key components for medium-scale steam production
Source: Open University, UK

be found at large wood-processing facilities such as pulp and paper mills in Scandinavia, where the heat can also be usefully employed in the process. Community district heating and power generation plants are also fairly common at this larger scale. Adoption of this technology commercially has been assisted by the recognition that waste organic materials would otherwise have a disposal cost. Instead they can be converted into useful process heat or electricity, thereby reducing both dependence on fossil fuel sources and greenhouse gas emissions.

Biomass combustion plant design can be either fixed-bed or fluidized-bed.

Fixed-bed combustion

Fixed-bed combustion (or stoker firing) uses mechanical devices to feed the comminuted or whole-bale biomass fuels into a bed at the bottom of the furnace, where they burn. The combustion air passes through the grate on which the fuel sits. Air flow pressure and rate are restricted so that the fuel is not stirred from the bed it sits on, and hence it remains in contact with other solids. Several methods are used to feed the fuel onto the grate:

- **Over feed stokers**   (or travelling grate stokers), originally designed for firing coal, feed the fuel by gravity onto the moving grate at one end (Figure 5.3). The depth of the fuel layer on the grate (and hence the heat output) is adjusted by the height of a gate located at the furnace entry. The grate travels slowly across the furnace as combustion takes place, carrying the fuel along as if on an open conveyer belt, with the residual ash and slag being continuously discharged at the opposite end of the furnace to the fuel inlet. The use of biomass with overfeed stokers can result in higher maintenance costs compared with coal, and the combustion efficiency can decrease owing to lower combustion air temperatures.

• **Spreader stokers**   consist of fuel distributors that force comminuted biomass fuel into the furnace above an ignited fuel bed on an air-cooled travelling grate. Either mechanical or pneumatic throwers can be used, depending on the fuel type. Some suspension firing also occurs before the fuel reaches the grate, depending on its moisture content and particle size. Fine particles tend to burn in suspension, while larger particles fall onto the travelling grate, where they are burned. Spreader stokers are responsive to heat load changes by adjusting grate travel speed, fuel feed rates and air intake, which can be achieved automatically. To gain maximum benefit from the suspension firing principle fuels need to be comminuted, and should be relatively dry with a moisture content less than 50% w.b. An ash layer needs to be maintained on the grate to protect it from thermal degradation, which can also result from frequent start-up due to intermittent operation if the heat load varies. Biomass ash can have a higher silica content than coal. It is abrasive, and can result in higher maintenance costs of the grate. A further disadvantage with this type of combustion design for biomass is that there can be a significant amount of fly ash and unburned carbon in the flue gas, resulting in lower combustion and boiler efficiencies and higher costs of emission controls.

• A **sloping-grate fur nace** is the most common design selected for biomass combustion systems. Pre-drying of the fuel occurs on the grate after the fuel is fed into the upper part of the furnace prior to its tumbling slowly under gravity onto a reciprocating grate lower in the furnace, where combustion takes place. The grate is either water-cooled or air-cooled, and so does not require an insulating layer of ash to protect it. This makes it suitable for biomass fuels with lower ash contents.

To achieve similar combustion rates with the same residence time as for coal in a fixed-bed combustor using biomass would require a significantly greater fuel bed depth. The hearth loading rate of coal-fired stokers is 3.6–4.3 GJ/h/m^2 of grate area when combusting fuel over a 30–35 minute period at a bed depth of 90–105 mm. By comparison, to achieve the desired hearth loading rate of 2.5 GJ/h/m^2 of grate area when using woody biomass at 30% m.c. would need a 550 mm bed depth. The ignition plane travels down through the bed at a greater rate than for coal, but the depth of bed will still be limited, leading to reduced combustion rates. Coal chain grate stoker designs have a fuel bed depth limit of 150 mm, which would give a hearth loading rate of only 1.8 GJ/h/m^2 when using biomass, thereby severely limiting boiler output. This leads to the derating of a coal-fired plant when converted to biomass, and indicates the need for purpose-designed grates of a larger area, and hence capital cost, than for coal. For some designs it may be possible to increase the grate speed to compensate for the limited bed depth, but in practice this would lead to ignition and excessive grate wear problems.

Fluidized-bed combustion

Fluidized-bed combustion requires finely comminuted biomass particles to be fed into the furnace onto a bed of sand and then to be subjected to an evenly distributed, upward flow of air passing through the bed from below. When the air velocity is steadily increased a point will be reached when the individual fuel particles lying on the bed will be forced upwards by the flow so as to be suspended in the air stream. When the air velocity is increased further, by turning up the fan speed, the bed becomes turbulent, and rapid mixing of the particles occurs. The surface of the sand is no longer well defined but appears diffuse, as bubbles of air similar to those in a boiling liquid rise up through the bed, which in this state is said to be fluidized (Figure 5.4). It behaves as a liquid by finding its own level and possessing a hydrostatic head. Further increases in air velocity cause progressively larger fuel particles to become entrained in the air stream, and smaller particles may then be carried off by it without being combusted if the air flow becomes excessive. For a particular bed design to suit a given particle size range there will be an upper and lower limit of air velocity within which satisfactory fluidization will be established.

Fluidized-bed combustion requires high air velocity throughout the entire area of the fuel bed to give it fluid properties. The bed will normally consist of coarse particles of sand, which assist the mixing of the fuel with the air and also increase the heat transfer to the fuel for initial drying and then ignition. The

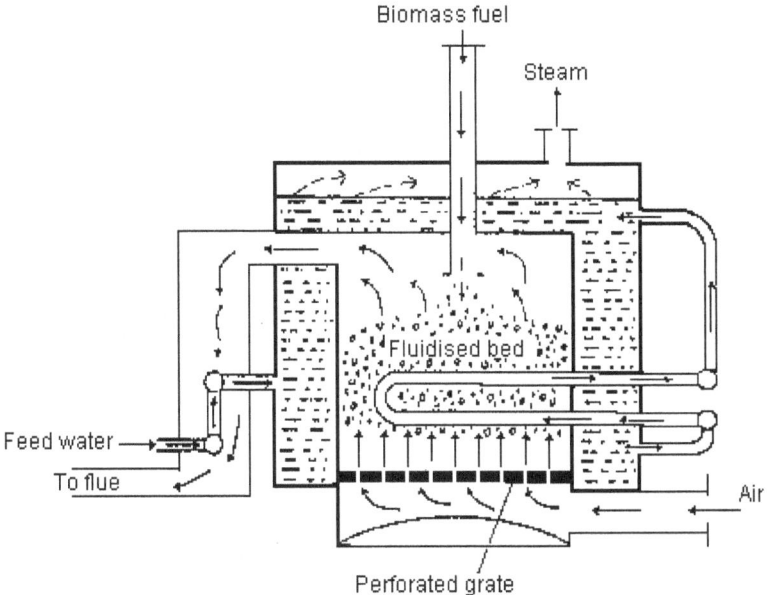

Figure 5.4 An example of a fluidized-bed boiler design suited to a range of fuels with higher moisture and ash contents

separation of the fuel from the other bed particles of sand occurs above the **minimum fluidization velocity** , this being a function of biomass particle size, density and the pressure drop through the bed from its base to the top surface. As air velocity is increased, the bed can change from bubbling, to turbulent, to circulating with increasing recycling rates of material in suspension.

Commercial designs are either bubbling fluidized-bed (BFB) or circulating fluidized-bed (CFB), and can be pressurized or not and use either air or oxygen. In BFB systems the air velocity used is typically 1–3 m/s, causing the sand bed to bubble and thus separating the bed particles from one another. Primary air supply is via nozzles beneath the bed, with secondary air flow entering the furnace above the bed. Bed temperature is maintained and controlled by altering the ratio of primary to secondary air, or by recirculating some of the hot flue gas in with the air.

In a CFB system the air velocity is typically a higher 4–9 m/s, which circulates the sand into the furnace, thereby assisting heat transfer. The flue gases and entrained solids leave the furnace together when they pass through cyclones that collect the biomass and sand particles and return these to the area immediately above the bed. There are both primary and secondary air supplies as for the BFB design. CFB systems tend to be more expensive to purchase and operate but reduce NO_x emissions significantly owing to their lower operating temperatures, below 900°C. Co-firing of biomass with coal is easier to accomplish than with fixed-bed combustors as both pulverized fuels can be fed into the furnace using separate feed systems, and hence the feed rates can be easily adjusted.

Fluidized-bed combustion is more tolerant of a wider range of fuel characteristics. Low-grade fuels with high moisture content, such as bark, can be interchanged with high-grade fuels with low moisture content, such as bituminous coal. Good control of bed combustion temperature to relatively low and even levels allows a range of fuels with varying ash properties to be burnt without the occurrence of slagging problems. Limestone can be added to the bed material to provide in-situ capture of SO_2, though this is not necessary for biomass fuels since they are low in sulphur. The burner can be quickly adjusted to meet changing heat loads, there being only a small quantity of fuel in the chamber at any one time compared with a fixed-bed design, and it can be fired successfully down to low loads of only 35% of full capacity. Conversion efficiency is relatively good, owing mainly to the low volumes of unburned fuel passing through.

However, fluidized-bed combustor designs are more technically intricate than fixed-bed systems, giving higher costs for design, construction and operation. Fluidized beds show economic advantages over fixed-bed combustors at around the 8–10 MW$_{th}$ scale, and so are not readily available at a smaller scale.

Chemistry of combustion

A comparative analysis of woody biomass and coal is given in Table 5.2. From this it can be seen why using coal as a source of energy produces atmospheric

Table 5.2 Ultimate and proximate analysis of typical samples of softwoods, hardwoods and bituminous coal (% of total weight)

	Softwood	Hardwood	Coal
Ultimate analysis (dry basis)			
C	53.2[a]	51.0[a]	75.6
H	6.1[a]	6.2[a]	4.6
O	38.9[a]	39.9[a]	9.2
N	0.1	0.2	1.6
S	0	0	1.0
Ash	1.7	2.7	8.0
Proximate analysis (wet basis)			
Volatiles	40.6	52.9	32.4
Fixed carbon	12.5	13.3	54.7
Ash	0.9	1.8	7.6
Moisture	46.0	32.0	5.3
Proximate analysis (dry basis)			
Volatiles	75.2	77.8	34.3
Fixed carbon	23.1	19.5	57.7
Ash	1.7	2.7	8.0
Moisture	0	0	0
Gross heat value (MJ/kg dry weight)	20.4	19.8	30.8

[a]Pure cellulose ($C_6H_{10}O_5$) for comparison, based on molecular weight, is 44.5% C, 6.1% H and 49.1% O

emission problems relating to CO_2, NO_x and SO_x, which are all formed during combustion.

Combustion of biomass occurs in four sequential phases. Of course in practice the whole combustion process is dynamic, and in an industrial burner, where fuelwood is continually fed into the combustion chamber to replace that being consumed, each of these four phases is carried out simultaneously.

Phase 1: Heating and drying

When the biomass is heated any surplus moisture in the plant cells evaporates. This moisture content varies between 10% and 50% of total weight (wet basis), which is higher than that of coal (normally between 3% and 10% when mined). It reduces the effective or useful heat available from the energy contained in the fuelwood. It also affects the temperature gradient, which can occur within the piece or particle of wood and in turn slows the heating and drying process. Wood is a poor heat conductor, which leads to high heat gradients across the wood when burning it of up to 30°C/mm. For this reason, to give an even burn and satisfactory combustion in an industrial plant the maximum distance from the centre of a particle of fuelwood to the surface should not exceed 20–30 mm.

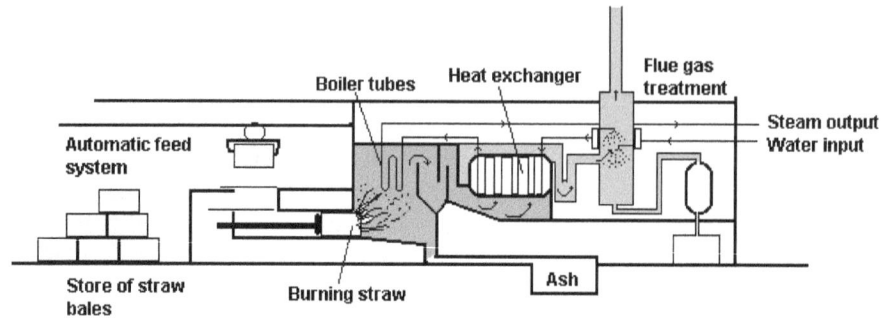

Figure 5.5 Straw-burning plant around 4–5 MW burns straw bales directly without shredding or chopping. Water vapour in the flue gases is condensed to improve the efficiency
Source: Based on a diagram from Open University, UK

Thus wood chips and sawdust, as well as shredded straw and pulverized biomass fuels such as bagasse, are preferred. However, in an attempt to save the cost and energy involved in the comminution process, burners have been designed especially for burning whole trees and whole bales of straw (Figure 5.5).

Phase 2: Distillation of volatiles

After evaporation of the water molecules is complete, further heat is absorbed and the volatile gases are released when temperatures of between 180°C and 530°C are reached and distillation occurs (similar to the distillation of crude oil products at an oil refinery). These gases comprise complex saturated and unsaturated organic compounds, including paraffins, phenols, esters and fatty acids, which all have separate distillation temperatures and hence could theoretically be processed in a 'biorefinery' to produce a range of chemicals. In fact biomass can provide a similar range of chemicals used for manufacturing plastics, etc., as is now sourced from the petrochemical industry.

Phase 3: Combustion of volatiles

Ignition takes place between 630°C and 730°C, when the gases produced from the biomass combine exothermally with oxygen to produce heat and mainly CO_2 and water vapour. The flame temperature reached in the burner during this gas phase will be dependent upon the level of excess air present and upon the amount of moisture in the wood at the outset, since this will have been evaporated into this zone where the gas phase reaction occurs.

In the gas phase part of the process it is important to provide excess oxygen by means of a secondary air supply, otherwise too low a temperature will result, giving poor combustion and high emissions of particulates. Some of this unburnt carbonaceous material, commonly known as *soot*, absorbs the volatiles after they have condensed in the cooler parts of the plant, forming an oily product or tar. While some of the soot could be subsequently burnt before it reaches the exit to the combustion chamber, it would be necessary to maintain the particles at relatively high temperatures in the presence of oxygen to do so, and this is not always practical in combustion appliances. So it is better to avoid soot production in the first instance by using dry wood and controlling the air intake to match the moisture content of the fuel.

Phase 4: Combustion of fixed carbons

Once the moisture and volatiles have been driven off, the fixed carbon component of the fuel remains as char, which then begins to burn as more oxygen becomes available. In a typical fixed-bed combustion appliance the oxygen reacts with the carbon to form mainly CO. This gas burns to form CO_2, but this is then reduced by more carbon back to CO. These three processes overlap. Hence the main gaseous product leaving the fire bed at this stage is CO. When it reaches the secondary air, it burns (along with the volatiles) and forms CO_2, which is then emitted to the atmosphere.

In a fixed-bed combustor the efficiency of these reactions depends upon the intimate contact between the solid carbon and the reactant gases. The smaller the particle size, the greater will be the surface area exposed to the oxygen for a given volume of fuel bed. Conversely, the greater the size of the wood pieces, the greater the bed depth that will be necessary for efficient combustion of the fuel.

In theory about 6 kg of air (with 20% oxygen) is necessary to burn 1 kg of dry wood. Owing both to the main product from a bed of burning char being CO (which is not oxidized to CO_2 in the bed) and to the high volatile content of wood, only half the air necessary to fully burn the fixed carbon component needs to be passed through the fixed bed. Hence approximately 85–90% of the theoretical total of the calculated combustion air should be supplied *above* the burning bed as secondary air. In practice, however, a large proportion of the primary air passes through the fuel bed (depending on wood particle size and bed depth) without reacting, and this then acts as secondary air when it reaches the volatile combustion zone above the bed. The actual volume of air passing through varies with burner design. For example, for mechanical stokers considerable quantities of excess air are free to pass through the relatively exposed burn-out section of the grate.

Emissions

At the small domestic scale biomass combustion can have a health impact owing to respiratory problems created by the products of incomplete combustion. This is particularly the case in developing countries, where open fires are used for cooking and heating, often indoors without flues or adequate ventilation. Even when using more efficient wood-burning stoves, atmospheric emissions, especially particulates, can be a problem where poor-quality fuels with high moisture contents are used. Also, if the stove is refuelled and then damped down and left to simmer overnight without first burning off the volatiles from the new load of firewood, higher atmospheric emissions can result, along with tars that condense in the flue.

At the commercial scale more stringent emission regulations, such as the USA's Clean Air Act, quite rightly now need to be met. This tends to increase the overall cost of biomass plants, as well as that of fossil-fuel-fired plants of course. Standards are being increased in many countries for emissions of SO_x, NO_x, CO, particulate matter <2.5 μm and other hazardous air pollutants. Biomass plants can have particular problems in meeting the NO_x minimum targets and particulate levels, especially at the sub-micron size. However, good plant design using up-to-date technologies can satisfactorily reduce emissions to within accepted standards.

Atmospheric emissions from solid waste incineration have been drastically reduced since the dioxin debate of the 1980s, but at a cost. Particulates in the form of fly ash can be easily controlled by cyclone filters, scrubbers or electrostatic precipitators. Toxic gas emissions can be controlled where necessary by wet or dry gas scrubbers or, for smaller quantities, by adjusting the waste feed rates and using fossil fuels to maintain the temperatures within the incinerator. Emissions of SO_x and NO_x are still present, but have been reduced to half what they used to be.

Environmental problems associated with the incineration of MSW usually result from incorrect operation, assuming the system was adequately designed in the first place. The main problem areas are those associated with the products of incomplete combustion. Incomplete combustion commonly occurs because insufficiently high temperatures are reached, and a wide range and variety of different chemical compounds result. Some of these, such as dioxin, are toxic and persist in the environment, but they are usually produced in very low absolute quantities. Manufacturers of the latest incinerator designs claim to have eliminated these chemicals altogether. Heavy metal emissions can also be avoided almost entirely by good plant design and improved sorting of the waste. Batteries are generally excluded from the waste combustion stream, which helps, but toxic heavy metals and compounds can also be removed in the flue. The collection and transport of large volumes of MSW to a central processing site create some negative environmental impacts, and would need to be carefully evaluated as part of a waste management system.

Co-firing of biomass

Co-firing involves burning biomass together with conventional fossil fuels such as coal or natural gas. It represents a promising technology for immediate increased usage of biomass fuels to reduce carbon dioxide emissions from fossil fuels. Where the biomass material is in a suitable form, such as dry sawdust, it may be possible to add it for minimal capital cost in low proportions to the coal fuel supply of an existing thermal plant. However, if this results in greater fouling and corrosion in the boiler or changes to the slag and fly ash, additional treatment of solid and flue discharges may be necessary and add to the maintenance costs.

Some biomass fuels such as MSW when co-fired produce greater risks from air emissions, poorer combustion efficiencies and reduced system performance due to slagging and fouling. A proper evaluation of each specific fuel blend is therefore usually required. Co-incineration of waste with wood chip fuels is being investigated in Sweden for use by rural towns located near the forests but away from an urban waste incineration plant. In spite of increasingly clean fuel fractions being obtained by better separating techniques, it is likely that the European Union environmental regulations on waste incineration will still apply. These are tight, and impose strict demands on flue gas clean-up, but also state that 40% of plastic packaging can be reused through incineration.

Within the general field of energy recovery from waste, packaging and waste plastic incineration remain particularly controversial topics. There is currently no commercial technology for the combustion of separated mixed plastics. The debate centres essentially on the merits of co-firing plastic waste with other products. For example, the Association of Plastics Manufacturers in Europe conducted tests at the 20 t/h incinerator in Wurzburg, Germany, by blending various amounts of plastic to the household waste processed at this facility. As might be expected, the more plastic incinerated, the more heat and hence electricity produced, giving reduced airborne pollution from unburnt particles and compounds. Where the biomass fuel is free on-site or has a disposal cost, and environmental benefits are factored into the economic equation, it is likely that co-firing will be viable even at current low fossil fuel prices. Under these conditions, in the USA alone co-firing of biomass has the potential to generate 50–100 TWh/yr of electricity by 2015.

Combustion technology for co-firing has not been fully evaluated for the many biomass fuel types that could be used, but testing continues, and an extensive range of test programmes – particularly on the co-firing of woody biomass and coal – have been implemented in recent years. These tests have been undertaken at low biomass content (<3%) to higher levels up to 20% on a full range of combustion plant types and also in several gasification plants. The tests have generally indicated:

- the need for a separate feed for the woody biomass

- a boiler efficiency penalty, in the order of 2% reduction for a 20% (by mass) biomass contribution

- a reduction in overall NO_x emissions due to increased fuel volatility

- that SO_x emissions are nearly always reduced

- that corrosion risk of the plant components is greater owing to the greater chlorine and alkali contents of the biomass, though this is fuel dependent

- that boiler damage can occur if the design is not properly suited for the fuel mix.

A summary of co-firing experience in Australia was presented by Delta Electricity at the 2000 Bioenergy Australia conference held in Brisbane. Key points were as follows:

- Co-firing is being investigated by a number of coal-fired power station owners to feed in around 3% of biomass onto the top of the coal on the standard open conveyor.

- Co-firing in large plants gives high efficiencies (33–35%) due to scale, which are not possible in small, stand-alone wood-fired plants at around 18–22% thermal efficiency.

- Generating costs would be high, at around 10–15 c/kWh for the additional co-firing component of a plant.

- If biomass is 3% by weight at 10 MJ/kg versus coal at 25 MJ/kg, the biomass accounts for only 1.2% of the energy output of the plant and requires about 1 t of biomass per MWh generated from it.

- Pulverizing the biomass is an option for use in some power stations, but not all of them are suitable for co-firing.

- Overall, in Australia biomass has a technical potential of 650 GWh/yr if added at low levels to coal to generate 1% of the electricity from only the coal-fired plants suitable. This equates to 600,000 t biomass per year.

- Little is known about char burnout, ash deposition, storage issues, handling, fuel processing, safety issues or emissions.

- Metering the biomass onto the coal conveyor is the best way to ensure a consistent blend. Mixing occurs during travel over the chute gate drops.

- Particle size is important but pulverizing 50 mm × 10 mm softwood chips produced thin fibres 5–10 mm long, like paintbrush hairs, which were undesirable.

- Tests showed that the plant temperature dropped soon after *P. radiata* sawdust was added to the coal; this temperature drop then had to be compensated for by increasing the feeder speed.

- It is not difficult to quantify the 'green' electricity output from the biomass component. The biomass needs to be characterized on delivery, and the individual weights of coal and biomass used have to be recorded. If the total MWh generated are recorded, then the 'green' component can be calculated:

- Green MWh =
$$\frac{\text{tonnes biomass} \times \text{specific energy biomass} \times \text{calculated plant efficiency}}{3600}$$

- The gaseous emissions did not change significantly when the biomass was added.

An alternative to co-firing is to run a dedicated combustion system, installed for a biomass fuel, in parallel with a conventional combustion system using coal, gas or other fuels, both providing heat which is then pooled for a common purpose. This method allows special biomass materials to be burnt in a smaller biofuel combustor and boiler while avoiding corrosion and emission problems that could otherwise be experienced with co-firing in the larger, more expensive fossil-fuel-fired boilers. A further advantage is seen in the secondary conversion process, where economies of scale are often associated with larger steam plants.

The cogeneration facilities at the New Zealand Kinleith pulp and paper mill (Case Study 7) are an example of parallel firing. The heating plant consists of a new dedicated wood-waste-fired boiler working in parallel with existing black liquor and gas-fired boilers to produce steam at different pressures for process supply and electricity generation. Constructing the dedicated wood-waste-fired boiler independent of existing plant also allowed continuation of operations during construction, with minimal interruption to pulp production.

Novel biomass combustion and waste-to-energy systems

Waterwide close-coupled gasifier

This is more correctly classified as a combustion system than as a gasification system, since full combustion occurs in the plant. Many of these systems have been installed for heating plants around the world. The basic design has been used extensively for heat production (Figure 5.6), fired by a wide range of wood wastes and crop residues with the heat used for such varied applications as drying tea and heating greenhouses. Plant performance has at times been disappointing, owing partly to the relatively high moisture content (>65%) of the fuels used (such as freshly harvested *Pinus radiata* wood waste), but when biomass with lower moisture content is used few problems have resulted. The

Figure 5.6 An early version of the Waterwide burner installed to heat a greenhouse complex

plant has a high combustion temperature (1200°C), but very low emissions are claimed (typically particulates are ~80 mg/Nm³, SO$_x$ ~3 mg/Nm³, NO$_x$ ~60 mg/Nm³). Large-scale plants up to 30 MW$_{th}$ are under development, as is the integration with power generation.

Offset direct-fired biomass combustion

This concept has been under trial for several years, having been pioneered by the University of Wisconsin-Madison in the 1980s. A pressurized, wood chip-fired, downdraught, gravel bed combustor is directly coupled to a modified Allison gas turbine. The initial trials resulted in significant ash deposition and blade erosion on the leading edge of the gas turbine in the early stages. However, the process has continued to move towards commercialization.

Sawdust

Bioten of Knoxville, Tennessee has successfully used sawdust to fuel a full-scale combustion LM-1500 gas turbine. Gas clean-up is achieved using a cyclone, and availability is now 85%. The unit produces 6 MW$_e$ at a heat rate of 15.6 MJ/kWh, with the electricity being sold to the Tennessee Valley Authority. Capital cost of the plant was $7.25 million or $1250/kW. The cost of electricity generated is 3.9 c/kWh, of which fuel plus operations and maintenance costs contribute 1.8 c/kWh. The plant will be used for testing other biomass fuels including prunings, rice hulls and bagasse. Particular problems that have been addressed over the original prototype include the following:

- Sawdust feed was improved by the use of a dual-lock vessel pneumatic injection system.

- Combustion and air temperatures were reduced to minimize turbine blade fouling problems associated with alkaline depositions.

- Blade fouling was reduced by the use of additives while by-passing 50% of the gas generator air around the combustor, which also helped to reduce particulate carry-over and increased cyclone efficiency.

- Thermal efficiency was improved by reducing the pressure drop of the combustor/cyclone system from 41 kPa to 13 kPa.

- Fuel drying was improved by using air-swept pulverizers that comminuted, dried and classified the fuel in one operation.

Bioten is now marketing the arrangement as a package that can be constructed from eight shippable modules.

Municipal solid waste to RDF (refuse-derived fuel)

The EnerTech process currently under demonstration in the USA is based on a pre-treatment of MSW in water slurry form to facilitate the removal of recyclables. The slurry is then subjected to high pressure and temperature conditions and partial dewatering to turn it into a higher calorific value RDF amenable to combustion in a high-pressure steam boiler or gasification to power a gas turbine. If successfully demonstrated this process, albeit expensive, will have very low pollution levels and significantly higher thermal efficiency than mass burns.

Biorefinery

The Convertech biorefinery technology developed in New Zealand aimed to process biomass into valuable products such as chemicals, reconstituted wood products such as panel boards, heat and power. The concept potentially offered a solution to monomer recovery that could prove more competitive than other overseas approaches. In essence part of the core process involved venting the volatiles produced by the preheating of the biomass. Through steam-entrained distillation it was possible to recover the volatile products either as a fuel for process heat or as valuable products such as essential oils from *Eucalyptus* biomass. Biomass feedstock fed into the plant as particulate matter was entrained at about 20 m/s in a steam atmosphere at pressures of up to 30 bar and temperatures in the range 200–300°C. After screening for metals and glass, finely shredded MSW, being 75% wet biomass, could be easily processed in the same way. By raising the temperature in one of the process modules to about 500°C, the same basic volatile venting process as described above could be used to extract gases and vapours coming from the organic polymer fraction of MSW. In this way monomer recovery could be achieved without extensive preliminary sorting of the organic MSW stream. The steam-entrained gases could then be processed to recover the naphtha-like wax, while the shredded MSW could be processed further in the superheated steam multiple-effect drying stage of the system to be dried into a stable RDF.

It was claimed that this process could be implemented at relatively small-scale levels down to 2 t/h or about 15,000 t/yr of MSW or 1000 t/yr of polymer waste contained in such MSW. On the basis of preliminary estimates, a larger-scale biorefinery processing MSW at the rate of 20 t/h and equipped to recover about 9000 t/yr of monomers, while also generating power from the RDF produced, would have the following financial characteristics:

- investment cost: $27 million

- gate fee avoided for the MSW: $10/wet tonne

- average cost of power produced: 2–3 c/kWh

- average price of power produced: 8 c/kWh (over 20-year life of the plant)

- sale price of naphtha: $360/t.

A claimed internal rate of return of 18% (over the 20-year lifespan of the plant, with 75% borrowed capital at 9% interest), even though indicative, looked sufficiently promising to warrant further investigation. However, after several years of successful progress, the company was wound up. Perhaps it was ahead of its time.

The steam cycle

In a typical factory processing say food or fibre products, the heat from the biomass fuel (or from coal, natural gas or oil) is used in the boiler to turn water into steam, just as a kettle does. This steam under pressure can be used 'live' on-site for a variety of heating needs such as kiln drying, as a stripping agent in a distillation column, for indirect heating through a heat exchanger, or for water heating. It can also be used to drive a steam turbine (Figure 5.7).

Figure 5.7 Steam turbine driving a 25 MW$_e$ generator in a wood-fired power plant in Canada

Table 5.3 lists some terms and definitions relating to combustion and steam generation that it will be useful to know.

A steam system usually comprises:

- a feedwater treatment plant
- a steam distribution system
- a steam boiler
- a steam turbine.

Feedwater treatment

The water feeding into the boiler needs to be treated in order to remove suspended solids by filtration, dissolved solids and dissolved salts by ion

Table 5.3 Terms and definitions relating to combustion and steam generation

Term	Definition
Back-pressure steam turbine	A simple non-extraction, non-condensing steam turbine. High-pressure steam is expanded through the turbine to generate electricity, and is exhausted from it at the required steam conditions for the site.
Boiler drum	The drum or upper cylinder of a boiler where the steam generated is separated from the circulating boiler water.
Deaerator	Used to preheat feedwater before entering the boiler, and to drive off any non-condensable and potentially corrosive dissolved gases.
Economizer	A counterflow heat exchanger for recovering energy from the exhaust gas to increase the temperature of the water before entering the boiler drum. It uses otherwise wasted exhaust heat and hence increases the steam-raising ability of the plant.
Isentropic	Constant entropy, or in effect 'zero loss', is used in the context of a steam turbine's ability to convert steam energy into mechanical energy, as opposed to leaving the turbine as exhaust heat energy at the exhaust steam pressure.
Make-up	Treated raw water that is added to the system to replace any steam and water lost to the system from site requirements, blowdown, evaporation, sampling or venting.
Steam turbine	An engine in which a multi-vaned wheel is made to revolve by the impingement of steam to convert steam energy to mechanical energy.
Superheater	A heat exchanger part of a boiler for increasing the temperature of saturated steam to superheated steam (that is, above the saturation temperature at that pressure). Generally, steam admitted into a steam turbine must be superheated.
Topping cycle	High-pressure steam is raised in an auxiliary boiler, expanded through a back-pressure steam turbine generating electricity, and then released at the steam conditions required on the site to provide process heat.
Waste heat boiler	A boiler that uses waste heat (such as from a gas turbine or from the exhaust gas of a reciprocating engine) to produce steam or hot water.

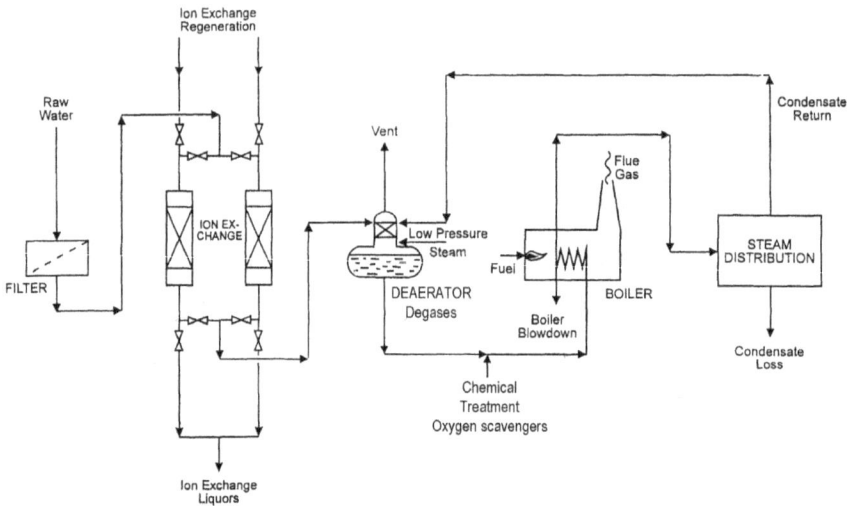

Figure 5.8 Boiler feedwater treatment system to remove impurities

exchange, and dissolved gases (mainly O_2 and CO_2) by deaeration before steam can be generated without problems occurring (Figure 5.8). A build-up of solids in the boiler can cause fouling and corrosion, and the solids can be carried over to the steam distribution system and cause blocking of the pipes. The de-aerator physically de-gases the water. A pair of ion exchangers are used so that one can be regenerated whilst the other continues to operate. Chemicals such as phosphates can be added to act as oxygen scavengers. The boiler blowdown purges out any dissolved solids such as silica (SiO_2) not removed by the filter, and they are discarded.

The volume of condensate returned to the system after steam distribution around the factory for process use varies between 0% and 90% of the flow, depending on the economics and any concerns about the possibility of returning contaminated water to the boiler. Normally it is not justifiable to waste water by discarding it. The volume of condensate lost determines the rest of the treatment plant design. If there is virtually zero loss, then less incoming water needs to be treated, and that part of the plant can be downsized.

Steam distribution system

A steam distribution system for any specific processing factory has several different levels of temperature and pressure, which can lead to complex interactions with steam being required at various high, medium and low temperatures and pressures. It is not necessary to study the engineering design details of such a system here, only to note that the steam from one back-pressure turbine could be at medium pressure while at low pressure from another. The interactions are such that the steam from each turbine is used for different heating duties within the processing plant.

Steam boilers

There are many different types of steam boiler depending on the fuel type (biomass, coal, oil or gas), output capacity and desired steam pressure:

- Very high pressure (10–14 MPa) is used only for power generation to drive a steam turbine.

- High pressure (4–5 MPa) is the normal maximum for steam distribution around the plant, but can also drive turbines.

- Medium pressure (1–4 MPa) is the conventional distribution pressure.

- Low pressure (0.1–1 MPa) is for applications such as water heating, and usually recoverable from waste heat from the plant.

A **fire-tube** or **shell boiler** is one of the more common designs. A large cylindrical shell holds the water supply, with heat from the furnace passing through tubes immersed in the water to generate the steam (Figure 5.9). The cylindrical shell structure restricts the maximum capacity of this boiler design and its operating pressure, which is normally only up to 20 bar. Hence it has a relatively small steam output.

A **water-tube boiler** design has the feedwater passing inside a network of tubes within the heated chamber (Figure 5.10). The water boils inside the tubes to produce the steam, which is at a higher pressure and output than that of a fire-tube boiler design.

A **fluidized-bed boiler** is particularly suited to biomass fuels with relatively high moisture and ash contents (Figure 5.3). Solid fuel particles such as sawdust

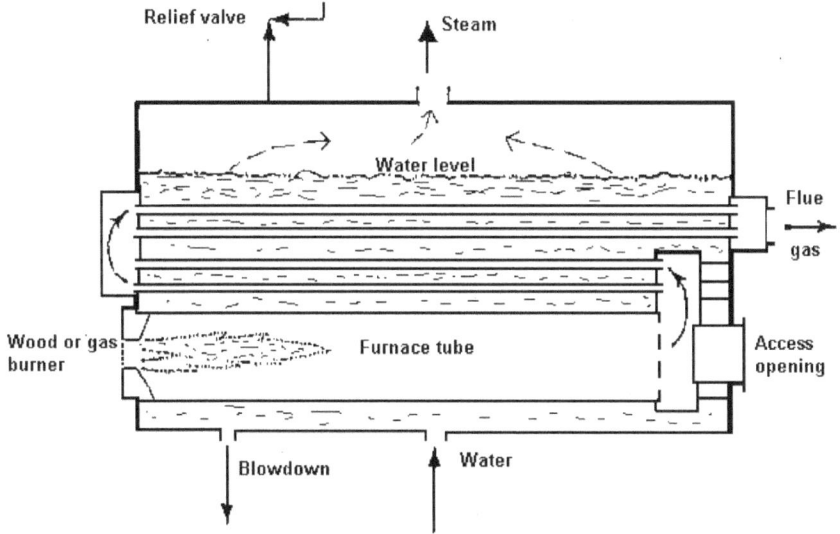

Figure 5.9 A fire-tube boiler using wood or gas fuel to produce hot gases. These pass through a series of tubes submerged in water, which is thereby heated to produce steam

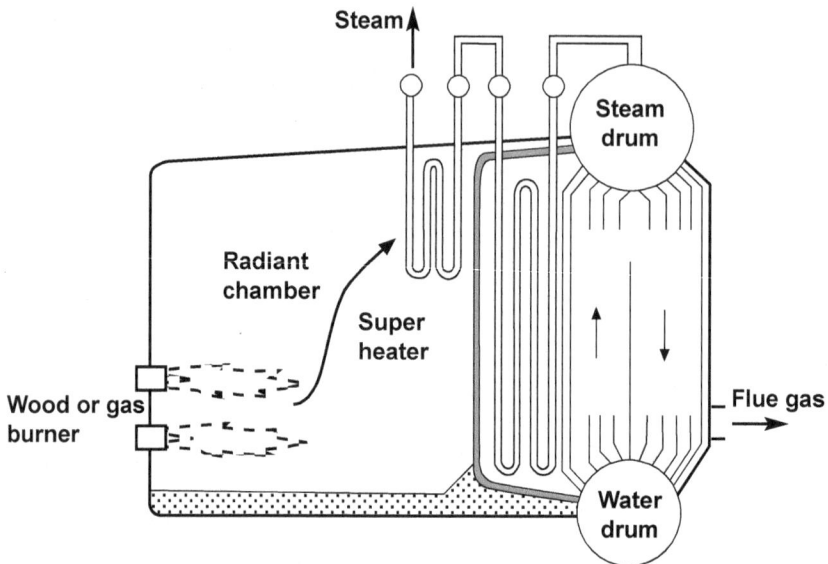

Figure 5.10 A simple water-tube boiler with the heat from combustion passing between and around the vertical tubes

enter the combustion chamber from above, meet the rising hot air and combust at relatively low temperatures to heat the water in the tubes in the chamber.

Steam turbines

There are basically two turbine designs, each with many blades around a shaft. The action of the steam on each blade varies with the design (Figure 5.11).

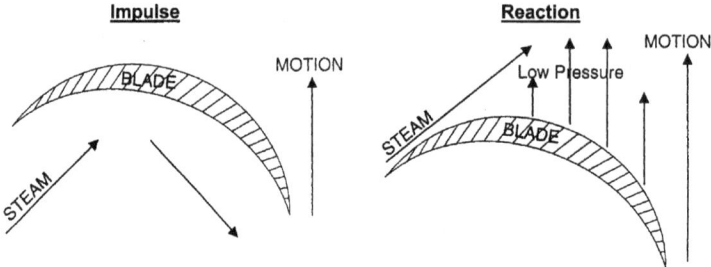

Impulse	Reaction
• Steam plays directly onto the blade and pushes it	• Steam passes past the outside edge of the blade and pulls it
• Steam actually strikes the blade	• Steam induces low pressure behind blade
• 75% efficiency	• 85% efficiency
• Large clearances during manufacture	• Finer tolerances so more costly to manufacture
• Used in low-power back-pressure turbines	• Used in medium- and low-pressure stages of large turbines
• Used in high-pressure section of large turbine	

Figure 5.11 The power from a steam turbine design arises from the action of the steam on the multi-blades of the turbine using either impulse or reaction to impart rotational energy

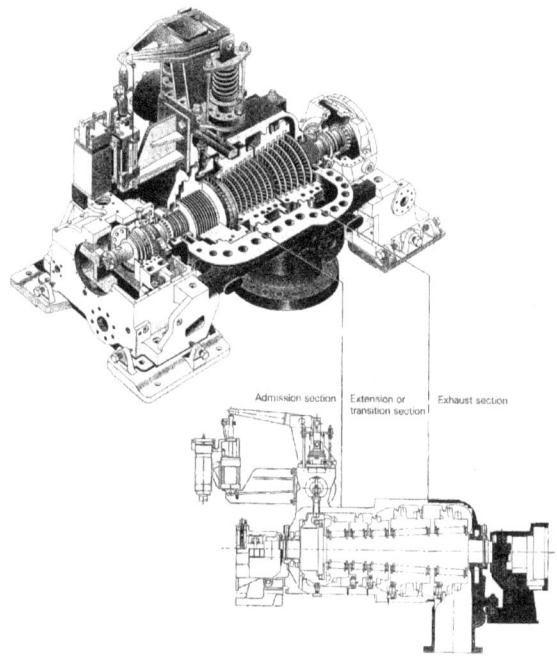

Figure 5.12 Cutaway section of a back-pressure steam turbine
Source: Siemens

Each individual turbine has to be matched closely with the steam parameters available at the site. Turbine manufacturers therefore 'make to measure' rather than mass-produce. To reduce costs the turbine is constructed from different modular building blocks: an admission section, where the steam enters; an extension or transition section, where the work is produced; and an exhaust section, where the steam is released at various pressures (Figure 5.12). These are available in various sizes and designs from any one manufacturer, and can be 'mixed and matched'. In addition the turbine consists of a baseplate, integrated oil supply system, gearbox, condensing plant, and monitoring and protection equipment.

The power outputs and efficiencies of turbines vary, so selection depends largely on the application and steam conditions available in the plant (see Table 5.4).

Table 5.4 Power outputs and efficiencies of steam turbines of various scale

	Power output (MW$_e$)	Turbine efficiency (%)	Steam conditions
Small	<1	35–45 single wheel	1–3 MPa 260–400°C
Medium	1–4	55–60 small multi-wheel	2–4 MPa 400–440°C
Large	4–40	75–85 multi-wheel	4–6 MPa 400–480°C
Very large	10–90	80–87 multi-wheel	6–14 MPa 440–540°C

The chosen configuration of a steam turbine varies with the installation method and the demands for heat on the site.

• **Back-pr essur e turbines**   (Figure 5.13) let down the steam from high to lower pressures, taking out the energy in the process of generating electricity.

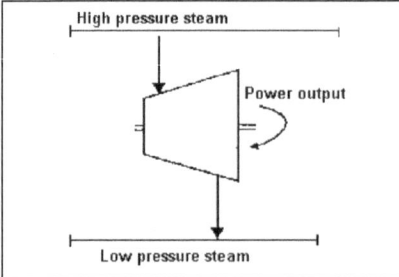

Figur e 5.13   Back-pressure turbine

• **Condensing back-pr essur e turbines**   (Figure 5.14) use cooling water to cause the greatest differential between input and output temperatures in order to maximize the shaft output power. Some useful heat is lost in the cooling, and the condensate is released at very low pressures: around 10 kPa.

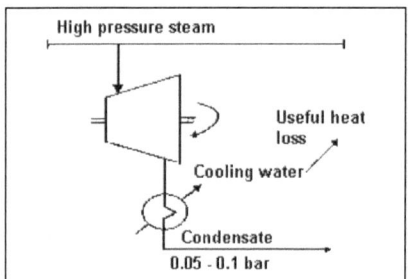

Figur e 5.14   Condensing back-pressure turbine

• **Extraction or pass-out turbines**   (Figure 5.15) let down some of the high-pressure steam intake through the turbine to provide medium- and/or low-pressure steam for use on the site. They can be condensing or not.

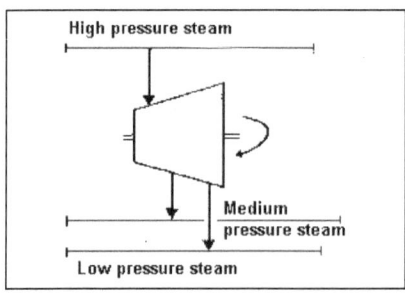

Figur e 5.15   Extraction or pass-out turbine

The steam extraction can be controlled (throttled) so that up to 95% can be extracted from the high-pressure steam intake to supply medium-pressure steam, or any proportion below that level to supply low-pressure steam (Figure 5.16). Uncontrolled extraction is an alternative design that can give up to 15% extraction, or alternatively up to 15% induction whereby steam is drawn from the medium-pressure steam mains to drive the second stage of the turbine and lower the steam pressure, which then supplies the low-pressure steam mains.

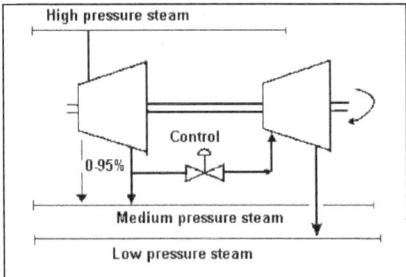

Figure 5.16 Controlled steam extraction

Combustion of wastes and MSW

Wastes have traditionally been disposed of by burning in the open air, dumping into landfills or discharging into waterways. These disposal solutions are no longer acceptable in a more environmentally aware world (though they are still all common practices, especially in developing countries). Alternative solutions to minimize and then treat the wastes are being sought. Meanwhile 'waste-to-energy' is one short-term solution, since most waste products have an energy value, which can be utilized and thereby help to offset the costs of disposal. For example, if the costs that would have been incurred in disposing of the waste material into a landfill for say $25/t for tipping fees and transport are avoided, then this saving becomes a negative cost to offset against the energy project costs.

There are three main options for recovering the energy in solid refuse:

- by mass burn (combustion or direct incineration) of MSW without pre-treatment

- by the production of more or less refined fuels out of the main waste stream, either partially processed or more highly processed RDF for later combustion in incinerators (such as rotary kilns) or via new pyrolysis or gasification techniques

- by the development of new approaches involving the recovery of chemicals such as plastic monomers combined with gasification, pyrolysis, hydrogenation and/or reforming of the gases and oils produced.

Waste combustion is gaining support as a means of waste disposal after a period of stagnation due to high costs and the very negative environmental image of early incinerators characterized by very poor performance. This is due to:

- higher landfill costs due to long transport distances out of large urban areas and environmental and sanitation requirements reaching well above $80/t in some parts of the USA and elsewhere

- increasing scarcity of suitable landfill sites in urbanized regions for both geological and political reasons

- recovery of the energy otherwise lost in landfill

- cynicism with respect to the cost of some separation, collection and manual recycling schemes.

For each proponent, however, there seems to be at least one opponent of incineration as a generic approach. The concerns most often quoted are:

- increasing costs of incineration technology due to ever more stringent environmental standards. In Europe or Japan incineration costs are often above $80/t, and a typical incineration facility costs $4000 per installed kW of capacity, whereas a combined-cycle power generation plant of similar capacity would cost a lot less

- uncertainty and risks incurred with the large investments required in the face of frequently changing regulations

- environmental impacts, in particular concerns over dioxins and furans, acid gas, heavy metal emissions and CO_2 emissions resulting from the combustion of fossil fuel origin materials such as plastics

- waste of materials and resources that are lost just as much through some form of incineration as with landfill

- low energy efficiency and poor energy recovery.

For the existing plants, the waste is often delivered from transfer stations, which act as convenient locations for bulking up the waste collected from different areas within the catchment region. Ideally it would first be passed through a waste processing centre to sort and allow recycling of glass, metals, etc. The preparation of the waste stream to remove incombustible materials and produce a more homogeneous and easily handled dry fuel can allow the use of more conventional combustion and other conversion equipment without modifications (Figure 5.17). MSW with the glass and aluminium components previously removed or excluded at source is delivered by truck or train to a waste reception and covered storage area, where the delivery trucks or wagons are unloaded and the MSW is pulverized. The rotary screen separates out the fines and oversize fractions that would not burn well. These are delivered to a nearby landfill for disposal. Magnets are used to separate any ferrous components for recycling, and the remaining material or **fuel fraction** is compacted and formed into pellets, dried, and conveyed to the energy conversion plant. The RDF pellets have a heat value of around 60% that of coal, and can be burnt in conventional solid fuel appliances. To avoid high corrosion

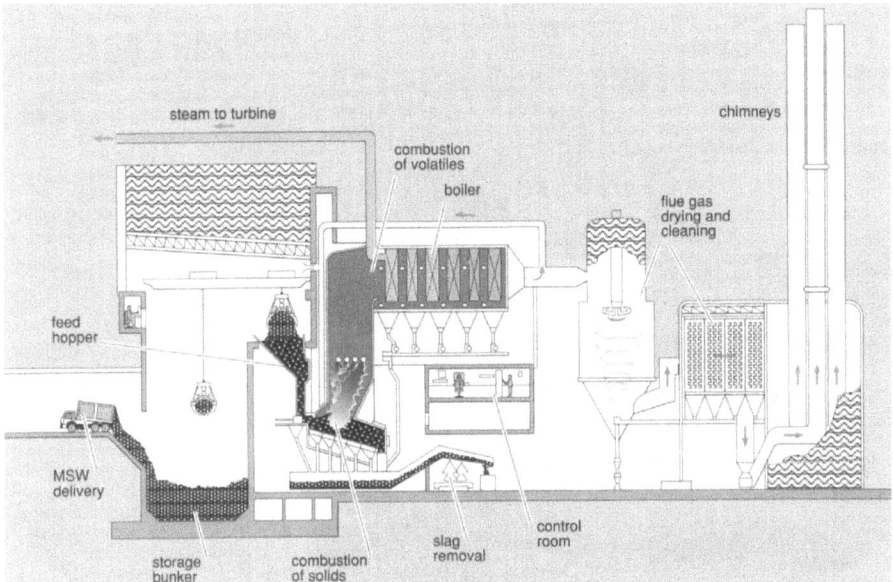

Figure 5.17 A typical large-scale MSW combustion plant
Source: Open University, UK

the steam temperature in some designs of these plants is restricted to less than
400°C. As a result, their total system efficiency is usually only between 12% and
24%.

Gas-cleaning equipment, a flue stack, and ash storage and loading area are
the other key components of the plant. The trade-off between fuel preparation
costs (requiring considerable manual labour input) and the need for more
elaborate combustion equipment often leads to the most cost-effective
compromise solution.

Incineration   is a generic term that encompasses a wide range of options
that differ markedly in technology, economics and environmental impact. In the
Netherlands a 40 MW mass-burn incinerating plant has recently been developed
and is apparently operating successfully. In countries such as the USA and
Australia, where there is adequate land available, while a number of
incineration schemes have been considered over the last decade, few so far
have been found to be economic relative to landfill. However, the debate
continues. Present trends indicate a move away from single solutions (such as
mass-burn incineration) towards the integration of more advanced incineration
technology within overall waste management strategies based on setting
priorities for waste treatment methods. These include waste minimization,
recycling, materials recovery, composting, biogas production, energy recovery
through RDF and residual landfill.

Many mass-burn combustion schemes that were started decades ago could
not be initiated nowadays because of far more stringent environmental
regulations. In many countries most older facilities are simple incinerators
without any substantial heat recovery. In Japan, for example, only about 35%

of schemes are truly waste-to-energy. In the USA, 75% of existing facilities recover the heat, but 80% are of low-efficiency mass-burn or modular types. In most countries that use incineration schemes the high capital investment cost is subsidized by the government.

The incinerator designs for mass burn differ from those for RDF, and so do their costs and environmental impacts. Mass burn is typically a low-efficiency approach to eliminate large amounts of refuse but with little energy recovery. Typically, MSW has an average heat value of 8–12 MJ/kg and mixed plastics of 33 MJ/kg as compared with 20 MJ/kg for dry wood. Most dry wastes can be used as fuels and burned successfully to produce heat for direct use as process heat in industry or space heating in buildings. Municipal and industrial wastes tend to:

- be even less homogeneous than wood and crop residues

- have a lower energy density

- differ widely in chemical composition

- contain a considerable proportion of incombustible material, which ends up as a high ash content

- contain heavy metals, and result in toxic emissions if not combusted properly.

This means that to burn these dry wastes effectively, and in an environmentally acceptable way, conventional combustion equipment, usually designed primarily for coal combustion, has to be adapted or more expensive specialized

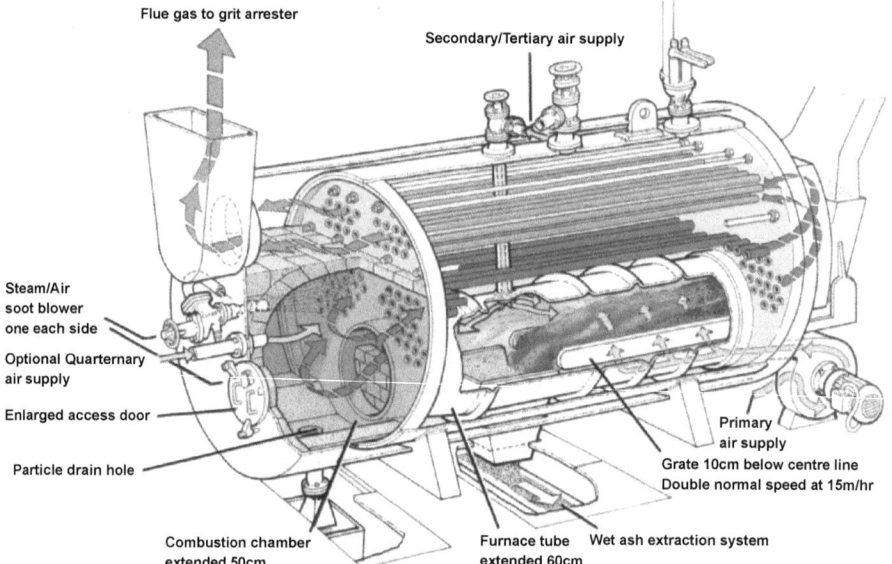

Figure 5.18 A 5.5 MW chain grate stoker shell boiler, modified to be able to burn RDF
Source: ETSU, UK

equipment used. The design of suitable systems that have low capital and operating costs, and which therefore can compete better with fossil fuels, presents a significant engineering challenge.

The use of raw, unprocessed MSW as a fuel causes problems owing to the heterogeneous nature of the material, which varies from suburb to suburb and from season to season. It also has a low heat value and high ash and moisture content. This makes it difficult for plant designers and operators always to provide acceptable, pollution-free levels of combustion. Processing of the waste to RDF partially overcomes these problems, and the fuel can then be used more successfully in either chain grate water-tube boilers or in circulating fluidised beds. Substitution of RDF pellets for coal in conventional equipment is not always straightforward, as some fuel characteristic variations such as moisture content can still occur. Fuel characterization of RDF is therefore an important tool for engineers in order to be able to understand the fundamental properties of the fuel and match it with the combustion system.

Net GHG emissions from waste-to-energy facilities are usually low, and comparable to those from biomass energy systems, because the energy generated is largely from originally photosynthetically produced materials such as paper, MGW and organic wastes, rather than from fossil fuels. Only the combustion of fossil-fuel-based wastes such as plastics and synthetic fabrics contributes to net GHG releases, but increased recycling of these materials will generally produce even lower emissions.

Case study 7

The Carter Holt Harvey 39 MW$_e$ plant using wood process residues, Kinleith, New Zealand

The Carter Holt Harvey (CHH) Kraft pulp mill at Kinleith, New Zealand, has been in operation for over 45 years, and was built when electricity prices were low and private generation was actively discouraged by the state power-generating authority. Deregulation of the power industry in the early 1990s reversed this policy. A variety of pulp and paper products are manufactured from plantation *Pinus radiata* logs, which are first debarked, giving a considerable quantity of wood waste. Historically some of this was used on-site to produce steam in two small boilers, but the majority of waste was traditionally landfilled.

The current total energy demand of the mill when in full operation is 470 t/h of steam and 60 MW of electricity. In the past steam was generated at high pressure in two black-liquor-fired recovery boilers, with additional steam from the two small wood-fired boilers and a natural gas-fired boiler as there was insufficient energy from the black liquor and wood wastes alone. The high-pressure steam produced was then somewhat wastefully let down to medium and low pressures for process use, and all electricity was bought in. In the early 1990s the two wood-fired boilers were due to be decommissioned as they were inefficient and had unacceptably high emission levels.

CHH, the site owner, carried out a number of studies over many years to solve these problems, and a new wood-fired cogeneration plant, as proposed by the then New Zealand state-owned power generator, ECNZ, was the most attractive arrangement. CHH would use the steam; ECNZ would take the power and sell some of it back to CHH. In 1994 CHH entered an agreement with ECNZ to set up a joint venture company, Kinleith Cogeneration Project Ltd, and to develop a cogeneration plant on the mill site. The plant was originally developed by ECNZ, but under the recent electricity reforms ECNZ was broken up into three companies. Genesis Power Company Ltd inherited and operates the plant.

The power plant consists of a new wood waste and natural gas co-fired boiler and a 40 MW_e passout, back-pressure steam turbine generator (Figure 5.19). Construction began in early 1995, and the plant was commissioned in late 1997. Any steam generated from the wood waste will replace an equivalent amount of steam generated by relatively expensive natural gas and therefore also reduce greenhouse gas emissions. The original small natural gas boiler was less efficient than the new plant but will remain as back-up. Other atmospheric emissions have also been reduced.

The function of the plant is to provide steam to the pulp and paper mill as required for process needs, primarily from the combustion of wood waste, and

Figure 5.19   General view of the 40 MW_e cogeneration plant at the Carter Holt Harvey pulp and papermaking mill at Kinleith, New Zealand

to generate electricity as a consequence of the pressure reduction from the steam generation pressure to the required process pressure. As well as 39.6 MW$_e$ it can supply up to 370 t/h of medium-pressure steam at 1250 kPa and up to 197 t/h of low pressure steam at 450 kPa.

The plant burns over 200,000 t/yr of *Pinus radiata* bark, wood chip fines (residues that are generated on-site from process operations), and some imported sawdust. The boiler is designed to handle double this volume if needs be in the future, as considerable quantities of wood residues are available from other plants and forests in the region. For 35 years or more before the plant was built, most of the bark and wood waste had been dumped into landfills. It is estimated that just one landfill site has 3 Mt of the biomass waste. Samples were recently taken by CHH from 2 m depth and it was found that there had been little change in the heating value of the material during this time, so there is the possibility that this material can be extracted (mined!) and also used as fuel for the plant. But further tests need to be done to 20 m depths before any decision will be made.

From a greenhouse gas point of view it is probably beneficial to burn this material rather then leave it to decompose slowly and release methane (which has a global warming potential 21 times that of carbon dioxide per molecule). Fossil fuels will be used to extract and transport this biomass fuel, but this would be only a relatively small amount of energy by proportion. Natural gas is available as a back-up and can be burnt alone or co-fired with the wood waste. The equipment supplier was Rolls-Royce Industrial Power Ltd, the boiler plant being designed and manufactured by subsidiary company John Thompson NZ Ltd.

Most of the wood waste is produced on-site from a dry drum debarker and a hydraulic debarker, giving a typical particle size range from sawdust to stringy bark material up to 500 mm long. A volume of around 200,000 wet t/yr is available, with moisture contents ranging from 43% to 67% w.b. The design moisture content was taken to be 59.5% with a gross heating value of 19.0 MJ/kg dry weight and 2.5% ash content (dry basis). The new plant was designed to burn around 53 t/h of fuel at this high moisture content, but for short periods it can cope with fuels of over 65% m.c.

The wood waste is delivered onto a reclaimer with capacity for 16 h of boiler feedstock, which is a reciprocating moving floor. It discharges the wood residues onto a slow-speed drag chain conveyor and then onto conventional inclined belts, which are covered to avoid dust nuisance. A self-cleaning electromagnet is installed over the first belt to remove any ferrous material, and at the end of this belt is a classifier to remove any oversize wood particles. The belt conveyors deliver the wood waste into an elevated storage bunker located at the front of the boiler and with 4 h fuel use capacity of 200 t.

At full load up to 80 t/h of fuel is consumed, requiring a bucket load per minute to be delivered to the conveyor from the stockpile by the yard loader. Depending on the moisture content of the incoming fuel, some gas is normally burnt to boost output above the 100 t/h steam possible on wood alone.

The plant was purchased by CHH through the developer, ECNZ, under a turnkey contract with the contractor, Rolls-Royce Industrial Power Ltd, for approximately $625/kW. The equipment is as listed below and identified by numbers on the schematic (Figure 5.20).

The two old recovery boilers continue to burn the spent 'black liquor' arising from the pulp-manufacturing process to generate some of the high-pressure steam. This is fed into a common steam header system, also used by the new back-pressure steam turbine, which therefore had to match the existing steam conditions of these boilers, being 4500 kPa and 400°C. These are well below the steam conditions normally encountered in a utility power station but provide a good balance between maximizing the efficient conversion of fuel energy to electrical power and providing useful process steam in an industrial environment. The common header delivers steam to the turbine generator at a flow rate of 477 t/h, which can then generate up to 39.6 MWe.

The plant was designed primarily to provide steam to the pulp and paper mill to meet process needs from the combustion of wood waste. Electricity is generated by the pass-out turbine as a consequence of the pressure reduction from the steam, which is initially at a higher pressure than needed for the process steam pressure requirements of the pulp plant.

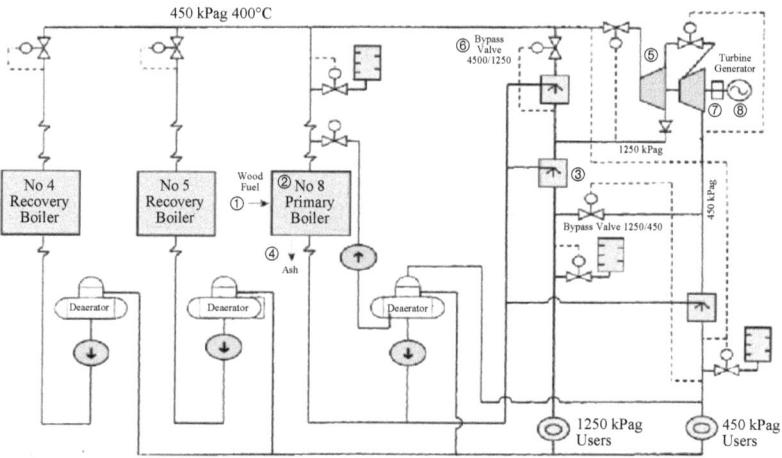

Figure 5.20 Steam system at Kinleith pulp and paper mill integrating a new No. 8 primary boiler and steam turbine generator with existing black liquor recovery boilers

1 Wood residue handling system including a reciprocating ladder-type reclaimer, a chain scraper conveyor, belt conveyors, a disc classifier and electromagnet.
2 No. 8 primary boiler: a John Thompson manufactured Eckrohrkessel, bottom-supported, corner-tube type, with Kablitz reciprocating grate wood-fired system. Also a natural gas burner, wood residue feeder, forced and induced-draught fans with variable-speed drives and grate cooling systems.
3 Electrostatic precipitator.

4   Ash-handling system.
5   Allen steam turbine, back-pressure, pass out, impulse type, rating 40 MW$_e$ at
    4932 rev/min.
    Maximum inlet steam flow is 477 t/h at 4500 kPa and 400°C.
    Maximum pass-out steam is 309 t/h at 1280 kPa and 260°C.
    Maximum exhaust flow is 170 t/h at 464 kPa and 172°C.
6   Steam pressure let-down stations from high- to medium-pressure steam.
7   Allen reduction gearbox 3.28 ratio (4932 to 1500 rev/min).
8   Peebles Electric Machines brushless generator.

Boiler plant

The boiler plant is based on a John Thompson wall tube boiler with sloping
grate. It consumes approximately 50 t/h of wet wood waste (mainly bark) to
generate the rated 100 t/h of steam. The moisture content can easily exceed
50% wet basis without problems occurring with this design. Slagging as a result
of contamination of the bark by the pumice soil in the region should not occur
owing to the design features. Natural gas is available as back-up supply, and
can also be used as a mix with the wood or alone at up to 5900 Nm3/h, when
up to a rated 180 t/h of steam can be produced. Gross thermal efficiency is 62%
on wood and 70% on natural gas.

The boiler is a single-drum, natural circulation, bottom-supported, water-
tube unit (Figure 5.21). The Kablitz sloping grate has a relatively large area to
handle the wet fuel, which can be conveyed into the boiler at the designed level
of around 59% m.c., but it can, if necessary, cope with extremely high moisture
levels up to 70%. The fuel is fed by a hydraulic feeder and enters at the top of
the grate before moving down it under gravity. This upper section of the grate,
designed to dry the fuel, is stationary and consists of spaced boiler tubes with
fins welded between them and preheated air introduced through small slots in
the fins. Thus the fuel meets primary hot air at around 200°C from beneath the
bed, and this dries the fuel ready for combustion.

Combustion does not occur till further down on the moving, mass-burning
part of the grate, which also contains a large number of holes for admitting
preheated primary combustion air. As the fuel burns, more moves down the
grate to replace it. The grate has eight sections that reciprocate to move the fuel
but also to break up any clinker that might form. If any of the pumice soil
enters, it has a low melting point, which could lead to clinker formation, but
steam jets are also aimed at the bottom of the grate to remove it as it forms.
The third section is a dumping grate where the combusted fuel is discharged
into the ash system.

Primary air is approximately 50% of the total air, with secondary air entering
towards the bottom of the grate and some colder tertiary air entering at the
throat of the combustor to combust the volatiles. The gas burners have their
own air supply. Air supply inlets can be controlled individually by the operator
to maximize thermal output for varying fuel types. The fuel can be too dry
(below around 30% m.c.) for this boiler design as this would limit the primary

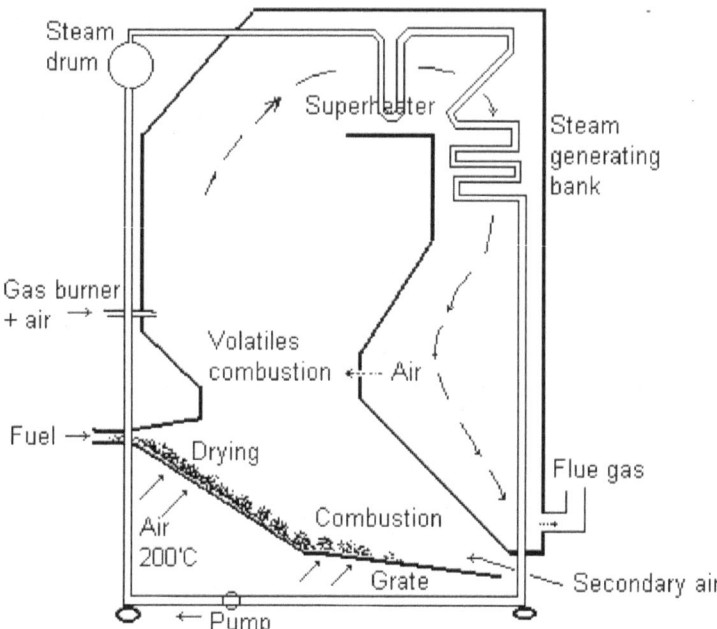

Figure 5.21 Simplified cross-section through the boiler, showing the sloping grate and fuel entry

air passing through the grate, leading to poor combustion taking place on a small part of the grate area, which would in turn lead to poor thermal efficiency overall.

The flue gases pass through the bank of steam tubes that form the main structure of the boiler (Figure 5.22), and the relatively high residual temperature is used to heat the incoming air to 200°C in a gas/air heat exchanger. Under normal operation the particulate emissions from the boiler should not exceed 70 mg/Nm³ as was approved by the environmental consents process. To meet this, the boiler has an electrostatic precipitator comprising a multitude of collecting plates, emitting electrodes, transformers/rectifiers and cleaning equipment mounted in a steel structure. It has three electrical zones in series, and is designed so that it can still operate at reduced efficiency when one zone is out of service.

Feedwater to the boiler is heated to 155°C in the de-aerator using the low-pressure process steam at 450 kPa, and then further heated to around 185°C in a feed heater using medium-pressure process steam at 1250 kPa. A combination of air preheaters and economizer for flue gas heat recovery achieves good thermal efficiency.

The boiler superheater has primary and secondary stages. The temperature of the steam leaving the boiler is controlled by an attemperator, which sprays

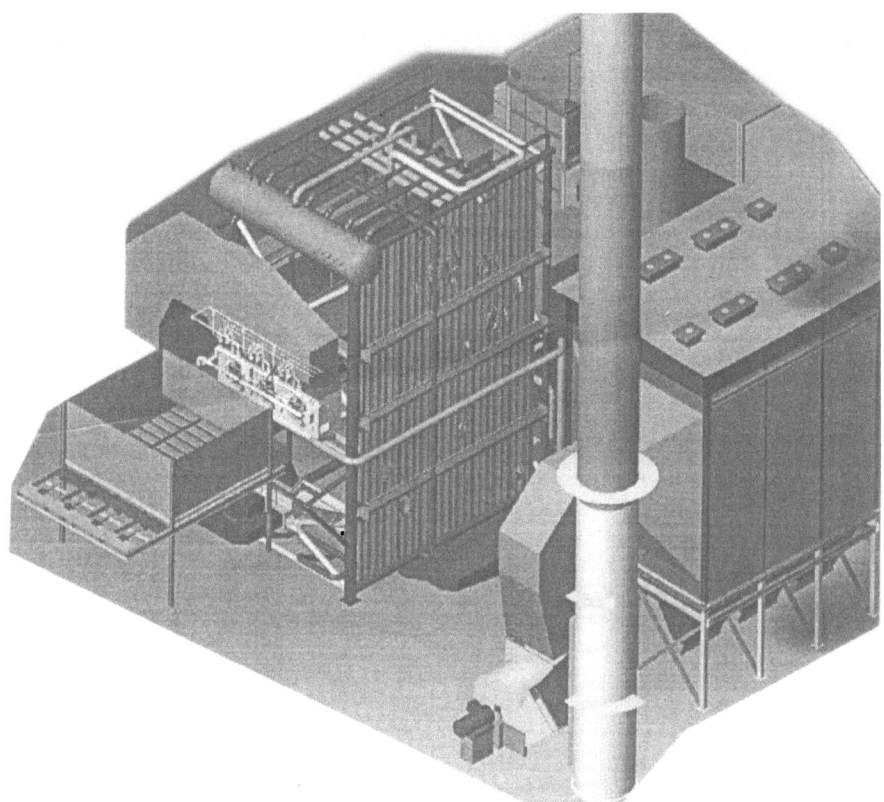

Figure 5.22 Isometric illustration of the boiler plant, showing corner tubes, steam cylinder and feed fuel system bin on left
Source: Rolls Royce/John Thompson NZ Ltd

water into the steam between the primary and secondary stages. The steam delivered to the turbine must be of high quality. The superheater spray purity is maintained by using the steam condensed in the feed heater as the source of water.

The gross thermal efficiency of the boiler under combined fuel firing is 70% but up to 86% when fired on natural gas alone. These high efficiencies are due partly to the combination of air heaters and use of the economizer.

Turbine

The turbine is a W H Allen H-5 single cylinder, pass-out, back-pressure steam turbine rated at 40 MW$_e$ and 4392 rev/min maximum (Figure 5.23). Under normal conditions, after high-pressure steam has been used for generation, the turbine can supply up to 309 t/h of steam at 1250 kPa and 170t/h at 450kPa to satisfy the mill processes. Under this condition the electrical power produced in the steam turbine generator is a function of the steam flow and is not controlled.

Figure 5.23 Steam turbine, gearbox and generator

Generator

The generator is a Peebles 39.6 MW$_e$ driven at around 1500 rev/min through an Allen gear box to provide 11 kV output. It is rated for an electrical output of 39.6 MW at 11 kV, 0.85 power factor, and 1500 rev/min at three-phase, four-wire and 50 Hz. Stator windings are star-connected, and the neutral point is high-resistance earthed through an earthing transformer and resistor.

Since the main business of the plant is to produce pulp, not electricity, the power generation is secondary to meeting the steam demand, and on average only around 25 MW$_e$ is generated. More power is purchased to make up the site demand of around 60 MW$_e$ capacity when under full operation.

Steam load at the plant varies as process machines are brought on line and others are shut down. The boiler and system was designed to cope with load changes of ±5 t/h in the high-pressure system and ±12 t/h in the medium-pressure system. There is an incentive therefore for CHH to keep up the steam load in order to be able to sell more power since the high-pressure steam for use in the process first has to pass through the turbine in this topping cycle.

In practice the plant has performed to expectations after initial problems with ash-handling conveyors were overcome. The environmental problems of waste disposal have been solved, and it is providing an economic return on investment. Plant designers have since stated that, with hindsight, a slightly larger boiler would have been an advantage.

Case study 8

MSW waste-to-energy plant, London, UK

The south-east London combined heat and power plant (SELCHP) is a new mass-burn incineration waste-to-energy plant generating around 30 MW$_e$ from MSW when running at its peak (Figure 5.24). Construction began in mid 1991 and took nearly 3 years.

Approximately 420,000 t of MSW from 400,000 homes, local businesses and commercial buildings is consumed annually. The raw waste is deposited by road trucks into a totally enclosed tipping hall, which can store 4 days' supply in bunkers. The waste is fed by crane into one of two identical incineration streams, each capable of burning 29 t/h. The two furnaces use reverse-acting stoker grates to keep the waste agitated during combustion. Energy recovery is achieved by integral three-pass membrane-wall boilers and economizers (which heat the incoming water). These produce nominally 152 t/h of superheated steam at 4.7 MPa and 395°C. The steam is fed to a single medium-pressure turbine driving a four-pole synchronous 31 MW$_e$ alternator with automatic regulators at a nominal 11 kV. Exhaust steam from the turbine is condensed by air-cooled condensers.

In the event of shutdown by one of the two incinerators, the full steam redundancy gives operating flexibility sufficient that the plant can keep operating. This gives reliability so that the plant availability exceeds 85%. The excess heat is sold to 7500 neighbouring domestic users and local schools. Over 6 km of double-insulated underground pipes will be used to connect the plant to both new and existing building heating networks via heat exchangers. Use of this heat will avoid the current use of several fossil-fuel fired boilers, which will then be retained for use only when the plant is being maintained or as top-up during particularly cold periods.

Pollution is minimized by using carefully controlled high-temperature combustion and advanced flue gas cleaning equipment, including acid gas

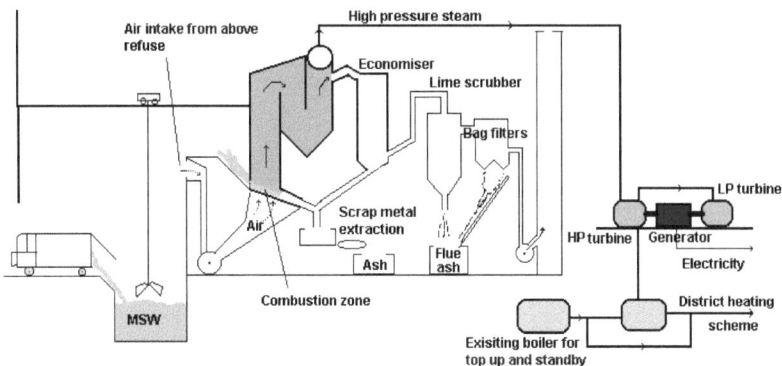

Figure 5.24 A typical modern waste-to-energy combustion plant for generating electricity and, in this case, also using the waste heat for district heating
Source: Based on a diagram from ETSU, UK

scrubbers and bag filters to remove particulates and dust. The site was selected for its convenience to the local electricity distribution system and the ease of waste delivery by road without adding disruption to the lives of those in the local community. This involved noise control down to 50 dBA around the plant at night by fitting sound-attenuating doors to the waste bunkers and tipping hall. On-line combustion chamber temperature monitoring equipment was provided to the local health authority so they could monitor the operation of the plant.

Costs were around $4000/kW in order to meet high European Union standards, which makes it non-competitive as an electricity generator compared with traditional forms without some form of government incentive or grant. Operating costs, including labour, consumables and ash disposal costs, are around $10 million per year. Offset against these costs must be the avoided costs of waste disposal. Revenue comes from the sale of the electricity (for around 10 c/kWh), the heat sales, the 20,000 t/yr of ferrous metals recovered from the fuel, and the ash for recycling. Around 50 full-time jobs have been created.

This facility illustrates the low efficiency of mass burn, and it is only economically viable because of very advantageous power purchase rates negotiated under the Non-Fossil Fuel Obligation[2] (NFFO) system. It has the prospect of extending into a cogeneration operation, which would be viable only in areas of very dense urbanization. Such a facility would not be viable in many countries, as the investment cost is too large, lower market prices exist for the electricity, and there is not always a viable market for the heat. An MSW mass-burn plant requires processing of at least 200,000 t/yr to be viable.

This case demonstrates the constraints and limits of most mass-burn technology. It was the first incineration facility to have been built for years in Europe. In the UK there are some 30 mass-burn incinerators, with only five designed to recover energy. Most were built in the 1960s and 1970s, and many will have to be closed because they no longer comply with EU environmental legislation. A similar situation also prevails on the European continent where some facilities, such as the Zindorf incineration plant in Bavaria, Germany, have to be retrofitted to improve combustion processes and cleaning of flue gases. The Zindorf retrofit will cost $4 million for a 28,000 t/yr capacity.

Notes

[1] Char and fixed carbon are used synonymously.
[2] The Non-Fossil Fuel Obligation was established to enable renewable energy (and initially nuclear) projects to enter the electricity market before becoming mature technologies by seeking tenders from developers of specific technologies such as landfill gas, wind or biomass gasification, and then awarding a fixed unit-price power generation contract to the most competitive of them. It has since been superseded by another mechanism to support renewable energy projects.

Chapter 6

Thermochemical conversion by gasification and pyrolysis

Converting solid or liquid biomass into a gaseous form can be achieved biologically (for example by using anaerobic bacteria to turn it into 'biogas' – Chapter 8) or thermochemically. Combustion, gasification, pyrolysis and liquefaction are all processes involving the thermal degradation of biomass fuels for which the available supply of air or oxygen (being the gasifying agent) determines the products. As discussed in Chapter 5, the combustion process and steam cycle is commercially available with minimum risk to investors but is limited by materials and technologies to around 30% conversion efficiency (from energy in the fuel to electricity) in the supercritical systems in large-scale units up to 600 MW$_{th}$. Most smaller-scale boilers have thermal efficiencies below 70%, and because of poor-quality steam limitations most small electricity generation plant designs have higher heat value efficiencies of around 25%, giving an overall conversion efficiency of only 15–18%. Gasification (Figure 6.1)

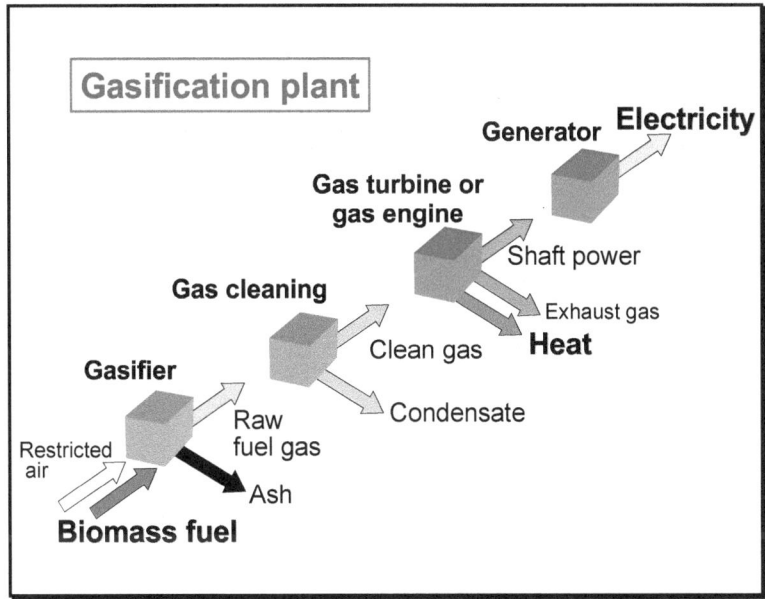

Figure 6.1 Gasification of biomass to generate electricity and heat

routes are therefore gaining interest as hot gas efficiencies for a gasifier can be over 95%, leading to around 35% overall conversion efficiency with 45–50% as a near-term possibility.

A biomass-fired gas turbine is similar in principle to a natural-gas-fired system, which is now well proven. It can be used in a stand-alone power generation system, which – assuming the gas can be readily produced or can be stored temporarily – is useful to meet peak power demands as it has a short start-up time of around 10 minutes. Alternatively the gas turbine can be integrated with a steam turbine to improve overall efficiencies, in which case it is termed a **biomass integrated gasification combined cycle** (BIGCC; Figure 6.2). Surplus heat can also provide process heat supply.

Pyrolysis is always the first stage of thermal decomposition in either combustion or gasification, and is considered as a separate process later in this chapter.

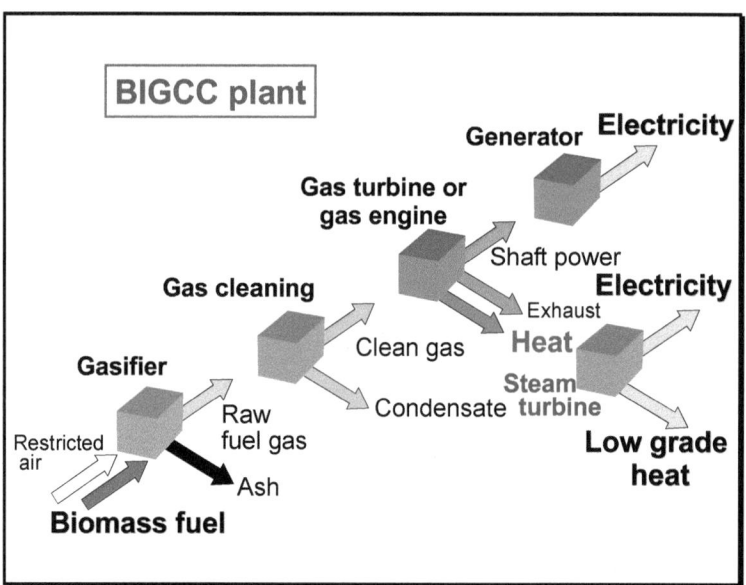

Figure 6.2 More efficient heat and power generation using gasification and steam combined cycles

Biomass gasification

Gasification of biomass is based on the partial combustion of material in a restricted supply of air or oxygen to give **producer gas**, which consists mainly of carbon monoxide, hydrogen and methane. Gasification in air produces a low-energy gas with a low heating value of around 3–5 MJ/Nm³. Gasification in oxygen yields a gas with a heating value around 15–20 MJ/Nm³, which is about half that of natural gas methane.

Gasification of biomass was first contemplated in 1801 by using the volatile gases driven off from the wood when producing the solid fuel 'charcoal'. By the 1850s, this 'wood gas' was being used to fuel the gas lights in London. In 1860 the first successful wood gas engine used for stationary applications was invented by Linoir, and by 1903 the first producer gas engine automobile was driven over 1000 miles (1600 km).

Earlier in the 19th century 'coal gas' from the gasification of coal had been used for direct heat applications, and wood gas was also used for this purpose. Coal gasification is well developed, with for example the Shell Buggenum 250 MW$_{th}$ plant operating successfully on 2000 t/day of coal for several years at 43% efficiency of coal to electricity. In such plants the gasifying agent is oxygen, giving a higher heat value gas than in biomass plants using air. Biomass plants are unlikely to exceed 100 MW$_{th}$ at any one location so it would be uneconomic to install an air separation unit to provide oxygen.

As well as its use for heat and power generation and as a transport fuel, producer gas can be used as **synthesis gas** to produce ammonia or methanol, which can in turn be used to produce synthetic petrol or as a source of hydrogen.

Indirect gasification to produce gas with a heat value range of 15–20 MJ/Nm^3 is also being demonstrated. This takes advantage of the thermochemical characteristic of biomass in that it will volatilize to about 80–85% gas, leaving a char of 15–20% of the input fuel energy. The char is separated and burnt with air to produce heat to drive the pyrolysis/volatilization/gasification process.

The temperature of reaction affects the type of volatile gases evolved. If the temperature is over 600°C then hydrocarbons are broken down to carbon monoxide and hydrogen. If the temperature in the reaction zones is lower than this, the resulting gas contains methane and other hydrocarbons, including ethane (C_2H_6) and ethylene (C_2H_4), together with oils and tars also present to a greater degree. Overall, the products are mostly gases of low to medium heat values.

The efficient operation of a gasifier involves careful management of the ratio of biomass to oxidant (air or oxygen) in order to obtain complete conversion of the char while avoiding excessive combustion, which would dilute the fuel gas with unreacted carbon dioxide. Gasifier operation is therefore more demanding than operation of a combustion system, and fuel specifications also need to be more controlled in terms of moisture content, particle size and other variables.

Gasification can have advantages over combustion if the plant is close to a natural gas line, which can then be used as a back-up supply if feedstock supply is limited, thus increasing the overall plant availability. In addition, natural gas can be used as a start-up fuel for pre-warming the system as it is easier to ignite. Biomass feedstocks suit gasification as they tend to have a relatively high inorganic content, which can build up on the walls of the boiler during combustion, causing scaling and associated heat transfer problems. In gasification plants these inorganic materials are removed as part of the gas-scrubbing process, so this problem does not arise.

The Texaco gasification process is an example of a proven large-scale gasification technology being actively marketed for a wide range of applications, including MSW processing. The core of the process is a pressurized gasifier operating at 2–8 MPa using an oxygen supply. In Germany, Veba-Oel uses a similar gasification approach to produce an oil substitute (40,000 t/yr) followed by hydrogenation at 30 MPa in its oil refinery. The process is apparently affected by a poor energy balance and negative public perception of it as an energy source rather than as a materials recovery operation. Texaco consider that a 100 t/day plant (that is, about 30,000 t/yr of pre-sorted waste) would cost about $40 million (without the ancillaries and downstream processing plant), which would be economic in the USA.

The gasification process of partial incineration with restricted air supply to create an air-deficient environment can also be used to convert MGW (municipal green waste) into synthesis gas. When integrated with electricity production at the medium to large scale it can prove economically and environmentally attractive, assuming that the alternative cost of waste disposal is treated as a negative cost. There have also been some interesting and innovative ideas demonstrated for using small-scale gasifiers to dispose of special wastes such as clinical waste from hospitals by mixing it with other biomass sources using an entrained-flow, downdraught gasifier.

The gasification process

Gasification of solid fuels involves a combination of decomposition and devolatilization reactions in a high-temperature environment of around 1200–1400°C with controlled oxygen or air supply to produce volatile compounds and a char matrix. In the devolatilization stage, also known as pyrolysis, the volatiles such as methane and higher hydrocarbons are driven off by the action of heat to leave a reactive carbon char. The volatile products are partially combusted at higher temperatures in secondary reactions while the char C is further gasified in the presence of restricted air, oxygen or steam to produce additional combustible gases giving the producer gas. This contains H, CO and CH_4, together with CO_2, H_2O, higher hydrocarbons and condensable tars.

Following the pyrolytic process in the oxidation zone:

$$[CH_2O] + O_2 + heat \longrightarrow C + CO_2 + CO + H_2O + hydrocarbons + heat$$
biomass

the following secondary reactions (simplified here under standard temperature and pressure) occur sequentially during this complex dynamic energy conversion process.

- The solid char and any hydrogen present are oxidized in the combustion air. These reactions, being exothermic, provide the source of heat energy to drive the other reactions:

$$C + O_2 \longrightarrow CO_2$$

$$H_2 + O_2 \longrightarrow 2H_2O + 393 \text{ MJ}$$

- The reversible water gas reaction or carbon/steam reaction is endothermic, and heat is absorbed:

$$C + H_2O \rightleftharpoons H_2 + CO - 131.4 \text{ MJ}$$

- The presence of water vapour in the gasifier results in the production of hydrogen as a secondary fuel component:

$$C_xH_{2y} + xH_2O \rightleftharpoons CO + (x + y)H_2 + heat$$

- The Boudouard reaction is highly endothermic. This reduction stage involves a reaction between heated carbon dioxide and the remaining char to form carbon monoxide fuel gas:

$$C + CO_2 \rightleftharpoons 2CO - 172.6 \text{ MJ}$$

- The water gas shift reaction then occurs, which is weakly exothermic:

$$CO_2 + H_2 \rightleftharpoons CO + H_2O + 41.2 \text{ MJ}$$

- The methane formation reactions are also exothermic:

$$C + 2H_2 \rightleftharpoons CH_4 + 75 \text{ MJ}$$

$$CO + 3H_2 \rightleftharpoons CH_4 + H_2O + 206 \text{ MJ}$$

So in summary the aim of the gasifier reactor is to produce volatile gases and carbon char, to convert the volatiles to CO, H and CH$_4$ gases, and to convert the char to CO. If the residence time in the hot zone of the gasifier reactor is too short, or if the temperature is too low, medium-sized molecules may escape and condense as undesirable tars and oils in the low-temperature reduction zone. Tar-cracking reactions involve the volatile products in the reactor, but these vary with the conditions and so cannot be easily predicted. Together with the water gas shift reaction, tar cracking generates CO, H and H$_2$O.

The use for the gas (direct heating, electricity generation or vehicle fuel, for example) determines its desired composition and the amount of clean-up of tars and particulates required. The proportions of the component gases are influenced by the reactor type, the operating conditions and the nature of the feedstock. The biomass fuel characteristics influence the thermochemical behaviour of the gasifier in terms of system performance and the consistent quality of the gas produced. The final fuel gas product can be burnt in a gas turbine, internal combustion engine or boiler. Since the gas has a lower heat value than natural gas, gas turbines usually require modification.

Thermal gasification needs process heat, which can either be generated in situ in the gasifier (autothermal) or be applied from external sources (allothermal). In an autothermal gasifier heating and drying of the fuel are preceded by pre-combustion reactions involving large quantities of heat generated from the in-situ combustion of part of the biomass charge, which provides the energy for subsequent endothermic reactions.

The process is complex owing to the varying behaviour of the biomass components cellulose, hemicellulose, lignin and volatile extractives when under high-temperature conditions. Gasification can be considered to be an incomplete combustion process manipulated by the interactions between:

- the design and physical configuration of the reactor

- the controlled volume of air admitted

- the temperatures necessary to yield a specific combination of products in the gas.

Key physical and chemical transformations occur at defined stages of the process (Figure 6.3). The supply of air is limited by the configuration of the gasifier, which is designed to constrain the primary gas-phase process and hence prevent the secondary oxidation reactions that occur during the combustion process. Heating and drying occur in phase (i), pyrolysis in phase (ii), secondary gas-phase reactions in (iii), char gasification in (iv), and tar-cracking reactions in (v). These phases correspond to varying temperature levels reached within the gasifier.

The heating and drying of biomass are physical processes, rather than chemical, and are affected by the physical size of the particles. The energy required for drying is increased by the presence of moisture in the feedstock, which therefore reduces the conversion efficiency since a significant amount of

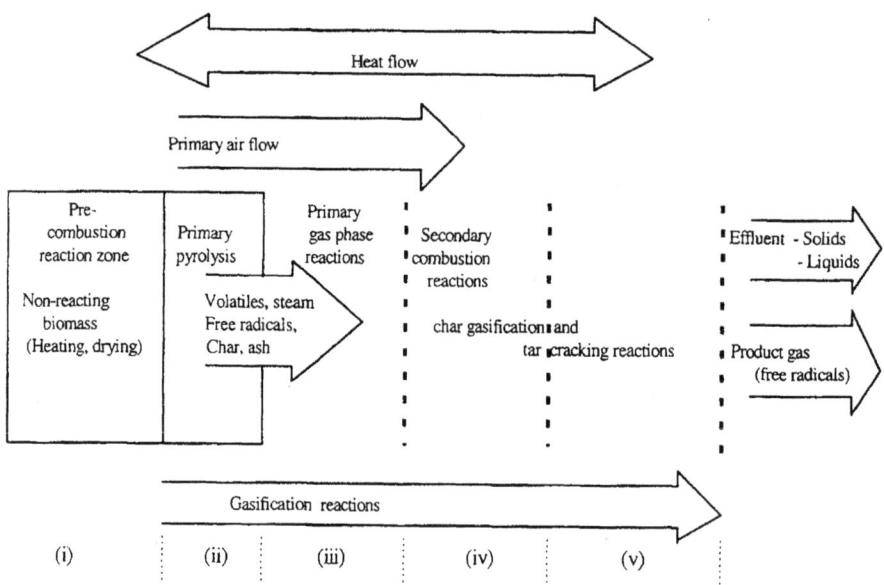

Figure 6.3 Physical and chemical transformations occurring during the biomass gasification process in a downdraught gasifier at temperatures of around (i) 100–200°C, (ii) 300–500°C, (iii) 1300–1400°C, (iv) 1200–1300°C and (v) 900–1000°C

Source: K Senelwa

heat is required to dry the feedstock prior to the pyrolysis and gasification reactions. Drying is also associated with the density characteristics through the porosity of the biomass, which accounts for differing rates of heating and drying between types of biomass, and even between species of tree or plant.

Biomass gasifier design

There are many types of gasifier including updraught (counter-current of air against fuel), downdraught (co-current), cross-current, and a full range of fluidized-bed gasifiers. A large variety of gasification systems are available commercially at the small scale (<100 kW$_{th}$), which produce gas of varying quality depending on the fuel and technology used. The broad type of gasifier selected for an application will be driven by economics as well as the choice being a function of size, fuel type and availability. There has been considerable investment in recent years in medium-scale plant (10–30 MW$_{th}$) also linked with cogeneration. At this scale biomass gasification is still of a prototype nature, at the demonstration stage. However, research is leading towards significant advances in the technologies, giving capital and operating cost reductions, performance improvements and life extensions. Cleaning the gas remains the most difficult challenge.

The typical thermal output capacity range of the various designs is:

- updraught: 20 kW to 1 MW

- downdraught: 1–15 MW

- bubbling fluidized bed: 2–50 MW

- circulating fluidized bed (CFB): 10–120 MW

- pressurized fluidized bed (PFB): 80–500 MW

Owing to fuel availability and the size of bioenergy opportunities, PFB gasifiers are unlikely to find application, and opportunities to use medium-scale CFB gasifiers may also be limited. Much research is focused on this design in Europe and the USA, but full commercial applications – especially for power generation – still carry high technical risks.

Larger-scale applications are normally efficient plants using integrated gasification systems with combined-cycle gas turbines and the surplus heat used where feasible. Such BIGCC technology has good potential at the 15–75 MW$_{th}$ scale, with economies of scale and improvements in generation efficiency. Several plants have been developed mainly for using woody biomass and bagasse feedstocks, and the concept is nearing full commercial realization as the pilot and demonstration projects are showing varying degrees of success.

Each gasifier design has a similar mode of operation regardless of scale, with broad zones within the gasifier where the biomass undergoes drying, pyrolysis, oxidation and reduction reactions (Figure 6.4).

- In the updraught, counter-current, moving-bed design the biomass feedstock flows in the opposite direction to the flow of air (Figure 6.4a). The resulting

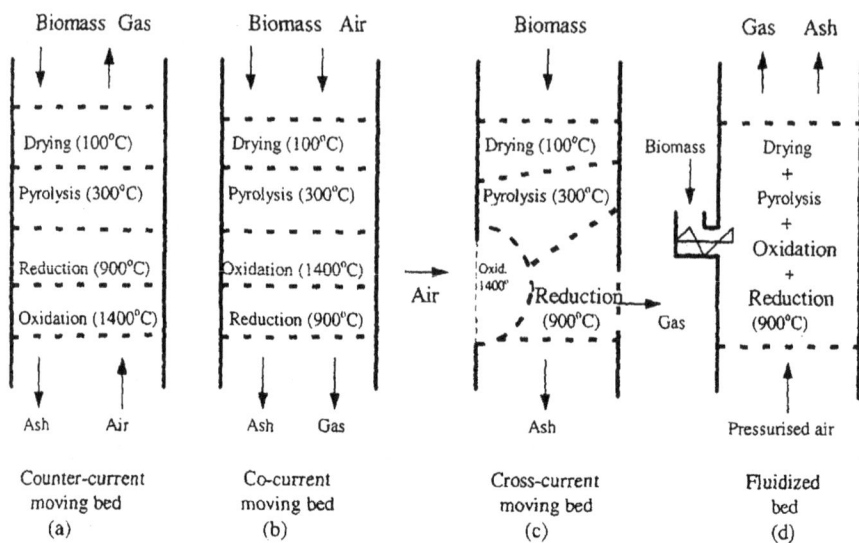

Figure 6.4 Common types of wood gasifier design and reaction zones: (a) counter-current moving bed, (b) co-current moving bed, (c) cross-current moving bed, (d) fluidized bed
Source: K Senelwa

ash and producer gas continue to flow in the same direction as the feedstock and air, respectively. The gas that leaves the reactor contains steam from the drying zone and a range of heavy hydrocarbons as pyrolysis products or 'tars', so it is more suitable for direct heating than for use in gas engines or turbines. A wide range of feedstock particulate size is possible.

• Feedstock and air flow in the same direction in the downdraught, co-current, moving-bed design, the bed of fuel moving under gravity (Figure 6.4b). All the heavy hydrocarbon decomposition products (or tars) from the pyrolysis zone have to pass through the high-temperature oxidation zone, where they tend to convert into simple hydrocarbons giving a gas with a relatively low tar content. This is advantageous where the gas is to be used in internal combustion engines or gas turbines, the tars causing many problems if still present. Hence it is critical to maintain the operating temperature, though with biomass fuels of varying moisture content and other variable characteristics this is not always straightforward to achieve.

• In the cross-current gasifier design the air and gas product move across the flow of feedstock so that it is difficult to control the gas to pass evenly through the zones (Figure 6.4c). This design is the least commonly used.

• Fluidized-bed designs are most common in the larger sizes of reactor (Figure 6.4d). A hot bed of sand particles is kept in constant motion by the air or oxygen, which is usually introduced through nozzles at the base of the reactor. The feedstock, having been previously reduced to a small particle size, is fed into the upper part of the reactor. The temperature throughout the reactor tends to be more even (800–1000°C), but the product gas often contains quantities of tars and ash: hence gas clean-up must be a key part of the process, and is often the most challenging.

Gas turbine systems

Standard gas turbines are engines that operate on what is known as the Brayton cycle, with continuous, steady-flow compression of air, constant-pressure combustion and expansion of the compressed heated gases through an expansion turbine. The working fluid is usually air, and fuels can be gaseous or premium liquid fuels such as distillate diesel.

- Aero-derivative gas turbines originate from the aviation industry, are made from advanced alloys, and are lightweight. They are twin shafted with two expanders and appropriate mechanical links, giving higher thermal efficiency but higher capital and maintenance costs than industrial gas turbines. Exhaust gas temperatures are generally lower than in industrial gas turbines. Performance is good under part loads but decreases dramatically at higher ambient temperatures.

- Industrial gas turbines are heavier, more robust and cheaper than aero-derivative gas turbines as they have a single shaft. Generally they have lower thermal efficiency and higher exhaust gas temperatures than aero-derivative gas turbines but can run for longer periods without overhaul.

A gas turbine consists of a **compr essor** , which consumes work (W_{comp}) to compress the air, and an **expander** , which produces work (W_{exp}) when the gaseous fuel is combusted with the compressed air. Since $W_{exp} > W_{comp}$ an export of the difference in work (W_{net}) is possible:

$$W_{net} = W_{exp} - W_{comp}$$

Combustion within the turbine occurs by one of two systems.

- Diffusion combustion occurs when the gaseous fuel and air are mixed in the combustion chamber. Local temperature peaks occur in the flame owing to improper mixing. When particularly high localized flame temperatures are reached, N in the fuel is oxidized by the air to produce NO_x.

- Pre-mixed combustion occurs when the fuel is mixed with the air prior to combustion. Initially, staged burning occurs in the fuel-rich zone of the combustion chamber nearest the flame. Then more air is added into the air-rich zone at the opposite end of the combustor. Flame temperatures can also be high enough here to cause NO_x to be formed.

The temperature at the expander inlet is typically between 800 and 1300°C, and is limited by the materials of the turbine blades. Not all the air goes through the combustion chamber, and excess air is mixed after combustion. The exhaust temperature after expansion is usually between 450 and 600°C.

A range of fuels can be used in gas turbines, including natural gas, methane, propane and light fuel oils, as well as producer gas from biomass gasification.

High-grade fuels are required, and they must be free of impurities, especially particulates (solid particles), to avoid damaging the turbine blades. The current major challenge for biomass gasifier designers is to produce gaseous fuel of consistent quality that does not contain tar and particulates. Ideally fuels are under high pressure when delivered to the turbine; otherwise they have to be compressed, which takes useful work away and hence decreases efficiency.

Gas turbines range in size from 250 kW$_e$ to 250 MW$_e$, and electrical efficiencies vary between 20% and 40%, though some manufacturers of newer designs claim even higher levels. There is a significant economy of scale for gas turbines, with costs ranging from $800/kW at the 250 kW$_e$ scale down to $200/kW at the 250 MW$_e$ scale (Figure 6.5). (This is just the cost of the turbine, and does not include installation costs, gasifier etc.)

Currently over 100,000 MW$_e$ of combined-cycle, natural gas turbine plants have been installed worldwide. It is currently the most cost-effective generation option in many regions of the world where natural gas is available. They were developed in the 1960s; uptake was initially slow, but as more experience was gained from the installation of increased capacity the costs declined rapidly and the demand went up. Typically, a unit cost reduction of 20% occurs for every doubling of installed capacity, and it is anticipated that biomass-fired gasification systems, currently at a very early stage of development, will follow a similar path.

Emissions

Normally complete combustion occurs within a gas turbine so there should be no CO remaining in the exhaust gas, which in theory is just 'hot air with high N'. If CO is present it would indicate poor operation and low excess oxygen. Other pollutants are:

- CO_2 from the fuel, but of no significance if from biomass since it is recycled

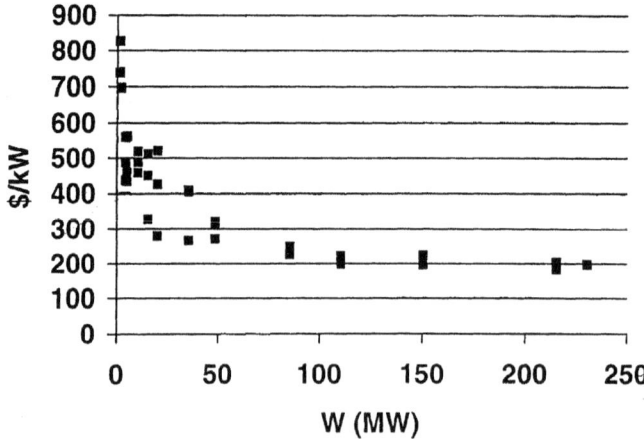

Figure 6.5 Comparative costs of gas turbines by output, based on quotes from a range of manufacturers

- SO₂ and SO₃ if S is present in the fuel, but biomass is usually low in S

- NOₓ from the fuel, especially if it contains leaf and bark material, and from the combustion air since N oxidizes at high temperatures

- unburnt hydrocarbons if incomplete combustion occurs.

It is interesting to compare emissions from a variety of fuels and conversion technologies (Table 6.1). The indicative differences shown are significant, although these will vary with specific fuel type. Switching from coal to natural gas will reduce emissions. Using gasification and gas turbines will give lower emissions than using the equivalent output capacity from steam boilers and back-pressure turbines. Emissions from biomass are comparatively low even when using boilers and steam turbines, and BIGCC (not shown) would be lower still.

A life cycle assessment of the production of electricity in a BIGCC plant was undertaken to determine how it would affect climate change and the environment in terms of CO_2 and energy use. Carbon closure (the percentage of the carbon in the biomass delivered to the power plant that is recycled through the system) was calculated to be approximately 95% assuming no change occurred in soil carbon levels. The energy ratio analysis showed that one unit of fossil fuel input produced approximately 16 units of carbon-neutral electricity exported to the grid.

Biomass can produce high NOₓ levels, especially when the flame temperature in the combustion chamber reaches high temperatures – above 1400°C – even for short periods.

There are several methods available to reduce NOₓ emissions:

- Water or steam can be injected into the combustion chamber at around 0.5 kg per kg of fuel to act as an inert material that absorbs some of the excess heat and hence lowers the flame temperature during combustion. The more water added, the lower the NOₓ emissions. Increased power output can result from the increased mass flow, but this does not compensate for the additional cost of the operation, nor the corrosion problems that may result in the burners and expanders.

Table 6.1 Atmospheric emissions from a range of fuels and conversion technologies

	Boiler + BPT Coal (2.5% S)	Boiler + BPT Natural gas	GT + HRSG Natural gas	GT + HRSG + BPT[a] Natural gas	Boiler + BPT Biomass
CO₂ (g/kWh)	2700	1010	610	510	15[b]
NOₓ (g/kWh)	5.2	1.5	1.1	0.9	3.0
SO₂ (g/kWh)	34.3	0	0	0	0

BPT = back-pressure turbine. GT = gas turbine. HRSG = heat recovery steam generator
[a] Combined cycle
[b] Carbon dioxide emissions from combustion are recycled

- Dry burners that give lean burning at the pre-mixed stage can be used to produce very low NO_x emissions of around 20 ppm. The technology is not available for all makes of gas turbine.

- Selective catalytic reduction is possible by injecting ammonia upstream of the catalyst into the exhaust from the turbine. The NO_x splits into N and water so that very low emissions can be achieved. High capital and operation costs are limitations, and a resulting increased back-pressure gives reduced turbine performance.

Examples of demonstration and commercial gasifier plants

Future Energy Resources Company, USA

An innovative indirect 40 MW_{th} gasifier developed by Battelle Columbus Laboratory was constructed and evaluated at the McNeil wood-fired power plant, Vermont, USA. It is a CFB design that is claimed to be one of the few specifically designed for biomass as opposed to coal. It uses 200 t/day of biomass, is steam blown as opposed to the more usual air blown, and so produces a gas with high heating value of around 16–18 MJ/Nm^3. The cold gas efficiency of the gasifier is about 70% (HHV), giving a 35–40% fuel to electricity conversion efficiency, compared with 20–25% for a conventional steam boiler plant. The plant uses two separate fluidized-bed reactors with circulating superheated sand applying the heat to the biomass particles to produce the gas. The design avoids the use of a hot gas clean-up system (often prone to technical difficulties), though problems were experienced with the novel gas scrubbing and cooling system, and modifications had to be made to improve reliability.

The project is hosted on the existing McNeil wood-fired power station-site, and a staged approach has been taken to development. It proved possible to develop the gasifier first, and use the gas directly in the existing station's boilers before testing the gas turbine as the next phase. The gasifier was completed in 1997 followed by start-up operations and testing, and has since been operating successfully to fuel the turbine.

Encouragingly, gasifier throughput of more than 400 t/day of forest and agricultural residues has exceeded the design estimates of 225 t/day. In addition the LHV of the gas has remained fairly constant regardless of fuel moisture content, so it can be used as a natural gas substitute.

The US Department of Energy estimated in its *Annual Energy Outlook 2000* that the capital investment for a high-pressure, direct gasification combined-cycle plant of this scale will fall from over $2000/kW_e$ at present to around $1100/kW_e$ by 2030. This will stem from experience gained, with the operating costs (including the cost of fuel supply) declining from 3.98 c/kWh to 3.12 c/kWh. By way of comparison, the capital costs for traditional combustion boiler/steam turbine technology were also predicted to fall from the present $1965/kW_e$ to $1100/kW_e$ in the same period. The current high operating costs of 5.50 c/kWh for steam systems reflect the poor fuel efficiency compared with gasification, but this may eventually reduce to 3.87 c/kWh. An OECD (1998) study of commercial-scale plants showed that at around 60 MW_e BIGCC would be more profitable than CFB combustion if the fuel could be delivered for $2/GJ.

Varnamo, Sweden

The world's first complete and longest-running pilot wood-fired gasification combined-cycle plant is the 6 MW_e and 9 MW_{th} cogeneration pressurized, circulating fluidized-bed plant in Varnamo, Sweden, constructed by Sydkraft AB with a Foster Wheeler gasifier and an Alstom gas turbine. The process is being marketed by Bioflow Ltd, a joint venture company between the participants. The decision to build was made in 1991, and plant start-up was completed by mid-1996. Operating experience of the gasifier at well over 8500 hours was gained using a range of fuels including wood, RDF, straw and bark, and with over 3600 hours operating on the product gas as a fully integrated BIGCC. The air-blown, 1.8 MPa pressurized gasifier gave an efficiency of 83%.

The low heating value of the gas (5 MJ/m^3) required the gas turbine fuel injectors and combustors to be redesigned to suit these conditions. The Alstom Power Typhoon turbine produced only about two thirds of its 6 MW_e rated output on this gas, and the Foster Wheeler heat recovery steam generator supplied the remaining 2 MW_e from a Nadrowski steam turbine. The heat produced was fed into the city district heating scheme in the winter. Initial problems included formation of deposits in the gas cooler and breakage of the ceramic hot gas filter, but these appear to have been overcome by changing to metal filters as the turbine is now being fired at 450°C without any other clean up. The N in the fuel was converted to ammonia in the product gas with around 70–80% yield, which was of concern, and NO_x emissions from the turbine required selective catalytic reduction to meet environmental limits.

Commercial applications of the Varnamo type of plant are likely to require a size probably around 60–70 MW_e to be competitive. Ultimately, when the system has been proven and experience has been gained from this pilot plant, then a larger 60 MW_e plant is expected to have a capital cost of around $1700/$kW_e$, and associated operating costs (excluding fuel) of 1.1–1.4 c/kWh.

Lahti Kymijarvi, Finland

This 60 MW_{th} Foster Wheeler CFB plant has completed several years' operation on paper, textiles, wood and peat fuels averaging around 50% m.c. It provides a hot but low calorific value gas around 2 MJ/Nm^3 used solely for heating purposes.

Bioneer

A number of gasifier designs have reached commercial production for heating purposes in Europe. One is the Bioneer updraught gasifier marketed by Foster

Wheeler for district heating. At least 10 of these plants fuelled by wood or peat have been installed with a 6 MW$_{th}$ capacity and an impressive average plant availability of 95–97%. Each plant employs three or four people. Other makes of around this size and smaller models are also available.

Small-scale projects

Many examples exist of gasifier projects between 30 and 300 kW$_{th}$, as numbers have expanded rapidly in recent years (Chapter 10). They are mostly fixed-bed units (mainly downdraught) with gas engines used to drive generators. In India and China there are at least 100 rice-husk-fuelled plants with the gas used to run modified diesel engines driving generators of around 200 kW$_e$. Recently India has implemented several demonstration and subsidy programmes to encourage greater deployment of these 100–200 kW$_e$ gasifier units in rural villages in an endeavour to move towards distributed generation rather than the more expensive option of expanding the grid supply network to reach outlying regions.

Adapting the small-scale gasifier/engine technology to meet environmental standards requires post-catalytic treatment of the engine exhaust gases in addition to placing emphasis on the handling and disposal of the ash and the toxic effluent streams from the cooling condensate from the gasification process. The Swiss Xylowatt company has developed a 55 kW$_e$ and 110 kW$_{th}$ system that meets the Swiss emission limit regulations, and has a claimed overall efficiency of 70%. It uses an IISc-Dasag Indian gasifier and a Swiss Bulle aspirated spark ignition engine.

Two systems in Northern Ireland (supported by the Northern Irish Non-Fossil Fuel Obligation) have also been developed by Eniskillen Research Centre and John Gilliland on his arable farm and by the small company B-Nine. In both systems the gas produced is used in gas engines to produce power for use on-site. Possibly in the future it could be used to power micro-turbines and fuel cells (Chapter 10).

Fischer-Tropsch liquids

Conversion of syngas to Fischer-Tropsch liquid fuels at the small scale has potential in areas where cheap biomass supplies exist. In practice producer gas must normally be utilized on-site as it is costly to store and transport, but if it is converted to a liquid fuel it could have wider application. The well-established Fischer-Tropsch process takes producer gas (CO and H) and converts it over catalysts into a range of liquid fuels that can then be used for heating or as a compression ignition engine fuel. This is creating much interest at the R&D level, not only for biomass but also for use in remote natural gas fields where pipelines are too expensive to construct.

Fischer-Tropsch liquids could also be produced in developing countries at the village scale and at similar prices to petroleum-based diesel, which tend to be high owing to transport costs over long distances from the urban centres. Simplified systems linking gasification, Fischer-Tropsch and combined-cycle gas turbine plants have been suggested to produce electricity and liquid fuels for rural areas in developing countries such as China using rice husks.

At the current stage of development gasification technologies for biomass fuels are still not fully mature, even after decades of R&D and with significant advances having taken place. Advanced concepts at the smaller scale, such as pressurized downdraught systems and combined-cycle systems, linked with fuel cells as the secondary converter, are under development and could reach 60% efficiency at the 5–10 MW$_e$ scale. Other areas requiring particular research and development for both small and large scales include improved tolerance to feedstock quality, improved feeding systems, increased automation, the development of hot and cold gas-cleaning concepts for tar and dust removal, more efficient compression of the 'syngas' for turbine combustion, and improved turbine design to handle producer gases with their associated contaminants.

Case study 9

Arable Biomass Renewable Energy project 10 MW$_e$ BIGCC, Yorkshire, UK

The Arable Biomass Renewable Energy project 'ARBRE' in Yorkshire, England, is led by First Renewables Ltd and supported by a 15-year NFFO contract awarded in 1996. The project is a fully integrated development including:

- constructing a high-efficiency combined-cycle power station supplied by biomass fuels

- contracting biomass supplies from dedicated crops of SRC *Salix* from local growers

- contracting for other sources of woody biomass

- applying domestic treated sewage sludge for land treatment and for use as a fertilizer on SRC crops

- recycling nutrients from the ash to the land

- the generation and sale of electricity, which is the main business income.

Planning approval was obtained in 1997 following widespread consultation with local residents, the local district council, the Department of the Environment and many others, together with the production of an environmental statement. The plant requires 43,000 odt/yr of fuelwood, which has been sourced mainly from SRC willow and other woody biomass supplies, principally from forest residues. These are required especially in the early years of operation before the SRC crops reach maturity. Over 1300 ha of SRC has been established by local farmers (out of a total of 1500 ha) within a 60 km radius and contracted for long-term fuel supply up to 15 years. The ecological benefits of SRC willow have created much interest, and after a slow start the encouragment of woodland grants of around $2000/ha to plant SRC has led to a greater rate of

Figure 6.6 Yorkshire ARBRE project: general view
Source: First Renewables Ltd

uptake. Six willow varieties from the British breeding programme have been grown to reduce the risk of yield loss from pests and diseases, and also to increase visual diversity. Treated sewage sludge is applied both to give increased yields and to provide an environmentally acceptable method of sludge disposal.

The SRC is harvested on a 3-year rotation in late winter/early spring before leaf growth, either directly as chips or as stems, and stored on-farm. Stems are later chipped after they have dropped below 30% m.c.w.b. Nearby 10,000 t of forest residues sourced from forests up to 100 km away are stored as a strategic reserve stock, but this resource will also continue to provide 20% of the plant's ongoing fuelwood supply after the willow crops have matured. Fuel is delivered to the power plant by road and tipped into the reception area, which holds 3 days' supply and is where the residual heat in the exhaust flue gas is used to dry the fuel. Processes are in place to ensure a continued supply of fuelwood at the required levels of quality and cost.

The plant design is a BIGCC (Figure 6.7), which greatly improves the overall power generation efficiency compared with traditional combustion plants and gives reduced atmospheric emissions. A Swedish Termiska Processor AB (TPS) atmospheric circulating fluidized-bed (CFB) gasifier had been successfully tested using a range of fuels, including those sourced from forest residues and SRC *Salix*, and this was selected as the heart of the process. The TPS Swedish manufacturing company (www.tps.se/egen/egen_en) is commercially operating two similar RDF-fuelled plants in Greve-in Chianti, Italy, and is also evaluating bagasse opportunities in Brazil.

The CFB gasifier design is air-blown and allows for a catalytic tar-cracking system. Chipped fuelwood is fed continuously into the lower part of the gasifier along with secondary air. Primary air enters the bottom of the vessel at high pressure to fluidize the bed. Gas leaving the gasifier contains ash, wood char

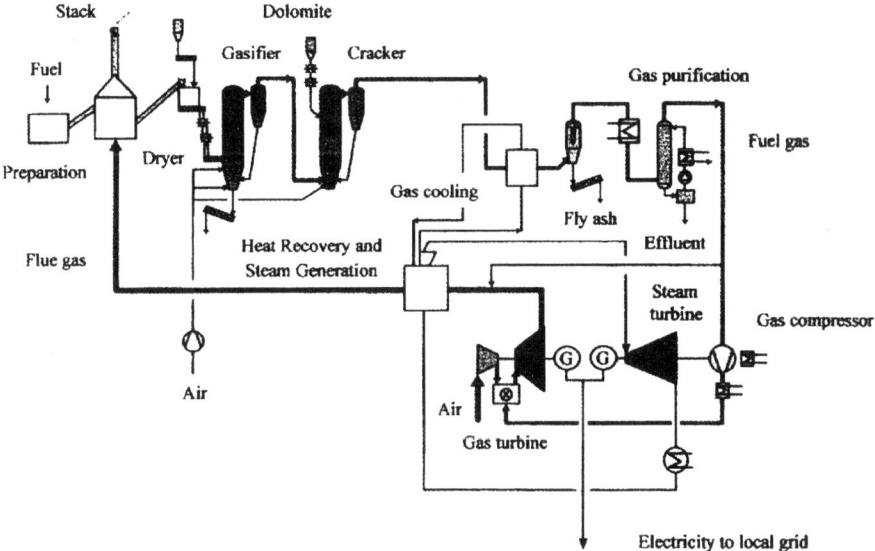

Figure 6.7 The Yorkshire ARBRE process for a gasification combined-cycle plant fuelled by willow coppice and wood process residue
Source: First Renewables Ltd

and sand particles, which are separated out in cyclones and returned to the bottom of the gasifier. The raw gas passes to the tar cracker, which is similar in design to the CFB combustor except with dolomite as the bed rather than sand. Here the hydrocarbons are broken down. The final gas stream consists of CO, H_2, CO_2, N_2, CH_4, H_2O and some traces of residual hydrocarbons and particulates that have escaped the cyclones. It then enters a gas-cooling process followed by cleaning using bag filters, a second cooling process and then to a wet scrubber to condense the water vapour and remove ammonia and traces of alkali compounds and tars.

Most of the cool, clean gas is then compressed ready for combustion in an Alstom Power Typhoon gas turbine designed to operate on low heat value gas. Heat is recovered from the exhaust leaving the gas turbine by the waste heat boiler. In addition a portion of the gas produced is fired into this exhaust stream to maintain temperature in order to produce superheated steam for a steam turbine. Any residual heat in the flue gases is used to dry the biomass fuel in the covered storage area. Generators coupled to the steam and gas turbines produce 10 MW_e of power, of which the plant requires 2 MW_e to run the process leaving 8 MW_e to be exported and sold to the grid.

Ash is removed from the bottom of the gasifier and from the filters and discharged into sealed containers for disposal or recycling. The plant components are modular, so future projects of similar design will have a relatively short construction period. Once the technology has been demonstrated and is considered proven, it is anticipated that rapid replication at a larger scale will then occur. This was one of the main reasons why the project also received support from the European Commission's THERMIE programme. Several projects are being developed at the 35 MW$_e$ scale to provide a significant contribution to the UK's renewable energy target of 10% electricity supply by 2010.

Based on the experience from this pilot plant, capital costs for a fully commercial 30–35 MW$_e$ plant are expected to be around $2700/kW, which it is anticipated can be reduced to $1400/kW with further project experience, and assuming future plants are expanded up to 50–55 MW$_e$. At the 35 MW$_e$ scale around 6000–7000 ha of short-rotation forest crops would be needed or the equivalent volume of biomass from sources such as the vegetative grass crop *Miscanthus* and wood process residues in order to supply the total biomass demand of 140,000–160,000 t/yr.

Pyrolysis

The goal of pyrolysis (Figure 6.8) is to produce from biomass a liquid fuel or **bio-oil** for blending as a substitute for crude oil, for direct use for heating or power generation, or for refining to produce a range of chemicals and fuels. The bio-oil or the refined fuel products from it can be combusted in a range of engine designs for either power generation or transport purposes. The key benefit of producing a liquid fuel is that it is easier to transport than either solid or gaseous fuels, as is exemplified by the petroleum industry. The biomass resource supply and the conversion plant can therefore be physically separated by a considerable distance from the end-use point for the bio-oil.

This complex thermochemical process converts solid biomass to a liquid form similar to crude oil by restricting the air available during the chemical conversion process to a greater degree than for gasification. Pyrolysis is the starting process for both combustion or gasification, which follow on if

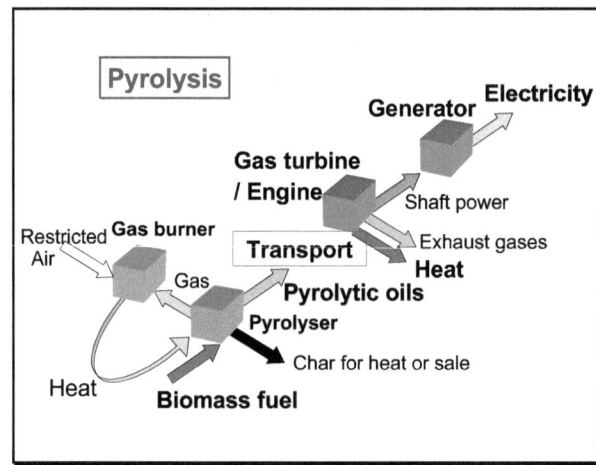

Figure 6.8 The process of producing pyrolytic oils from biomass then using the oil as fuel for an engine generating set

sufficient oxidizing agent is present. So careful control of the air intake is necessary to maximize the yields of bio-oil. Two mechanisms have been suggested for this pyrolytic conversion process based on using dry biomass. At temperatures below 280°C several sequential and competitive reactions predominate and are undertaken at slow heating rates. Above 280°C, and under faster heating rates, tars are produced through the rapid cleavage of the glycosidic bonds within the cellulose structure. These tars are combustible, and when heated to >500°C undergo exothermic reactions generating CO, CO_2 and H_2O.

The pyrolysis reactions and products are non-equilibrium and therefore hard to predict, but are dominated by pyrolytic oils, acids, water, solid char and a mixture of gases. They are dependent on temperature, the period of heating, the ambient conditions, the presence of oxygen, water and other gases, and the nature of the feedstock, especially with respect to any inorganic impurities that constitute the ash content. The nature of the oil product obtained is strongly dependent on the type of feedstock and on the conditions experienced by the products before they are cooled to prevent further reaction. The yield of volatiles increases with the heating rates, but at higher temperatures the tars (or condensibles) are cracked: this increases the total gas content, which in this process is undesirable.

The simplified pyrolytic reaction to convert the complex ligno-cellulose molecules of biomass to oils is

$$(C_xH_yO_z) + O_2 + heat \longrightarrow C + CO_2 + CO + H_2O + hydrocarbons + heat$$

Pyrolysis technology is less developed than combustion or gasification technologies and there are fewer commercial examples to date, although the Canadian company Dynamotive (www.dynamotive.com) and the US company Ensyn (www.ensyn.com) have been in this business for several years. Development costs are still particularly high and not well established because of the early state of development of the technology, so there is considerable scope for cost reduction. The production of pyrolytic oil decouples the source of biomass from the site of energy demand, which is not possible with combustion or gasification. Slow pyrolysis has also been used to turn carbonaceous material into charcoal, but this has limited application.

Current trends in R&D on both woody biomass and MSW processing show a renewed interest in fast pyrolysis and solvolysis approaches. Solvolysis refers to the use of organic solvents at 200–300°C to dissolve the solids into an oil-like product. Fast pyrolysis refers to the heat treatment of particulate organic matter at 300–1300°C under steam or other non-oxidizing gases at pressures ranging from atmospheric to 3 MPa to produce pyrolytic oils and/or medium to high energy value gases.

The current focus of commercialization is on fast pyrolysis, which produces varying quantities of char, gas and liquid but with the objective being to obtain as much liquid as possible. A particle of biomass has to be brought up to the operating process temperature within a few seconds, thereby minimizing the time during which it experiences the lower temperature range at which

Table 6.2 The typical product mix formed from different modes of pyrolysis to maximize the output of either bio-oil, charcoal or syngas (% by weight of dry feedstock)

	Liquid	Solid char	Gas
Fast pyrolysis	75	12	13
Moderate temperature, short residence time			
Carbonization	30	35	35
Low temperature, very long residence time			
Gasification	5	10	85
High temperature, medium residence time			

formation of charcoal occurs (Table 6.2). This is achieved either by using very small particles or by transferring heat rapidly to just the surface of the particle that contacts the heat source (known as an **ablative pr ocess**).

The bio-oil liquid as normally produced has a higher heating value of about 16–19 MJ/kg (compared with diesel at around 42 MJ/kg). About 25–30% by weight of the bio-oil is water, which cannot be easily removed by distillation or evaporation as on heating to 100°C it forms a part solid. The oil is not miscible with any hydrocarbon liquids but mixes with water, has a low pH of 2.5, has a relatively high density around 1.2 kg/l, and has an extremely wide range of viscosity, which tends to increase over time from 25 to 1000 cSt measured at 40°C. It is chemically unstable, and even more so at higher temperatures, so storage at room temperature is recommended.

Temperatures reached during the process typically range between 500 and 1300°C, while pressures range between 5 and 15 MPa. Since residence times must be quite short to maximize oil yield, thereby relying on effective heat transfer within the biomass particles, it must be finely ground, which normally requires dry material around 10% m.c.w.b. Under optimum conditions 60–75% ·of the dry weight of feed material can be converted into the liquid fuel. The residual gas and char are normally used as heat sources to drive the process.

The liquid fuel has the advantage of being able to substitute for or supplement transport fuels, in addition to its use for heating and power. The oil could be used directly in stationary diesel engines for power generation or in vehicles, but is normally refined or blended with diesel. Several pilot and demonstration units have been established for R&D purposes, and a wide variety of configurations have been tested, but no best design or method has yet been established. Commercial operation of pyrolysis plants remains limited, but interest continues to grow.

Areas for future development and process improvements include the following:

- Reactor. There is no best method or design, and because of the infancy of the technology radical innovations can be expected. The use of catalysts to improve oil quality is one possible area for investigation; scaling up the several pilot processes being investigated is another.

- Hot vapour filter for removal of char and ash carried over. Char can contribute to secondary cracking and to other instability problems (accelerating polymerization and increasing viscosity), so should be removed.

- Reduction of the vapour residence time will promote bio-oil stability and optimal yields.

- Methods for liquid collection. The liquids are mainly in the form of aerosols initially when first produced, so their efficient collection is needed.

- Modification of the bio-oil will be required for it to be more similar to current engine fuels.

- Operational aspects include problems with fuel handling and stability, engine start-up and shut-down, atomization, emissions, flame temperatures and patterns.

- The possibilities of co-firing with fuel oil warrant further evaluation, as do health and safety issues in handling and transport.

- The commercial opportunities for extracting chemicals, including furfural, resins, aldehydes, carboxylic acids and specific carbohydrates, need further assessment.

There is a need for further demonstration in terms of chemical and fuel production prior to coupling with engines for electricity production. Costs are not well established because of the immature state of development of the technology, but at present the bio-oil usually retails for 2 to 10 times the cost of fossil fuel oil. However, Ensyn has established a profitable business from fast pyrolysis in North America, with four commercial biorefinery plants producing 30 commercial chemicals and with the residual bio-oil sold as boiler fuel.

Pyrolysis is also another option for waste-to-energy that is being investigated. Pilot projects using pyrolysis for plastic wastes, and for mixed municipal solid waste, potentially have high energy efficiencies. Combined pyrolysis and gasification systems (such as the Thermoselect system) and combined pyrolysis and combustion (such as the Schwelbrenn-Verfahren system) have also been developed and implemented.

Incineration of waste can be a relatively expensive process, and gasification and pyrolysis even more so. Costs increase significantly to allow for more stringent pollution control measures and the safe disposal of the ash. Economies of scale are evident, which would inhibit development in many regions with relatively low and diverse population levels. British estimates by ETSU in 1996 for a plant processing 400,000 t/yr of MSW and generating 160 GWh/yr of electricity were over $80 million for investment costs, giving approximately 5 c/kWh electricity generating costs on average (at 8% discount rate). Against this must be offset the savings in gate fees for waste disposal by landfill and a comparison of transport distances and costs that would need to be included.

Case study 10

Pyrolytic oil from sewage sludge, Western Australian Water Corporation, Perth

An alternative to incineration or anaerobic digestion of sewage sludge (or dumping it out at sea, as is still often the case) is the innovative Enersludge™ process, which converts it into useful bio-oil (Figure 6.9).

The concept was first promulgated by Professor Bayer in Germany in the early 1980s, but it has only been more recently that environmental pressures and the increased costs of other sludge treatment options have made it competitive. The process has been commercialized by Environmental Solutions Ltd (www.environ.com.au), and the first plant has been installed at the Water Corporation's Subiaco water treatment plant in Perth, Western Australia. The main aim was to reduce the sludge disposal costs while keeping within environmental constraints. If income could be earned from sales of an energy product, or costs reduced by on-site power generation, this would be a bonus.

In essence the plant uses standard technology, fairly common in Europe, to produce dry pellets from the raw sludge; these have a soil fertilizer and conditioning value and are free of pathogens. The innovative part of the Enersludge™ process is the addition of a pyrolysis unit to produce gas, char and oil. The gas and char are used to heat the plant, leaving the bio-oil for revenue-earning activities, either for direct sale or for use on-site to fuel an internal combustion engine to produce electricity and thereby offset electricity purchases by the treatment plant.

The Water Corporation decided to invest in the process in 1995, and construction began in February 1997. The plant now treats 85,000 m³/day of

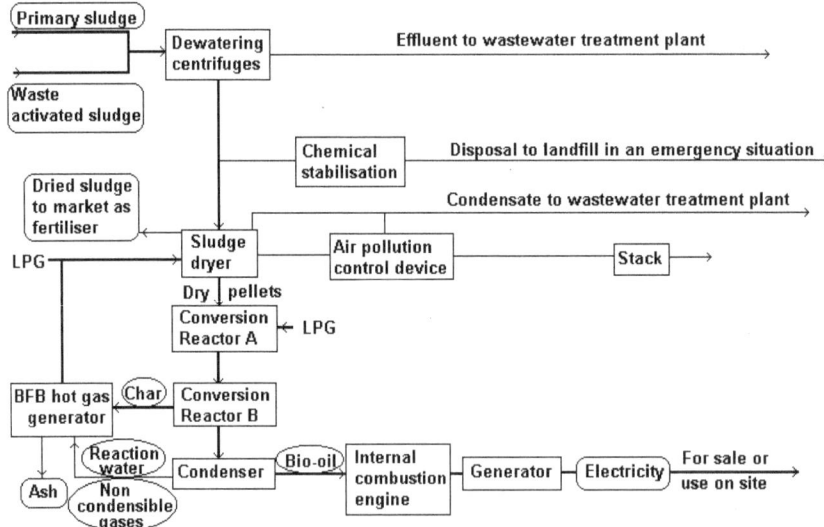

Figure 6.9 The pyrolysis Enersludge™ process for converting sewage sludge to bio-oil

sludge. It cost around $11 million, and is operated and supervised by one person only, requiring two 12-hour shifts a day. The wastewater enters the plant 24 hours a day, 365 days a year, so back-up systems have to be designed throughout the system in case problems arise.

After maceration, the raw primary sludge is mixed with active sludge at around 5% total solids (TS). The blend leaves the mixer tanks and, at around 4.5% TS, enters the de-watering centrifuges. Polymers are added to help settle out the solids, and a sticky cake material results with around 65% TS. The cake enters a triple-pass rotary dryer, which rotates at 3–4 rev/min forming 17 t/day of pellets each of around 1–4 mm diameter and at 95% TS. Drying heat comes from air having passed through a heat exchanger at around 800°C with heat entering direct in exhaust air from the fluidized-bed hot gas generator fuelled by the gas and char from the reaction, and leaving at 70°C.

The pellets are conveyed from the dryer directly to the pyrolysis conversion reactor, where the Enersludge process (Figure 6.10) converts these pellets into

Figure 6.10 The Enersludge pyrolysis section of the plant is physically separated from the dryer and hot gas generator

fuel, some of which is used to produce heat for the drying. The pyrolysis process is two-stage, with volatiles being driven off at conversion reactor A (Figure 6.11) and the char passing to conversion reactor B. The heavy metals present in the sludge act as catalysts. An LPG burner is installed to maintain the temperature at 400°C. From 1 t of pellets around 300 l of bio-oil are produced, which is currently sold for 10 c/l to several local companies for use as fuel in their oil-fired boilers. In the longer term it is hoped to produce this bio-oil to a sufficient standard in order to run the treatment plant's diesel engine/generator set and hence provide a portion of the on-site power demand.

Figure 6.11 First-stage conversion reactor of the pyrolysis unit

The hot gas generator is a fluidized-bed incinerator designed originally for burning coal dust but scaled down for this application. Getting it to run on the gas, char and reactor water (or on the pellets) proved difficult because the phosphorus concentration in the sludge caused slagging within the refractory area. The solution was to drop the temperature from 1050°C to around 800°C, which is below the phosphate flux temperature and therefore avoids slag

Figure 6.12 The bio-oil storage tank with ash hopper in the background outside the Enersludge plant building.

formation. Ash from the fluidized-bed incinerator includes the heavy metals now immobilized in it. This is either landfilled or used in a concrete mix to make terracotta paving bricks.

Another issue to be considered in the plant design was fire risk from, for example, the pellets being subject to spontaneous combustion, or from flammable gas leaking back up the char conveyor. Consequently a supply of liquid N was installed as a fire retardant and distributed to key points within the process. Odours are also treated and extracted from various parts of the process equipment to air pollution control devices, where they are treated prior to entering the emissions stack. The plant is fully monitored, and all the data are stored for any historical analysis that may be needed.

The bio-oil produced is stored in tanks ready for collection, and the ash is stored in a hopper ready for transport off site to landfill (Figure 6.12).

Plant operators are pleased by the success of the project to date, although even after several years of operation it is not yet fully developed.

Chapter 7

Biochemical conversion of wet biomass

The commonest biochemical process used to convert high-moisture waste biomass resources to bioenergy is anaerobic digestion, whereby bacteria produce biogas and landfill gas. Other biochemical processes are also used specifically for the production of transport biofuels such as ethanol, and these will be covered in Chapter 9.

Organic waste materials have an energy value and hence can be used as a source of biomass. Large volumes are discarded every year, but using conversion systems and techniques now being demonstrated around 20% of this 'technical potential' could be economically exploited at current energy prices. However, wastes generally have a low energy density, can be difficult to handle and process, and are often produced some distance away from the energy market. Minimizing and recycling wastes should remain paramount, being preferred to using waste-to-energy solutions, which it has been argued could encourage the production of even more wastes from packaging and the like.

Treatment of organic wastes and their management can also affect the release of greenhouse gases (GHG) in several ways:

- reduction of emissions of methane from landfilling of solid wastes

- reduction in fossil fuel use by substituting energy recovery from waste combustion

- more recycling of materials, leading to a reduction in energy consumption and process gas releases in mining and manufacturing industries

- greater carbon sequestration in forests due to decreased demand for virgin paper as a result of recycling

- reduced energy used in the transport of waste for disposal or recycling, though this is generally of minor importance.

The development of waste-to-energy plants often experiences major difficulties, not so much in their engineering design but in the legal complexities of obtaining planning consents and securing the waste supplies, which are important factors for investors and developers to understand.

Various technologies have been developed to treat organic wastes using gasification or pyrolysis cracking processes to produce what are effectively gas or oil substitutes. In Oregon, USA, the Conrad process was used to process urban waste to recover material from chemical polymers. A small-scale unit processing 5000 t/yr through a rotary kiln and a liming stage produced a bio-oil product. However, the process was banned because the oil substitute was considered by the authorities to be an energy product, and as such the overall process was not achieving the required level of material recovery. The Toshiba pilot process plant in Japan has a capacity of 250 kg/h over an 11-hour working day. It processes mixed plastics from Toshiba's factories to produce a range of oil substitutes. A high-density alkaline solution is used to neutralize the chlorine (from PVC, for example) and some of the additives that resist heat cracking. A second high-pressure cracking unit boosts reclamation further. No economic data are available, but other Japanese companies are pursuing similar routes.

Waste-to-energy facilities offer very substantial opportunities for the potential long-term development and use of liquid and solid waste streams for materials and chemicals as well as for fuel. The trend towards integration of incineration of dry wastes within mixed waste processing to complement recycling is paralleled by a technological trend towards materials and energy recovery. Technology is moving fast in this area, with a number of new approaches. New technologies for biochemical conversion of wet waste are also wide ranging, and several near-commercial projects link thermochemical with biochemical treatments.

- The **Wabio** process developed by Ecoenergy Oy, Espoo, Finland, is a form of biothermal waste treatment. Waste is pre-treated and divided into organic and RDF (refuse-derived fuel) combustion fractions. From 1 t of municipal waste 535 kg RDF is produced, which is combusted in a speacially designed fluidized-bed boiler unit. The temperature is kept below 900°C to avoid the formation of NO_x and of dangerous slagging compounds that could threaten the life of the boiler. The organic fraction is degraded biochemically into biogas and compost matter.

- The **Valor ga** process developed in France uses a similar approach. MSW is shredded and sorted mechanically (with manual input to check for omissions) to recover glass, metals, plastics and inerts such as sand and gravel, and to remove sources of toxic compounds such as batteries. The remaining fractions (including hospital waste) are separated into a dry RDF that is directed to a kiln for steam raising and base load power generation, while the fermentescibles are sent to a proprietary, high solids (above 45% solids), computer-controlled, high-yield methane anaerobic digester. The methane is used to produce peak load power. The organic residues are composted to produce a sterile high-quality soil conditioner. A plant processing 120,000 t/yr of fermentescibles could generate 31 GWh of power from the methane produced and 57,000 t of soil conditioner.

- The trend in favour of such new bioenergy technologies integrated within an overall waste management strategy focusing on materials and energy recovery is illustrated further by the French government's aiming to phase down landfills and develop up to 150 new MSW waste-to-energy conversion facilities.

This chapter concentrates on the biochemical treatment of this biomass resource, the combustion and gasification thermochemical conversion having been covered in Chapters 5 and 6.

Anaerobic digestion

Biogas production facilities intentionally convert high-moisture organic wastes to methane and then use it as a substitute for fossil fuels, thereby reducing GHG emissions (Figure 7.1). Under some circumstances the high ammonia content (for example from pig manure) can inhibit the conversion process, but this can be avoided by mixing the animal wastes with other, lower nitrogenous wastes to maintain the optimum C:N ratio. Wastes with high fat content, on the other hand, can enhance and increase methane output. The residue left after digestion, if free from contaminants, can be used as a soil conditioner and fertilizer. Woody wastes with high lignin content cannot easily be converted to methane, and MGW and wood process residues are therefore better handled by composting or combustion.

Small-scale digesting of organic wastes has become widespread in several developing countries. For example, many village communities in China and India digest food and human waste at the household scale using small, low-technology conversion facilities that are inexpensive and generally problem-free. Some European and North American communities have encountered greater difficulties when implementing larger-scale, mixed domestic, commercial and industrial bio-waste collection and composting schemes, ranging from odour complaints to heavy metal contamination of the decomposed residues. The alternative of large-scale composting requires mechanical aeration, which can be energy intensive, needing inputs of 40–70 kWh/t of waste. Waste treatment facilities that combine anaerobic and aerobic digestion are able to provide the energy for aerobic digestion from self-supplied methane if 25% or more of the waste is digested anaerobically.

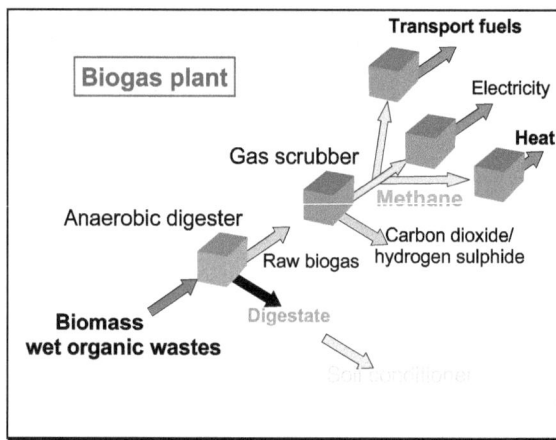

Figure 7.1 Biogas plant schematic

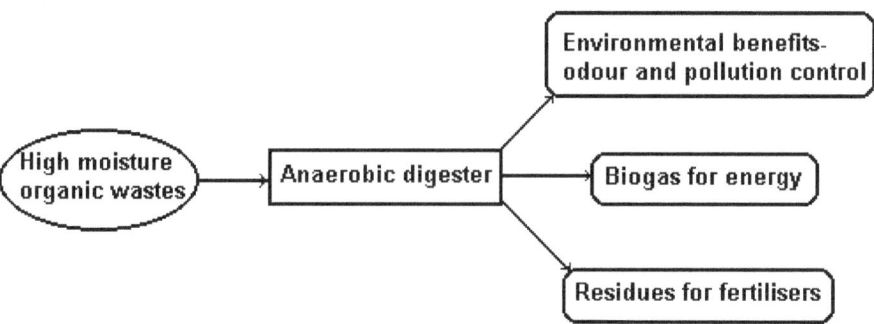

Figure 7.2 Key system elements of a biogas plant

The key system elements of a biogas plant (Figure 7.2) are

- procurement of the fuel, either organic wastes generally used on-site or, less commonly, green crops grown specifically to supply the plant

- biogasification of the comminuted or slurry feedstock, which is digested in a reactor, collected and stored (Figure 7.3)

- biogas utilization after scrubbing (removal of CO_2 and H_2S) to supply gas reticulation systems, to provide heat and power or for transport fuel after compression

- sludge disposal by returning it to the farmland, thus closing the nutrient cycle.

Figure 7.3 Biogas plant on a British pig farm with slurry storage tank on right, digester behind the small shed and, in the garage structure, two 100 kW$_e$ gas engines, which drive generators to power the farm and provide heat to dry lucerne for sale as high-protein animal feed

Anaer obic decomposition is the bacterial degradation of organic wastes or green crops in the absence of oxygen, resulting in the production of a methane-rich gas. The process occurs naturally on the bottom of lakes and wetlands, as shown by gas bubbles rising. Anthropogenic methane is one of the major greenhouse gases. It is produced from rice paddy fields, and also results from new hydropower plants when any land area upstream is first flooded, after which the vegetation decomposes over fairly long periods of time. Methane is also produced by bacteria living in the digestive gut of ruminant animals, and is an important part of their food digestion process. Any means of avoiding the generation of methane, or of capturing it and then converting it into carbon dioxide with a lower global warming potential, should be encouraged.

Decomposition is a natural bacterial decaying process that breaks down organic materials, and involves a number of biological steps. Well-defined classes of bacteria absorb energy for their survival from the gradually decomposing biomass, which is ultimately converted to methane, carbon dioxide and water. This process can be encouraged by placing the organic material and bacteria in large enclosed tanks or digesters maintained at set temperatures to maximize bacterial activity. The gas produced is then easily captured for use. Such biogas installations are commonly used for treating sewage, animal wastes and food process residues. Odours are removed as a result, and the pollution potential of the waste is reduced compared with its being dumped untreated into waterways as a cheap means of disposal.

The proportions of CH_4, CO_2 and impurities such as H_2S in the gas mixture depend on the feedstock material, the design and size of the plant, the retention time in the digester and the process temperature, around 35°C being preferred by the mesophilic bacteria. Typically a methane content of between 50% and 90% of the gas is achieved, depending on the feedstock (Table 7.1). It should be noted that the bacteria have to be nurtured for successful digestion to occur. If the temperature drops, or if antibiotics, detergents or toxic chemicals have been added into the wastes, the bacterial process will be inhibited and may even stop completely. At times the ammonia or sulphur concentrations in the feedstock are also sufficient to inhibit operation. The bacterial action generates some heat, but additional heating is often required by burning some of the biogas to maintain the digester contents at the optimum temperature.

In a continuous process (which is the usual case in commercial-scale biogas plants) the feedstock material is fed into the digester as a slurry, which then displaces the same volume of that already present. For a batch process, usually at the small scale, the digester would need to be almost emptied on completion of the digestion process before adding the next lot of feedstock, though it is important that some digested feedstock remains as this contains the bacteria. The average hydraulic retention time for the feedstock to be contained in the digester must be as short as possible for the process to be economic, otherwise larger digester reactors will be required to produce the same amount of gas. As an example, consider a pig farm producing 100 m³/day of waste for processing. If the retention time is, say, 20 days in order to optimize the volume of biogas produced, a 2000 m³ digester will be required. If under good conditions the RT

Table 7.1 Typical volumes and composition of biogas produced by digestion of a range of feedstocks at 35°C and with typical optimum retention times (RT) at a loading rate of 5% solids into a commercial farm-scale tank digester

Material	Biogas produced (1/kg of total solids)	Methane in biogas (%)	Suitable retention time (days)
Sheep manure	180–220	55–57	20–25
Cattle manure	190–220	56–60	20–25
Pig manure	170–450	60–65	20–30
Poultry manure	300–450	60–70	15–20
Waste newspaper	235–245	51–53	28–30
Aquatic weeds	370–390	55–57	20–25
MGW	370–390	46–49	25–25
Sugar beet leaves	375–385	65–68	20–25
Cereal straw (ground)	350–450	54–58	25–30
Hay	350–460	54–65	20–25
Whole green oats	450–480	51–55	20–25
Whole green maize	350–500	50–55	20–25
Grass	450–530	55–57	20–25
Kale (brassica)	440–560	47–58	20–25
Lucerne	450–600	56–64	20–25
Meat process wastes	590–610	58–60	25–30
Sugar beet root	610–630	64–66	15–20
Potatoes	860–890	53–55	15–20
Banana bunch	930–950	52–55	15–20

could be reduced to 10 days with minimal loss of gas produced, then the digester would need to be only half the size to ensure that the process has time to complete. In the last 20 years research has helped to shorten the retention time significantly.

The type of feedstock, its volume and characteristics need to be matched to the design of reactor and the infrastructural requirements, which is difficult in practice owing to the number of possible combinations. The volumes and strengths of the wastes determine the size of reactor. The bacteria present in the reactor break down the organic matter into sugars and then into several acids by a fairly complex process. The acids are decomposed to produce the final gas product, leaving an inert and odourless residue. This residue is mainly from the inorganic inerts in the feedstock and the compounds that are difficult to digest, but it also contains the bodies of the exhausted bacteria.

It is well understood that fibrous solids such as cellulosic materials in the feedstock require a longer retention time in the digester under mesophillic conditions (30–35°C) in order to break them down into acids initially. Undergoing methanogenesis at a higher temperature was thought to speed up the process. Therefore a laboratory-scale combination two-stage thermophilic digester operating at 55–70°C was developed at Manchester University in the

1980s in an attempt to speed up the limiting characteristic of solubilization. This followed an investigation of key digester designs and operating principles. However, the extra conversion efficiency gained from the fermentation being under thermophilic conditions was more than offset by the additional gas used to supply the heat energy required to raise the temperature of the feedstock. For on-farm wastes there was therefore little perceivable benefit. This was confirmed by subsequent work at University College, Cardiff, where it was found that thermophilic digesters produced far less gas and took longer to reach stability than mesophilic digesters did. Where the feedstock is already warm, such as from industrial food-processing wastes, the concept may have some application. However, confirmation will be required that greater gas production per cubic metre of slurry is possible than from a cheaper tank digester if it is to be an economic alternative.

The design and development of a biogas plant for biomethanation need consideration to be taken of such factors as the feedstock volumes and its characteristics, biogas production rates, seasonal variations, fertilizer requirements, the quality of the gas, compression and pumping technology, digester design and mixing and fermentation kinetics. In addition, planning consents will be needed. Technical problems such as blockage of feeding systems, scum formation, rheology and partitioning of solids in digesters can largely be avoided by careful design.

Digester design

There are many designs of digester reactor throughout the world, ranging from 1 to 15,000 m^3 capacity, some being more efficient than others. In Vermont, USA, for example, an evaluation of digester designs to suit best their dairy farm industry was conducted (www.state.vt.us/psd/ee/methane) as part of a broader evaluation. Virtually any biomass substrate within a wide range of moisture contents can be biomethanized, and purpose-designed digesters may mobilize a greater proportion of the organic feedstock for energy production. Typically, however, the process remains inefficient, with only 40% of the total energy potential being converted into methane gas.

Continuous stirred, tank reactor, anaerobic fermentation technology has been used for over 100 years. Today there are special automated feeding mechanisms and a degree of automatic control, which enables the solids concentrations of the feedstock to be doubled.

The semi-continuous, completely mixed, mesophilic single-stage tank digestion process (without recycling) is the most popular system on farms, particularly where feedstock is in the form of a slurry with 2–10% total solids (TS) rather than in a more solid form. Other designs such as plug flow, batch, two-stage and filter digesters are more costly and complex than a stirred tank.

Common construction materials for digester tanks of 100–800 m^3 are vitreous coated steel or concrete (often below ground). Smaller tanks (say less than 50 m^3) can also be manufactured in fibreglass. Typically a 30 m^3 digester would be suitable for 60 cattle or specifically grown green-crop feedstock from 6 ha of land, whereas a 500 m³ digester would be required for wastes from 1000 cattle or from 110 ha of green crops.

Four digester designs were constructed and evaluated by University College, Cardiff, for a range of feedstocks. A 30 m³ plug flow, 30 m³ conventional tank, 2 m³ experimental hydraulic and 1.5 m³ anaerobic filter were compared. A wide variation in gas yields resulted from the same pig waste feedstock. Generally the hydraulic and filter systems gave higher than anticipated efficiencies of conversion under short retention times and hence produced the highest gas yields. The plug flow was comparable to the conventional stirred tank but easier and cheaper to construct. The digester design and operation of two British-made farm-scale tank digesters were monitored by ETSU in the 1980s. They both proved satisfactory, though one of the plants experienced several problems with ancillary equipment. The usual method of effluent discharge by gravity flow through a gas trap or pump did not effectively remove the heavy, settled sludges formed of grit and inert materials that built up in the digester over a period of time. This reduced the digester volume and hence in effect reduced the RT. Hence there was a need to shut down the plant occasionally and manually clean out the digester. Should a design failure of any biogas plant prevent production of the biogas, depending on the application a standby fuel should be available.

Heating

Heating the digester contents can be achieved by internal heating coils and heat exchangers or external heating jackets. A range of heat sources can be used, including burning some of the biogas or using waste heat from an associated internal combustion engine. Insulation of the digester to reduce the amount of biogas required to heat the uninsulated digester tank to operating temperature can be internal, but owing to problems of its detaching as a result of thermal expansion and distortion of steel tank walls, external foam is sometimes preferred.

Mesophilic digesters need a relatively stable temperature around 35°C, but this often fluctuates (typically by ±4–7°C) within the tank owing to inefficiencies in heating and mixing systems. Since feedstock usually enters the digester at ambient temperature, and the effluent leaves at the higher process temperature, the feasibility of incorporating a heat recovery device has been investigated. Since the temperature difference is relatively low at around 20–30°C, a counter-current flow exchanger is preferred. Alternatively a heat pump could be employed to heat the slurry, either in the reception pit or in the digester.

The feedstock slurry should be of high total solids content for best results (around 5–10%, but even up to 40% if it can be handled without pipe blockages, etc.). Several farm installations that were closely monitored in the UK proved to produce only 60–65% of their design projections for TS, one reason being the dilution of the manure by rainwater from the roof, which lowered the TS concentration. This resulted in poorer plant efficiency, but could be overcome by better manure storage design.

Mixing

Mixing of the digester contents is required to:

- maintain an even temperature gradient
- maintain a homogeneous and even supply of feedstock
- prevent settlement of solids
- avoid crust formation.

Mixing by recirculating the gas generated through the contents of the digester by means of a compressor is a common means of avoiding crust formation or settling, for full-scale digesters. The length of mixing period each day may be a critical factor to obtain adequate mixing, which affects the associated power demand. Recirculation of the digester contents is a more popular mixing system, and only a few plant designs use a mechanical agitator system. Typically, at least 30 times the digester volume needs to be circulated daily.

Biogas plant operation

To optimize the economic returns per unit cost of the digester, gas yields of 70–80% of the maximum theoretical yields per kg of feedstock are needed. The optimum RT is a compromise between maximizing the gas yield per cubic metre of feedstock and limiting the size of digester and hence the capital cost of the installation. This depends on the organic loading rate (kg TS/m^3 digester volume), and therefore varies with the feedstock in question. Sudden variations in normal loading rates (for example, disposing of wash-down water into the digester effluent) can significantly reduce gas yields and also cause possible washout of bacteria. A European survey of digesters showed that most plants had an RT two to three times longer than the optimum values suggested at the laboratory and pilot scales, often as a result of oversizing of the digester and giving increased capital costs.

As a rough guide, in order to be economically viable a digester should be capable of producing 1 m^3 of gas per m^3 of digester volume per day throughout the year. In practice 0.5–0.8 m^3/m^3 day is typically achieved at a commercial plant, with higher rates obtainable at a pilot scale where operating conditions can be more carefully controlled and the feedstock is more consistent. To maintain flexibility in management it is also important that a digester can cope with a range of feedstocks.

Gas production depends on the digestibility of the solids in the feedstocks (m^3 gas yield per kg TS) and the amount of solids present. A higher concentration of solids in the feedstock results in higher gas yields per m^3 of feedstock fed. Typically cattle slurry can produce 0.2 m^3 gas/kg TS, pig slurry 0.3 m^3/kg TS and poultry slurry 0.35 m^3/kg TS. In practice it is very difficult to maintain a high constant solids concentration in the feedstock owing to poor feed reliability resulting from settling in the reception tank, feed blockages and a wide range of dilution rates resulting from poor design of slurry channels and collection systems. As a result, daily gas yields fluctuate widely within a typical range from e.g., 20 to 200 m^3/day, which makes matching supply of gas with energy demand difficult without building extra and costly buffer storage.

To obtain continuous and efficient biogas production, a properly designed feedstock supply is essential. Ideally 20 m^3 of biogas per m^3 of slurry should be achieved to satisfy both economic and energetic requirements. This requires a constant TS content of 6% for pig slurry, whereas in practice 2% TS is normal owing to dilution problems. Designing farm buildings to suit the digestion plant before construction is therefore preferable to retrofitting a biogas plant to an existing animal production unit.

Greencrop feedstocks produce a range of gas yields varying from 0.4 to 0.5 m^3 gas per kg total solids from fodder radish to 0.05–0.2 m^3 gas per kg total solids from the scrub weed *Bromus*. Variations in gas yields occur with time of harvest, methods of storage (including ensiling) and whether the crop is macerated prior to digestion, which increases the availability of biogas and hence lowers the RT. However, the gas yields tend to remain low (0.1–0.6 m^3/kg total solids), as does the methane content of the gas at around 55%. So the cost effectiveness of maceration and the energy demand for the process must make the practice questionable at this stage.

It is worth noting that the biogas composition can vary with the plant species, fertility level and, for animal wastes, with the animal species and diet fed. The methane content of the gas can vary from 55% to 75%, which affects the inherent energy value, ranging from approximately 20 to 26 MJ per m^3 of biogas. Gas from pig and poultry manures tends to be higher in calorific value than that from cattle.

The methane gas present has an energy content of 38 MJ/Nm^3, whereas the CO_2 component has no effective energy content and so should be removed before storing the gas, to conserve space and to reduce storage costs. Typically for each tonne of dry matter added in the feedstock, 200–400 m^3 of methane is produced containing 8–16 GJ of available energy. A typical biogas analysis is CH_4 65%, CO_2 30%, N_2 4%, H_2S 0.2%, H_2 undetectable. The raw biogas produced can be used directly for heating, but for other applications it will need 'scrubbing' to:

- improve the energy content by removing the inert diluents CO_2 and nitrogen, particularly if it is to be compressed or liquefied for a vehicle fuel

- reduce corrosion of process equipment, boilers, etc., by removing H_2S and water vapour, especially if the gas is to be used as fuel in internal combustion engines

- reduce potential health hazards by removing toxic sulphur compounds and the asphyxiant carbon dioxide

- reduce environmental pollution if sulphur dioxide is produced after combustion.

Several types of scrubbing technique have been developed, but they all tend to be expensive. The relatively simple techniques of passing the gas through a bed of iron filings (to remove H_2S) and a pressure water scrubber (to remove CO_2 and some H_2S) have been well tested. Although the equipment is capable of producing methane gas of 95-98% purity, its capital cost is high (around $0.1/m^3$ gas), as are the running costs ($0.08/m^3$ gas). Closed-loop water scrubbers show promise on a farm scale, as do membrane systems. An American system uses wood chips impregnated with iron oxide, but again the cost is high.

Biogas utilization

Matching gas supply to energy demand is one of the most difficult operations within the decision-making process when studying the feasibility of an anaerobic digestion project. A site survey is necessary to ascertain the volume and variability of the feedstock and to appraise the local energy requirements. It is essential that all the gas is used at the time it is produced if the economics are to be viable, since storage costs are high.

Agricultural uses for the gas (such as grain drying, space heating for greenhouses and intensive livestock production) and its use for domestic purposes (such as space and water heating and cooking) tend to be highly seasonal. This results in very inefficient use of plant, particularly if any gas excess to demand has to be flared or vented to waste. The problem tends to be further compounded since more gas becomes available in the summer months (when heating demand is often lower) as a result of less gas being required for heating the digester contents.

For direct use of biogas in gas boilers, only minor modifications are required (such as enlarging the gas jet to compensate for the lower heat value of biogas compared with natural gas). Greater difficulties can occur when using a commercial atmospheric, gas-fired boiler with a gas holder owing to the pressure fluctuations. Therefore a pressure jet-fired boiler would be preferable, giving less sensitivity to gas pressure fluctuations, reduced corrosion from sulphur deposition and improved air control, and making it easier to include a dual-fuel operation such as using oil as a back-up fuel.

It is doubtful whether a small-scale biogas plant would be economically viable if low-grade heat was the principle or sole use for the gas. The most popular use at this scale is to fuel a converted stationary gas engine, which drives a generator to produce electricity for use on-site. Where the waste heat from the engine can be utilized, a cogeneration (combined heat and power system) is warranted. Typically $1 m^3$ of biogas can produce 15–20 kWh of electricity. The low wholesale electricity purchase price in many places means that it is often not economically viable to export electricity to the grid. Hence as much power as possible should be utilized on-site to supplement power demand and save retail power purchase costs.

It is difficult to match generator capacity closely with demand, particularly since some reserve is needed to cope with short peaks such as motor start-ups.

Gas engines generate more electric power per m³ of gas when working at higher loads. It can therefore be a viable option to operate the alternator in parallel with the mains using a self-excited system so that the engine constantly runs near full load.

In many rural areas there is unlikely to be a demand from industry for piped gas from a nearby biogas plant. Introducing clean, scrubbed gas into the natural gas pipeline grid is another option, though only where the grid exists nearby, as extension would be costly. Another barrier may be the need to gain a suitable contract with the owners of the gas grid. The use of the scrubbed gas as a vehicle fuel is technically possible, and well proven in the form of vehicles powered by compressed natural gas. However, limiting factors are the costs of scrubbing and compressing the biogas, the cost of vehicle engine conversions and the need to match the gas supply with a suitable fleet of vehicles.

The use of biogas in spark ignition engines (either dedicated or converted) has produced several problems of operation:

- Bronze bushes need replacing with aluminium/tin bearings.

- Compatible lubrication oils must be used, and changed frequently.

- Carburettor adjustment is very sensitive.

- Variations in gas quality give lower engine efficiency and power outputs.

Even some specialist gas engines require further developments if they are to be fully reliable when running on biogas. All copper/bronze components need to be avoided and, where a basic compression ignition engine is converted to spark ignition, the compression ratio must also be reduced.

Environmental benefits

The sludge leaving the digester after processing contains similar nutrients to the raw organic waste, so it has potential value as a fertilizer or plant growth medium (after separation of the liquid fraction), or as an animal feed supplement, being rich in protein and odourless. If sold for such purposes it should be included in the economic assessment of the plant, but only if there is a guaranteed long-term market available, which in practice is not always the case.

The fertilizer value of digested effluent may be higher than raw slurry as, although the concentrations of total N, P and K remain similar, the nitrogen is concentrated into a more available form as ammonia ions. For this same reason, however, there may be greater losses of nitrogen by volatilization from digested effluents when stored for long periods. The added effects of digested effluents on crop growth are difficult to measure, and the results of field trials have proved insignificant, though one series of trials showed that effluent and raw slurry from a digester, applied at 22.5 m³/ha, was equivalent to 100 kg/ha of commercial nitrogenous fertilizers.

The recommended pasture rejection time after spreading raw sewage slurry is 3–4 weeks, whereas only a few days are necessary after the application of digested effluent, since the potency of any pathogens would have been greatly reduced during the process.

The increase in nitrate pollution levels in groundwater resulting from fertilizer addition and disposal of wastes from intensive livestock husbandry is causing public concern in some countries. Anaerobic digestion is one possible method of control, which also happens to produce valuable bioenergy and thereby can help to offset the costs involved. To be acceptable, any organic waste treatment process must:

- be cost-effective within the economics of the enterprise involved

- work all year round

- achieve the levels as set by the regulations

- be easy to operate and reliable

- stabilize soluble organic matter to prevent groundwater pollution

- control the spread of viruses, bacteria and pathogens.

The digestion process can in theory

- reduce the total solids by 30–35%

- reduce the volatile solids from around 60% of TS in the input slurry to 50% of TS in the output effluent (that is, by about 40%)

- reduce the volatile fatty acid content from 2000 mg/l to 800 mg/l

- reduce the biological oxygen demand (BOD) by 75–90%

- reduce the chemical oxygen demand (COD) by 15–60%.

Where pollution and odour control are the main objectives of anaerobic digestion, the digester should be operated differently from the method required to maximize production of the gas, for example by adjusting the RT.

If stricter environmental controls are applied, other means of organic waste treatment will compete with anaerobic digestion. So unless the economics are favourable it does not automatically follow that this method of treatment will be selected. Where more stringent environmental controls are likely to be generally applied, this will undoubtedly stimulate greater interest in biogas production, but in many countries there is no incentive to invest in a waste effluent treatment installation.

International experience

Anaerobic digestion to produce methane for fuel has been technically successful at a variety of scales in developed and developing countries. For many developing countries, the low cost and simplicity of small-scale digestion and the high organic content of the waste stream make it a promising solution, and millions of small digesters are operating in China and India alone. It is also realized that increased utilization of the organic fraction of MSW for biogas energy can also reduce waste management costs and emissions, while creating employment and other public health benefits.

Livestock manure management accounts for around 10% of the USA's methane emissions. Capture of about 70% of this methane appears technologically feasible for housed livestock, of which some 20% would be profitable under existing economic conditions (with zero carbon price). This would rise to nearer 30% if the GHG emissions avoided were to be valued at $20/tC$_{equiv}$, which would hence provide a greater incentive for farmers to collect and use the gas.

The development of the biogas industry is slowly advancing (though there have been no significant technological breakthroughs in digester designs and performance in spite of much R&D investment). For example, in Denmark a number of community-scale biogas facilities have been running successfully, accepting livestock manure as well as wastes from local food-processing industries, restaurants and the like. In Germany and Switzerland pilot projects compress the methane from biogas plants and supply it to natural gas vehicles. Canadian engineers have completed a pilot project using a mixture of waste-activated sludge, food waste, industrial sludge from potato processing and municipal waste paper. Methane production reached 50 l/kg total solids, and heavy metal contamination was found to be far below regulatory levels.

Anaerobic digestion has the advantage of generating methane that can be used as a fuel, yet many sewage treatment plants simply flare it, so the potential for energy generation is clearly large. New York City's 14 sewage plants, for example, generate 10.2 billion m³ of methane every year, most of which is flared. Cities such as Los Angeles sell methane to the local gas utility, and one New York plant in the Boston Harbor facility has been equipped with biogas-fed fuel cells that successfully provide electricity and heat, but this is still very expensive at this stage.

Concerns about contamination of sewage sludge by heavy metals have led to policies in many countries that now encourage incineration rather than soil application. Owing to the energy needed to dry the sludge for incineration, there is a net overall increase in GHGs. Alternatives to sludge incineration that yield improved energy efficiencies and reduce GHG emissions include gasification, wet oxidation, co-incineration with coal and anaerobic digestion.

Anaerobic digestion of MGW has also been successfully achieved. The process produces around 125 m³ biogas from 1 t of refuse of 50% organic matter content digested in a batch process in a 2000 m³ solids system. A fully integrated MSW recycling and processing plant coupled with anaerobic digestion of the organic fraction is technically feasible. The solid refuse from the digester cannot be applied directly to land, but can be used as fuel for power generation (Figure 7.4).

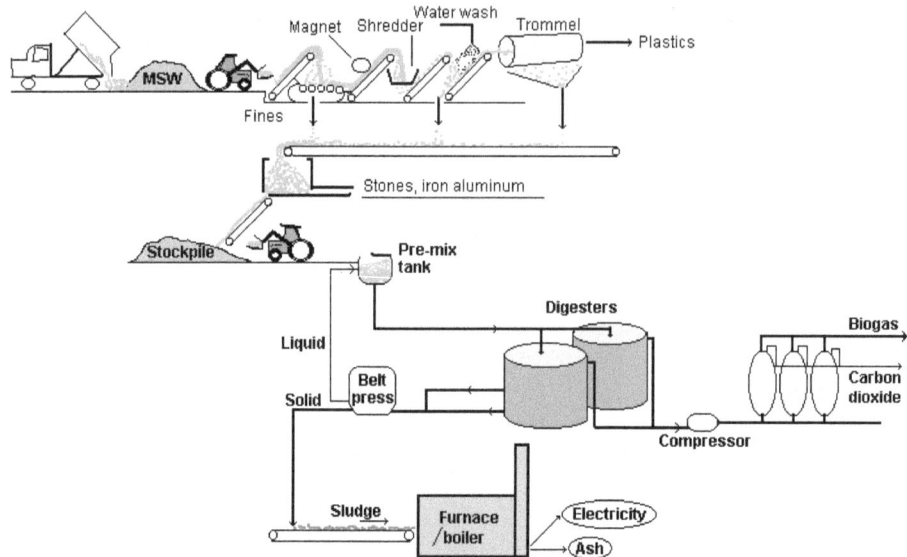

Figure 7.4 An integrated municipal solid waste materials plant in the USA recovers metals and removes plastics prior to digesting the remaining organic fraction and incinerating the solid digestate
Source: Based on a diagram from Open University, UK

Economic analyses

In the UK only sewage plants can produce biogas economically without subsidies. Electric power from farm wastes was estimated by British Biogen in 1997 to cost three times that from sewage at around 20 c/kWh at 8% discount rate. However, where the costs of disposal of the waste material are high, such as paying to transport and dump food process wastes to a landfill, payback periods of only 2–3 years can result from investing in a biogas plant for treatment on-site.

Using purpose-grown green crop feedstock for biogas production and ensiling for use in the off-season was analysed under New Zealand conditions. Each 1000 ha dedicated to crop feedstock production for the biogas plant would yield 10 GWh/yr of electricity. Biogas could be produced in farm-scale plants for around $3/GJ, but transport costs for the crop could add up to another $1/GJ, so it is not usually a viable proposition.

In Denmark an analysis of the 20 or so large-scale community plants running on farm and industrial food-processing wastes showed they were profitable. The gas was sold at a price comparable to that of natural gas, credit was given for the disposal of the wastes and the waste heat from the biogas plant was also utilized. Conversely a community biogas plant in India providing methane to run a dual-fuel diesel/gas engine/genset producing 5 kW$_e$ for lighting and pumping water was generating for over 35c/kWh, which was far more expensive than grid power. However, in this case the load factor was only 4 hours per day owing to the limited availability of the animal manure feedstock. Bringing in more manure to operate for 15 hours per day halved the generating costs, and then the system proved not only cheaper but more reliable than the grid power.

The capital costs for reactors to produce the biogas and power generation plants to convert it work out at around $1300/kW for a medium-scale plant of around 500kW$_e$. Total system operation and maintenance costs are relatively high at 6 c/kWh, so life cycle costs of over 20 c/kWh can be expected. With digestion of wastes the biogas can be considered as a free fuel if the alternative to digestion is the more expensive option of aerobic treatment. Furthermore digesting the wastes instead saves the energy that would have been used for the aerobic treatment to power the pumps, etc.

A wide range of digester sizes used to produce biogas from crops have been evaluated, ranging from 100 m^3 to 10,000 m^3. The smaller range is suited to on-farm use, 800 m^3 requiring feedstock crops from just over 100 ha of land, whereas the 10,000 m^3 facility would need feedstock from nearly 1400 ha or would need to be associated with a large sewage treatment plant or food-processing factory. Capital costs vary from $2000/kW for the smaller scale to $1000/kW for the large facility. Life cycle costs at 10% discount rate range from 6 c/kWh for the small unit to just under 4 c/kWh for the 10,000 m^3 facility.

Recent developments and greater experience in treating a wide range of waste organic products including MGW, high-protein wastes, high-lipids wastes and cellulosic wastes will lead to reductions in plant capital and operating costs, which should encourage more biogas plants to be developed. Indicative life cycle cost analyses can be summarized as follows:

- Waste feedstock: $0/GJ (with possible negative costs)

- Green crop feedstock: $5/GJ

- Transport: $1/GJ

- Capital cost of plant: $1000/kW for wastes, $2500/kW for crops

- Operation and maintenance: Low, High

- Electricity generating costs: 2–5 c/kWh, 10–18 c/kWh

Lack of greater adoption of the technology by many owners of organic waste streams is due to:

- poor comprehension

- the risk of the biological process failing

- high transport costs of feedstock

- multiple end-use decision options needed for what is a finite resource

- the inconvenience of running and maintaining a biogas plant.

Environmental legislation will provide opportunities for new plants, but research breakthroughs to enhance the gas yields, for instance by sourcing or engineering bacteria that will also digest the lignin component of green crop plant feedstocks, are needed if biogas plants are to become widespread.[1]

Case study 11

Anaerobic digestion of pig waste, Masterton, New Zealand

A pig farmer in the south-east of the North Island of New Zealand installed an innovative biogas plant in 1995 to overcome the problems of meeting the stringent environmental requirements of the Resource Management Act (1991) regarding disposal of pig effluent. The waste from the 6000 animals is separated into liquid and solid streams before being treated separately in order to maximize the gas yield (Figure 7.5).

The higher solids fraction of the effluent enters the first of two tank digesters heated to 35°C. After a 14-day residence time it is transferred to a second smaller tank and from there, after a few more days of digestion, the residual sludge is applied to nearby pasture, providing some useful nutrient components. The lower solids effluent stream is sent to a sludge blanket digester (Figure 7.6), and after a residence time of only 8–10 days is pumped to an irrigation scheme covering 340 ha of dairy farm pasture. In a dedicated energy system this effluent stream could be irrigated onto green crops, which would be harvested throughout the year and used as more feedstock for the digester.

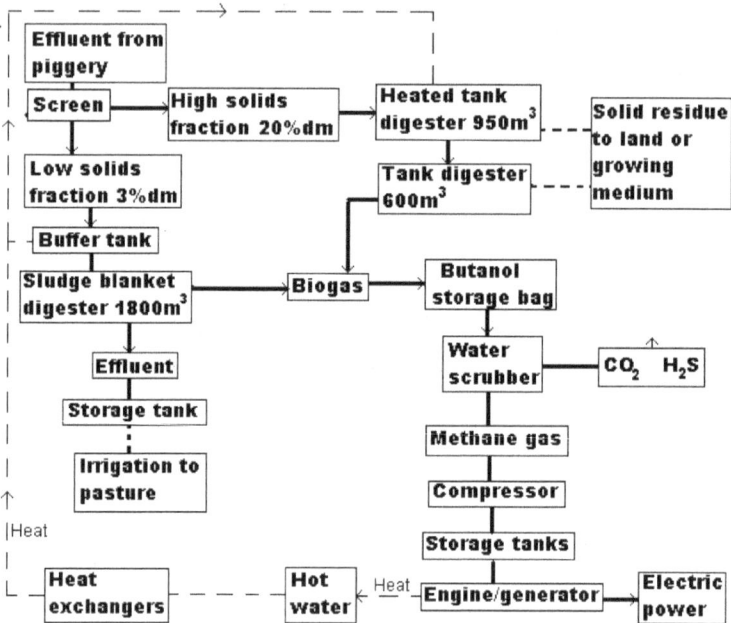

Figure 7.5 Outline of an innovative anaerobic digestion plant design and pig waste treatment process using two digesters, one for the high solids fraction and one for the low

Figure 7.6 Part of the heated and insulated tank 950 m³ digester for the high solids fraction (on the left) and the larger 1800 m³ sludge blanket digester for the low solids component (on the right) with sludge formed on the surface to prevent gas escaping. The beams visible are the tops of inverted V-shape gas collection troughs hidden just below the surface. The white horizontal pipe collects the biogas from the two digesters. The white tower in the foreground is the water scrubber

The biogas produced is stored initally in a flexible butyl rubber bag as a buffer store prior to scrubbing; then the clean gas is periodically transferred to a pressurized cylinder via a compressor. The compressed gas is then piped to fuel a Caterpillar gas engine driving a 190 kW$_e$ generator, which provides around half the electricity demand of the farm, mainly for feed conveyors, lights and underfloor heating of the farrowing pens. The waste heat from the engine is used to heat the tank digester, and any surplus is then used to warm the solid effluent stream. The economics show a negative return on investment based purely on the amount of purchased power saved, but the savings serve to offset the cost of disposing of the slurry, which would be required by other means in order to adhere to the environmental legislation.

Case study 12

Woodman Point sewage treatment and anaerobic digestion plant, Perth, Western Australia

The Woodman Point sewage treatment plant (Figure 7.7), operated by the Water Corporation of Western Australia, treats sewage and waste water from a population of around 600,000 people living in the southern parts of Perth. The 125,000 m³/day of intake fluid is initially screened to remove large objects and

Figure 7.7 Woodman Point sewage treatment plant

rags, and the organic solids are then allowed to settle out in ponds. During this primary treatment around 50–60% of the solids are deposited as sludge. The effluent containing the remaining solids is then piped 20 km south and discharged to the ocean through a 4.2 km outfall pipe. This is located close to a natural ocean trench in order to reduce the chances of any recontamination occurring along the shoreline. However, environmental concerns remain. Therefore a secondary treatment process has been developed at the plant that will remove 90–95% of the solids before discharge. This involved greater treatment of the solids on-site, and hence the anaerobic digestion plant had to be upgraded in 2000. An overview outline of the plant shows the liquid sewage entering at the Munster pump station and passing through the plant (Figure 7.8).

The settled solid sludge is dredged from the settling pond and skimmed from the surface before pumping it into two 38 m tall, 8000 m³ anaerobic digesters (known as the 'eggs') (Figure 7.9). The raw sludge contains 3–4% total

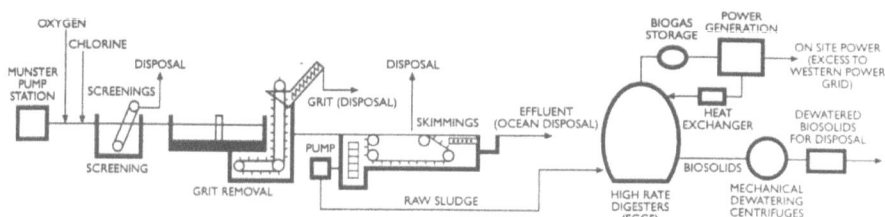

Figure 7.8 Concept of the sewage treatment process and anaerobic digestion plant at Woodman Point

solids, of which around 89% are volatiles. Pumps introduce the raw sludge from the settling pond via a below-ground pipe up through the floor of the biogas plant to the macerator on route to the digesters. It is fed in small spurts (around 3 m³ every 5 min) into the sludge already present in the digesters in order not to overload the system. The digesters are maintained at a temperature of 36°C, and the contents are frequently mixed by recirculation in order for the mesophilic bacteria present to break down the volatile solids in the sludge and produce biogas, in theory without crust formation. The egg-shape design of the digester, by Oswald Schultz Ltd in Germany, avoids any unmixed pockets where scum and crusts can form and which would inhibit the biogas production. The design also helps to maintain a uniform temperature and good contact of the sludge with the bacteria.

The designed residence time is around 26 days, though a minimum of 20 days would be acceptable. Previously the old digesters were three primary cylindrical tanks and one secondary digester giving a total capacity of only

Figure 7.9 One of the two 'egg' anaerobic digesters: the gas rises up through the sludge inside (which is circulated through the external pipes to avoid scum formation) and is then collected at the top

around 9000 m³ and hence a residence time of only 13 days to digest the same volume of raw sludge, resulting in lower biogas yields.

The output of digested sludge from the eggs is reduced down to around 1.7% total solids, of which the original volatile content is reduced by 65%. The sludge is dewatered by centrifuges after a high-charge polymer has been added to encourage flocculation (binding together) of the solid particles followed by a mechanical extraction process, which results in a 28% moisture content residual solids material. The extracted water is returned to the sewage settling ponds and then eventually piped to the ocean outfall with the sewage effluent. The 180 t/week of residual solids produced is stored in bins, and collected every few days by trucks and delivered free to local farmers for use as a fertilizer and soil conditioner. (The Water Corporation pays the $10/t to transport the material.) The pH is about 7.2, and some heavy metals are present (mainly Cu and Zn but also Cd and Cr), but these are considered to be too low in concentration to cause a soil accumulation problem at the rates applied. The fertilizer nutrient value per kg of sludge is: N, 2 g; P, 12 g; K, 21 g.

The biogas produced is around 61% CH_4 and 36% CO_2 with some H_2S contaminant. It is stored in a 3000 m³ buffer store (Figure 7.10) at constant pressure under a floating cover and then scrubbed to remove the corrosive H_2S prior to its being burnt as a fuel in two Waukesha V16 internal combustion gas engines, each driving a 600 kW$_e$ alternator (Figure 7.11). Electricity is generated at 415 V, which is stepped up to 22 kV for connection and export to the plant's ring main and grid, although most power is used on-site to meet the plant's 500 kW$_e$ load (excluding the 2.4 MW pumping station). This saves the Water Corporation around $9000/month in electricity purchases. In addition around

Figure 7.10 Buffer storage of the biogas following production in the digesters and before use as fuel in the two gas engines. The four H_2S scrubbing tanks are evident

Figure 7.11 One of the two generating sets, the 600 kW alternator being powered by a Waukesha gas engine

7000 kWh/day of excess power is exported to Western Power, the local utility, for around 3 c/kWh. The overall cost of generation is around 2 c/kWh, based on the capital invested in the digestion plant and the amount of power generated daily.

The combined heat output of the process is around 1500 kW$_{th}$, arising mainly from the cooling water for the two gas engines. This is first cooled to 68°C by passing it through a forced ventilation radiator system and then piping it to two heat exchangers near the digesters and through which the digester contents are recirculated to maintain the 36°C operating temperature as required by the bacteria. If the sludge temperature exceeds 36°C the hot water is automatically bypassed around the heat exchanger to prevent overheating of the sludge, which could kill the bacteria.

Problems experienced since the plant was completed include:

- pipe blockages, since the screening system proved to be less than 100% efficient

- a pipe rupture, causing sludge to be sprayed all over the pump room

- erratic gas formation, which still occurs, with the daily volumes of gas produced varying widely. This is thought to result from a build-up of scum on the surface of the sludge in the eggs, in spite of all the precautions taken and the special design of the digesters specifically to avoid it. Rotating scum shovels were installed to clear the surface of the digester contents periodically but have proved to be ineffective. The addition of a mechanical agitator could help to solve the problem.

- The H_2S scrubbers were not originally included in the design but proved to be needed as the H_2S exceeded 3000 ppm and the gas engine manufacturers would not maintain their warranty at these levels. Scrubbers were installed but have also caused problems. The gas is passed through a media blend of wood chips impregnated with iron oxide in the four tanks to physically absorb the H_2S and produce iron sulphide (Figure 7.10). The idea is that, after saturation by the H_2S, the medium in each of the four scrubbers is regenerated on a rotational basis by simply opening the hatch, allowing reoxidation of the iron sulphide back to iron oxide to occur. This has not proved totally successful, since of the 2500–3000 ppm H_2S entering the scrubber in the biogas, there was still 700 ppm coming out and passing through into the engines, which consequently had to be stripped down annually for cleaning and checking for corrosion. In addition the media, imported from Illinois, has to be replaced every 280 days or so at a cost of around $40,000 a time.

- The capacity of the engines was not matched closely to the gas production, which is a common problem of anaerobic digestion plants. In this case insufficient total gas is available to supply both engines simultaneously in order to produce maximum power. Therefore one engine is run for 24 hours a day but the second is run only intermittently when both convenient and depending on the amount of surplus biogas available. One engine running at full load generates 600 kW$_e$ of electricity output. However, when both engines are running simultaneously their total output is only around 800 kW$_e$ as there are no gas flow regulators on the engines. Restarting the second engine the next day also often leads to problems as the gas then becomes too rich. When only one engine is running the buffer gas storage volume increases over time. When both are operating the buffer store decreases quickly but there are insufficient reserves to last more than a few hours. This means that during the day both engines can be run simultaneously on only limited occasions. At night, without an operator in attendance, only one is run, so the gas storage level slowly builds up and, when the buffer store becomes full, excess gas is flared off.

The plant is fully monitored and instrumented. However, the information provided for the operator is not being used appropriately and some gas is flared. Extra investment may therefore need to be made to maximize the returns from the use and sale of the power. If at times both engines are not running yet excess gas is being flared off, then there must be room for improving the system.

Landfill gas

A large proportion of MSW is biological material that, once disposed of in landfills, experiences conditions suitable for anaerobic digestion. Methane is produced as a result: this vents to the atmosphere, but can be hazardous as it is explosive. Hydrogen sulphide and mercaptans are also produced during the

M S W   <u>**Landfill gas**</u>

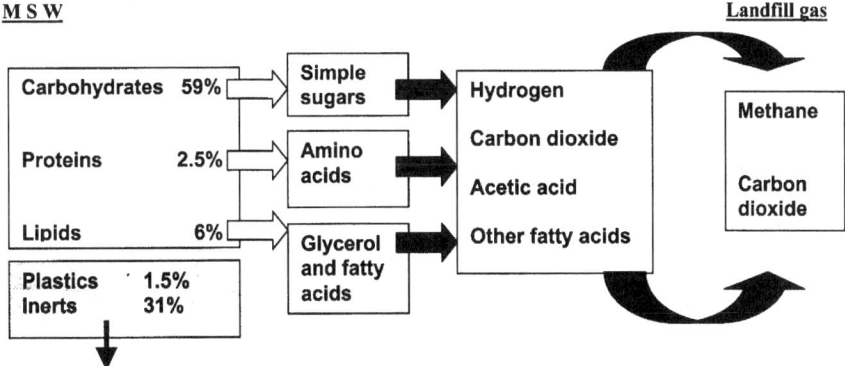

Figure 7.12 The decomposition process to convert the organic matter component of MSW to landfill gas

process, which cause the odours associated with traditional landfills. The concept of capturing this gas and utilizing it for heat and power in order to reduce the danger of explosion, reduce odours and reduce its global warming potential began in the 1970s in the USA and UK.

The process is slower than in a biogas digester owing to the lower temperatures and drier conditions, but the end-product gas is similar. In theory, for every tonne of MSW 150–200 m³ of gas can be produced with a heat value of 5–6 MJ/kg, though these yields are rarely obtained in practice. The process is illustrated in Figure 7.12 for a typical landfill MSW composition.

The gas is collected by an array of perforated pipes laid horizontally in the refuse as the landfill is constructed, and/or put in as vertical wells drilled into the buried refuse of an existing site. The landfill surface area is sealed (or capped) at intervals using clay or plastic sheets during construction, and again finally once full, to ensure that no gas escapes other than through the collection system. It is more efficient to plan for gas collection as the landfill is being filled and manage the site accordingly, rather than attempt to collect gas from established landfills. The gas disperses through the refuse seeking out the easiest path, and hence moves through into the perforated pipe network (Figure 7.13). Any leachate that is collected from a well-designed landfill (as opposed to leaching into the subsoil beneath if the landfill is not adequately lined) can be irrigated over the top of the landfill in order to treat it, as well as to encourage more gas production.

The landfill gas process is similar to anaerobic digestion except that the slurry is replaced with solid waste and the enclosed tank reactor vessel is replaced with a hole in the ground! In a well-designed system it is usually divided into separate large cells, each lined and covered with plastic and/or clay soil to prevent the gas from escaping. These cells can be thought of as large individual batch anaerobic digesters.

The composition of the gas changes with time as the decomposition process begins after sealing of the landfill (see Figure 3.7 in Chapter 3). Oxygen soon

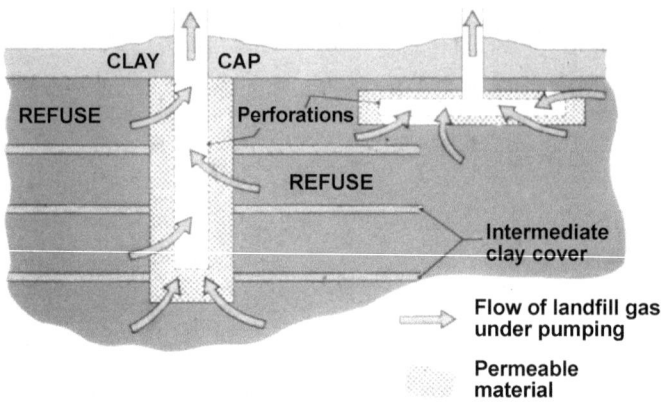

Figure 7.13 Horizontal and vertical gas abstraction wells and clay capping of a completed landfill gas system

declines, leading to anaerobic conditions when the methane production begins. It then stabilizes after some weeks to around 50% content along with the CO_2.

Landfill gas capture and energy recovery is by far the most desirable landfill management practice. There have been many initiatives during the past few years to capture and utilize the gas in internal combustion engines and gas turbines, and many such facilities are currently generating electricity. If using a gas engine to produce power the difficulty is to design a system so that all the gas can be used but without high investment in excess capacity just to meet the relatively short period when methane production is at its maximum. Technically, it may also be possible to use fuel cells in the future, but currently they are far too expensive.

USA regulations require capture of an average of 40% of all landfill methane nationwide in order to reduce emissions of this greenhouse gas. Even after compliance with this regulation it remains profitable (at a carbon price of zero) to capture a further 20% of the remaining methane. If a $20/tC$_{equiv}$ value was placed on the emissions as a tradable credit or subsidy, then up to 40% of the remaining methane could profitably be captured.

The most cost-effective use of landfill gas is for direct combustion in kilns, boilers and furnaces, provided the supply is in close proximity to the user. Electricity generation is a further option, or injection into an existing natural gas pipeline for reticulation. Purification of the gas depends on its utilization. To feed into the gas utility network, for example, the gas must be cleaned, scrubbed free of CO_2, and then blended with air to give the optimum gas blend. Alternatively it can be compressed or liquefied as a vehicle fuel, but this is even more complex and expensive. ·

If 60% of the gas on a landfill site can be captured and put through a gas engine to produce electricity, the overall system efficiency is only 18%. Using membrane collection systems to increase the collection efficiency to 90% and a cogeneration system using the heat to give 80% conversion efficiency would give an overall output system efficiency of 72%.

Most sites involve the passive collection of the gas as it is produced but without any attempt to adjust the landfilling practice to optimize the conditions to increase gas yields. Typically less than a third of the theoretical gas production is collected, and several sites have been abandoned since their gas output did not meet design expectations. As a result the plant proved to be uneconomic. Landfill gas production can be enhanced by

• improving the methods of abstracting, pumping and using the gas

• determining the factors that influence gas production rates using gas pumping trials and correlating the results with variables such as refuse age, site packing density, site geometry and local geology

• studying the effects of placement methods, and adding water, leachate and sewage effluent.

Little is known about the biochemical and microbiological processes involved in landfill gas production from cellulosic wastes, so there is room for improvement.

Economic analyses vary markedly with the site, scale and local gas market, as there is competition with the natural gas supplier if it is to be sold direct. Customers usually demand connection to the natural gas supply too, as they do not have confidence in the landfill gas technology being able to provide a secure supply over the long term. Many landfill sites are too small to develop for gas generation. In the long run, landfill sites will probably be superseded by recycling and more efficient recovery systems from MSW, and landfill gas will decline as a resource.

Municipal wastes and integration of treatments

City and town councils normally have the responsibility for dealing with the treatment and disposal of wastes generated in the area by domestic and industrial users. This includes solid wastes such as MSW, recycling and liquid wastes such as sewage effluent. They usually charge for the service through the rating system on properties, and at times let it out to contract rather than manage the business directly. Many specialist waste disposal firms exist.

Usually the treatment of solid wastes (the local landfill) is totally separate from that of liquid wastes (the sewage treatment plant). However, since both waste streams have an energy value, there may be merit in considering their treatment in an integrated fashion to minimize the costs and maximize the benefits to both the community and the environment (Figure 7.14).

As part of the treatment system an energy crop could be grown to treat the effluent and sludge components of the waste by land application and hence avoid dumping it out to sea or in waterways. The biomass would then be harvested to provide additional fuel for the combustion/incineration conversion plant and to ensure that clean burning of the MSW, MGW or RDF occurred by maintaining correct temperatures. An option exists not to produce RDF where transport distances for the MGW to the bioenergy conversion plant are short,

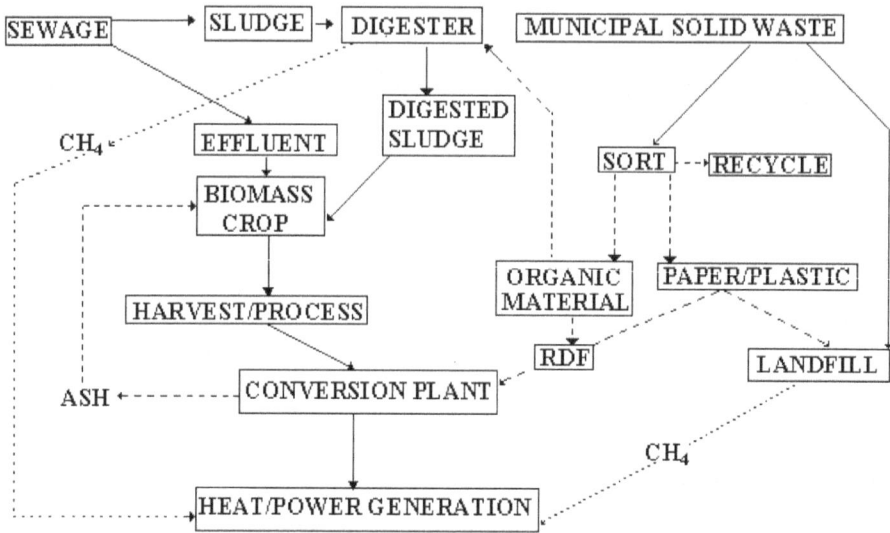

Figure 7.14 Integrated waste treatments for both solid and liquid municipal wastes leading to waste-to-energy solutions

but to simply feed the unprocessed MGW into the combustion plant after suitable comminution.

Case study 13

Nova Gas landfill plant, Porirua, New Zealand

The landfill site of this urban area was developed in 1996. It consists of a network of buried pipework laid approximately 2 m below the surface, which was then connected to a central feeder line going to the compressor (Figure 7.15). This creates a small suction pressure on the collection pipes to help draw the gas out of the tip site. A well-designed site should be able to collect around 60% of the gas in this way, but some sites have closed down after only a few years as the amount of gas produced proved to be far less than predicted by the design engineers.

A two-stage compressor is used, which causes water vapour in the gas to condense. The compressed gas is then treated in a water absorption column. High-pressure water is used to dissolve the CO_2 component to separate out the methane, which is then passed through a desiccant drier to remove residual moisture. The treated and dried gas is then metered, odorized to meet safety regulations and pressure regulated before being reticulated to local customers (Figure 7.16). After this treatment the gas is compatible with existing natural gas-fired equipment including boilers, driers and cogeneration plants, so it can be used in many applications. A decision was made to pipe the gas to customers for heating purposes rather than use it to generate electricity, thereby wasting two thirds of its energy content. A distribution network was therefore

Figure 7.15   The gas collection pipe network on the clay-capped landfill at Porirua

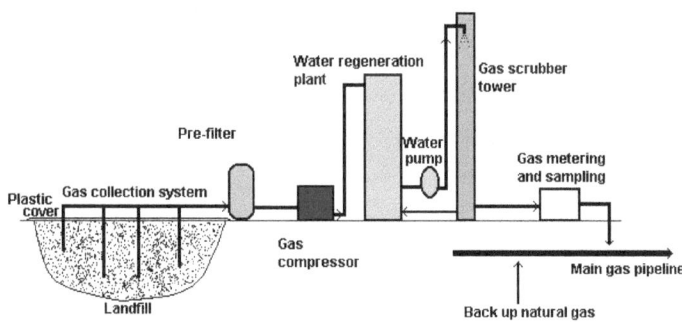

Figure 7.16   The landfill gas treatment process

constructed by the company to supply several suburbs of the city of Wellington, and major customers include Kenepuru and Porirua hospitals, a large abattoir, three public swimming pools and several commercial buildings. Later the network was connected into the main natural gas transmission network, and in total it supplied around 15% of the Wellington regional gas market, which increased to 40% when a second Wellington landfill was commissioned and also linked into the network.

Notes

[1] Useful Internet sites on biogas include http://www.eren.doe.gov/RE/bioenergy.html and http://www.rensselaer.edu/dept/chem-eng/BiotechEnviron/MISC/biotreat/anaerobic.html

Chapter 8

Cogeneration of combined heat and power

When biomass (or coal, oil or natural gas for that matter) is used as a fuel in a conversion plant to generate electricity, some heat is always produced. This relates to the laws of energy (thermodynamics), as covered in Chapter 1, and is unavoidable. However, if this heat can be usefully used and not 'wasted', as it normally is in traditional thermal power generation plants, then the overall system efficiency will be greatly improved. To be classed as 'cogeneration' the by-product heat must be put to good use in some way, normally as process heat in a nearby processing plant (Figure 8.1). Producing steam that only drives a condensing turbine (where the waste steam is condensed and the heat rejected into the environment) is not cogeneration.

In Europe cogeneration is also known as CHP or **combined heat and power** . 'Power' is normally assumed to be the generation of electricity, but it can also be used to describe work by a rotating shaft to drive a pump, compressor or other mechanical process machine.

So a definition of cogeneration is:

The generation of two energy products from a single fuel, usually being a combination of useful heat and electricity.

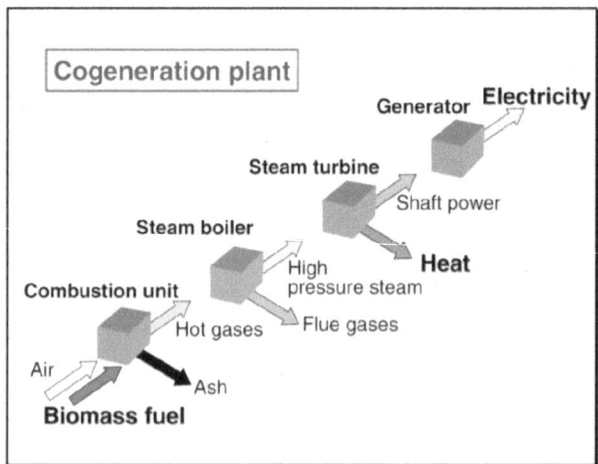

Figure 8.1 Cogeneration is the production of both electricity and useful heat

Large and small cogeneration applications provide heat (usually in the form of steam) and electricity for use in activities such as mineral processing or the manufacture of pulp and paper, petrochemicals, food and textiles, as well as in hospitals, hotels, office complexes, commercial buildings and swimming pools. A range of fuels can be used for cogeneration including biomass in the form of rice husks, straw and methane from landfill sites and biogas plants. A key growth area for cogeneration around the world is the increasing utilization of such biomass waste products.

Many successful examples of bioenergy cogeneration plants exist in Scandinavia, North America, Australasia and elsewhere. The global trend towards privatization of the power industry has created business opportunities for the owners of biomass material to set up new business ventures by becoming independent power producers (IPP). This is normally best achieved by the resource owner's partnering with a utility company and a third-party joint venture investor rather than trying suddenly to gain expertise in a totally new business area. A sugar company best knows how to produce sugar, not how to produce and market power! Sugar companies owning bagasse, farmers owning chicken litter, forestry companies owning wood residues and city councils owning MSW are all able to utilize their resource for generating heat and power, not only for use on-site but also for export to the grid or to neighbouring industry or dwellings.

Cogeneration technology

Cogeneration systems can be as small as a 3 kW$_e$ Stirling engine plant, or as large as a 450 MW$_{th}$ industrial on-site system. Typically cogeneration is two to three times more efficient than the major conventional forms of power generation, which in many countries are predominantly coal- or gas-fired centralized power stations of 20–40% efficiency (Figure 8.2). The range of thermal efficiencies and consequent CO_2 GHG emissions per unit of electricity produced from standard thermal power plants compared with combined cycle or cogeneration systems is wide (Table 8.1). The benefits from cogeneration, assuming the heat can be profitably utilized, are clear. For biomass fuels similar thermal efficiency benefits occur, but any additional carbon emissions avoided will be of little value since the carbon dioxide released during combustion is recycled through the next crop of plants anyway.

Cogeneration efficiency gains are achieved by harnessing the heat that would otherwise be wasted. When the thermal energy can also be recovered for cooling, this is termed **trigeneration** . As well as heating a commercial building in the winter, trigeneration can also be applied in the summer for cooling. In this regard the reject low-grade heat is used to drive refrigeration cycles in which the usual electricity-powered, compression refrigerator or heat pump is replaced by an absorption refrigeration process or vacuum steam jet ejector refrigeration. This is of particular relevance to developing countries in the tropics and subtropics where the waste heat from power plants has no direct application for water or space heating, but where refrigeration of

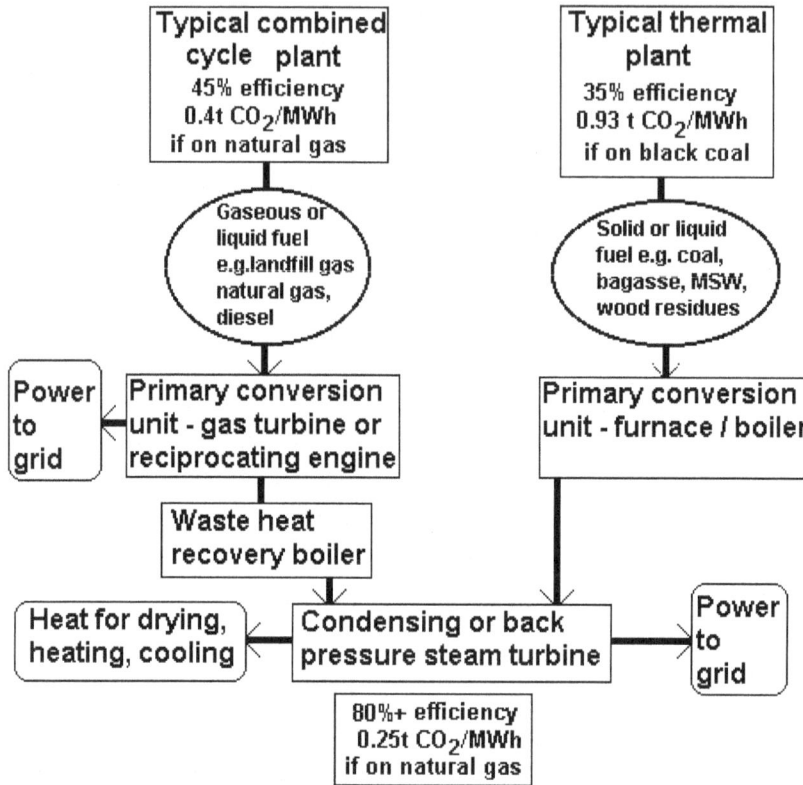

Figure 8.2 Comparison of efficiencies of a typical natural gas-fired cogeneration plant with a typical coal-fired standard power plant. Carbon emissions would be zero if using biomass fuels

agricultural products and fish intended for export or local consumption could result in major benefits to the economy. It also has potential for dairy farms in warm climates, where animal manure can be used in a biogas plant to produce electricity for use on-site but the waste heat can also be used for refrigeration to cool the milk down to 4°C for safe storage.

In Denmark a significant proportion of electricity generated is from biomass cogeneration plants using wood waste and straw, and in Finland it is around

Table 8.1 Carbon dioxide emissions and thermal efficiencies from traditional sources of electricity generation

	Typical CO₂ emissions (g/kWh)	Typical thermal efficiency (%, HHV)
Thermal (brown coal)	1230	29
Thermal (black coal)	930	35
Thermal (natural gas)	490	38
Combined cycle (natural gas)	390	48
Cogeneration (natural gas)	260	77

10% using sawdust, forest residues and pulp liquors. In other countries biomass cogeneration is utilized to a lesser degree as a result of unfavourable regulatory practices and constraining structures within the electricity industry.

Cogeneration can be linked with either a 'topping' or 'bottoming' cycle:

- **Topping cycle** : Steam is produced by a boiler and then passed first through a steam turbine to produce electricity, with the residual lower-pressure steam used for process heat.

- **Bottoming cycle** : Residual heat is recovered from a high-temperature industrial process (such as a furnace) and then used to generate steam through a boiler, which is finally used to generate electricity. Any residual lower-pressure steam from the turbine exhaust can also be used for process heat.

Cogeneration is the most efficient way to use a fuel by producing both heat and power, and hence it gives larger overall energy savings. As an example consider a factory that uses a gas boiler to produce process heat at around 80% thermal efficiency, and also buys in electricity supplied from a coal-fired power plant at around 30% efficiency feeding into the grid. The overall thermal efficiency is around 50% assuming the heat demand is double the electricity in terms of 'units' of energy consumed (GJ/yr) (Figure 8.3).

Now assume the same factory installs its own cogeneration plant on-site based on a wood-fired boiler and topping cycle back-pressure turbine. This would result in a thermal energy efficiency of over 80% (Figure 8.4) with considerable savings in greenhouse gas and other emissions, and the total energy bill for the factory should be lower.

If a cogeneration plant is fired by biomass rather than coal or natural gas:

- the efficiency would probably be similar depending on the coal type and biomass moisture content

- the cost (in terms of $/GJ of heat and c/kWh of electricity) could be more or less dependent on the relative fuel prices ($/GJ delivered), capacity of plant, etc.

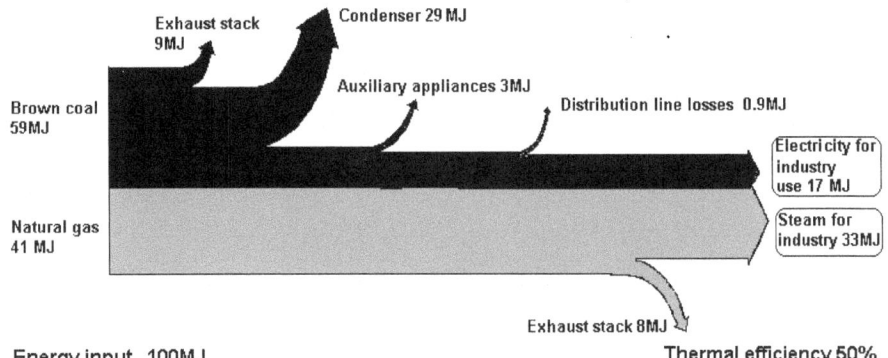

Energy input 100MJ   Thermal efficiency 50%

Figure 8.3 Example of the traditional use of heat and power in a factory

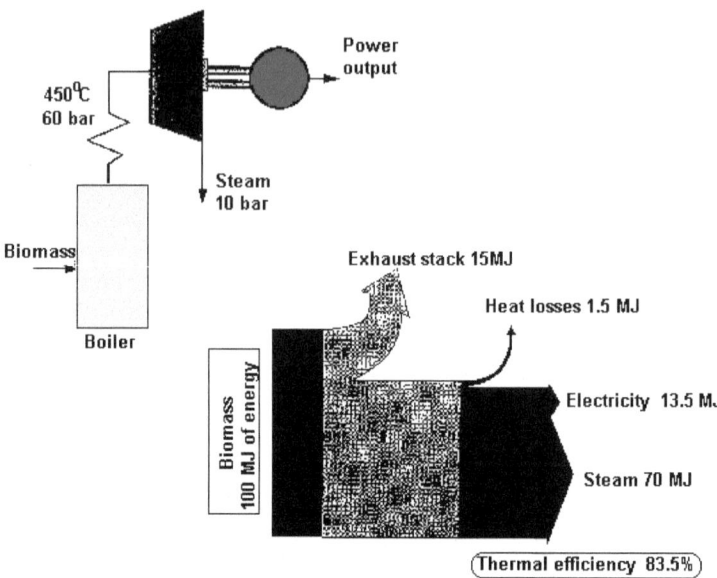

Figure 8.4 Typical biomass cogeneration installation based on topping cycle

- the greenhouse gas emissions would be far lower, which based on IPCC[1] figures is typically around 200–250 gC/kWh of electricity generated by coal, 90–110 gC/kWh by natural gas and around 5–10 gC/kWh by biomass, since in this case the CO_2 produced is recycled through the next forest or crop of say sugar cane. (Note: 1 gC/kWh is equivalent to 3.66 g CO_2/kWh).

In many countries the sugar industry has long used the residual bagasse (left over after extraction of the sucrose) to supply heat and power for processing the sugar. However, it has usually practised this inefficiently and with high stack emissions, mainly as a means of disposal of the large volumes of waste material. Since the worldwide trend towards privatization of the electricity industry, independent power producers have been established to generate electricity, and several have built modern cogeneration plants alongside sugar mills. Where a sugar plant traditionally produced say 3 MW_e of electricity and also heat for use on-site, it can now produce from the same volume of bagasse perhaps 20 MW_e of electricity, mostly for export, and still supply the process heat.

Similarly the pulp industry uses its waste biomass products, including bark and the non-useful lignin component of the wood, called black liquor, as fuel for on-site cogeneration plants. The dairy industry (and other food-processing industries) can benefit from cogeneration, as many processing plants need both heat to process the milk and electricity to refrigerate it and operate such things as lights and electric motors for pumps. However, unlike the sugar and forest industries, gas or coal is usually used to fuel an on-site cogeneration plant since there is no biomass by-product that can be used for energy (other than whey perhaps, which is already used to make ethanol in some dairy factories – but for gin and vodka production rather than as a transport fuel).

Environmental benefits

Since less fuel is used with cogeneration to produce a given amount of heat and power than if using separate heating and electricity-generating systems, then there must be environmental benefits, whether fired by fossil fuels or by biomass. From the carbon dioxide GHG emission perspective, its effectiveness as a mitigation option depends on which fuel is being replaced:

- If the cogeneration plant is fired by natural gas and displaces energy from separate heating and power plants, both coal-fired, then that is beneficial.

- If a new coal-fired cogeneration plant is installed to replace both a coal heating boiler and electricity from the grid from a coal-fired power station, overall emissions will probably be reduced slightly. But if the purchased electricity displaced came from a hydro station, then the new cogeneration plant might actually result in an increase of greenhouse gases.

- If the cogeneration plant is run on natural gas or coal and replaces an efficient natural gas boiler and grid power purchased from hydro or other renewables, it would produce more CO_2/MJ.

- If the new cogeneration plant is run on natural gas and replaces a wood-fired burner used for heat together with purchased electricity from hydro power, then it would definitely result in increased emissions and be a poor investment decision from an environmental perspective.

- If it is a biomass-fired plant using bagasse, wood chips or landfill gas to displace any fossil-fuel-fired plant, then carbon emissions would be reduced since the biomass carbon is recycled.

- If it is co-fired on some biomass and some fossil fuels, a detailed life cycle assessment would be needed in order to ascertain the benefits or otherwise.

Process energy

Consider a factory using heat for several different processes: A, B and C. The heat comes from a steam supply under different temperatures and pressures, having first passed through a steam turbine driving an alternator (Figure 8.5).

The high-pressure boilers produce steam to drive the turbine (1) and generate power either for use on-site or for export. Hence there is an interaction between the processes and the local power utility company (2). Steam from the turbine (3) is available at different pressures (4) for the various processes. In this example process A requires additional heat for high-temperature processing (5), any surplus being fed into the steam mains to supply the other processes. Process B requires high- and low-pressure steam (6), and process C high- and medium-pressure steam (7). So there are interactions between the processes through the steam mains (8). A cooling system is required for any excess heat rather than releasing it directly to the local environment (9).

So, overall, supplying energy to the plant is a complex problem for the mechanical engineer. The plant engineer has to figure out how best to operate

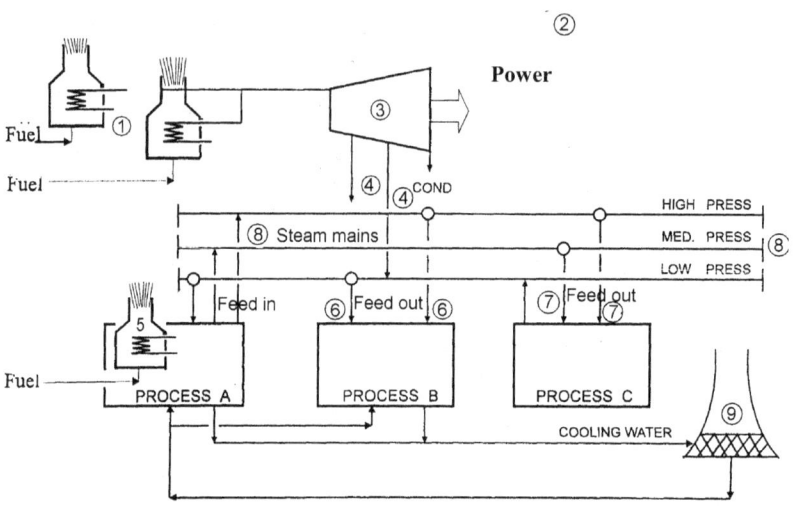

Figure 8.5 A typical industrial processing site using biomass fuel to generate steam for electricity generation from a condensing pass-out steam turbine together with process heat at high and low pressures

the system to save costs and, if using coal, gas or oil, how to minimize carbon dioxide emissions. Should the engineeer concentrate on saving medium-pressure steam in process C, generate more high-pressure steam for process A, shift load from high-pressure to medium-pressure levels, minimize fuel use or maximize electricity generation in order to export more power, redistribute the load on the steam turbine, or install a gas turbine instead? Overall there are usually many opportunities for improvement, and a structured approach is needed to identify the best ones.

Turbine operation

In Chapter 5 the characteristics and theory of back-pressure steam turbines used to generate power were discussed. If the steam leaving the turbine is not wasted, but used to supply useful heat for the process in a factory, then it becomes a cogeneration plant (Figure 8.6).

The biomass fuel is combusted in the boiler furnace to produce high-pressure steam, which powers the turbine. The resulting low-pressure steam is used for process heat, and as much condensate as is practical is fed back to the deaerator for treating before being returned to the boiler for reheating. The principle of cogeneration also applies when a reciprocating steam engine is used in place of a steam turbine to drive a generator (Figure 8.7).

A gas turbine can also be used in a similar manner (for example in place of the reciprocating engine in the schematic in Figure 8.7). However, when the gas turbine exhaust is used for the process heat, care must be taken that there are no flammable materials involved in the manufacturing process, as the exhaust is oxygen rich (up to 15% by volume) owing to some air being mixed in after

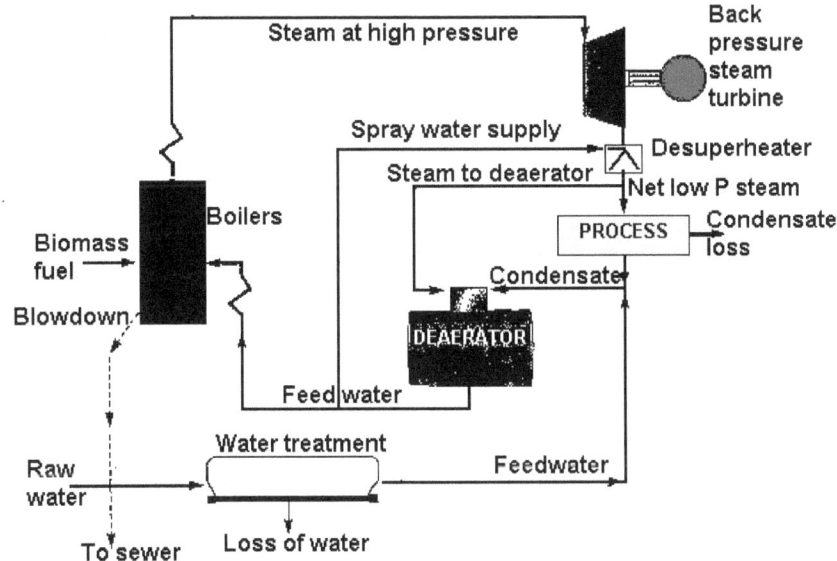

Figure 8.6 Back-pressure steam turbine (topping cycle), with residual low-pressure steam used to provide the heat supply for the process: hence this is a cogeneration system

combustion. An explosion is therefore possible. The exhaust heat can be safely used for drying indirectly through a heat exchanger or rotary dryer. It is also ideal as a source of combustion air for furnaces, again owing to its high oxygen content but also because it is preheated.

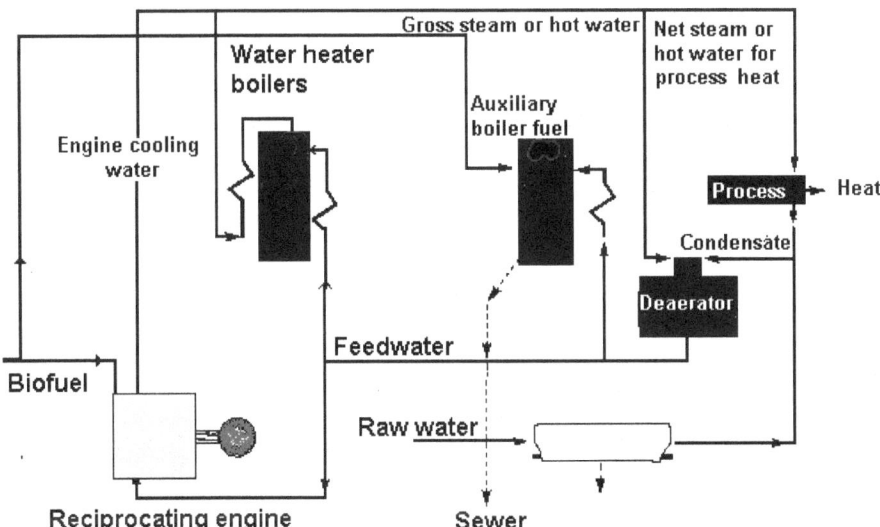

Figure 8.7 Cogeneration using a reciprocating steam engine to drive a generator, with surplus heat as steam or hot water used in the plant process

Assessing performance of a cogeneration plant

Pass-out steam turbines are similar to back-pressure turbines, but they are designed so that useful heat can be extracted as process steam at suitable pressures for use in the factory at the same time as the power is being generated (Chapter 5). A specific design of pass-out steam turbine can be operated within an envelope of boundaries, as shown by its operational characteristics (Figure 8.8). These are provided by the manufacturer to enable the correct turbine design to be selected for a given application. Mechanical limitations of any particular turbine design prevent operation outside the given set of conditions.

- Line AB shows the minimum pass-out conditions relating to the steam input flow (m) and power output (W). Line FE represents the maximum. Constant pass-out conditions are shown by the line 'Constant P', which can be moved up or down between AB and FE.

- Line GF shows the minimum exhaust conditions and line BC the maximum. The line of constant exhaust steam conditions, 'Constant E', can be moved across between GF and BC.

- Line ED is the maximum flow rate of steam acceptable by the specific design of turbine (kg/s as on the y axis).

- Line CD is the maximum power that can be generated (MW as on the x axis)

It is not easy to calculate the efficiencies and CO_2 emission avoidance levels that result from using cogeneration instead of traditional energy sources, and there is no standard method for doing so. The benchmark used is for a site that has its own thermal heat-raising facility to produce steam or hot water and also imports electricity from coal or gas plants, transported to the site via a transmission system with various line losses involved. Average system efficiencies and emissions can be calculated, allowing for the peak demand times of year when more than one plant is used and allowing for emissions when generating any imported electricity used.

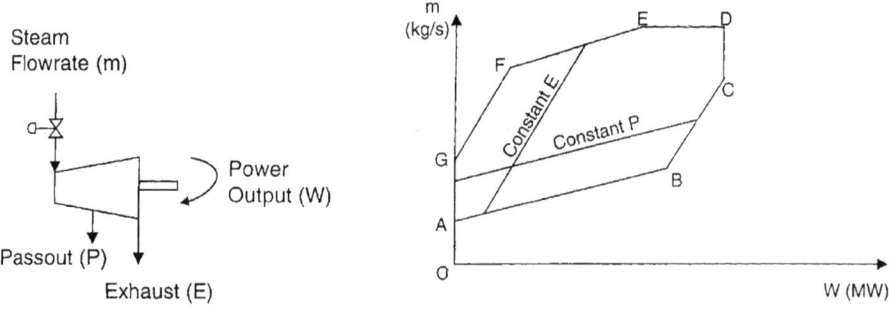

Figure 8.8 Representation of a pass-out steam turbine and operational characteristics

Net electrical efficiency =

$$\frac{\text{cogeneration gross electrical output + imported electricity - parasitic electricty consumed}}{\text{fuel to cogeneration unit + fuel used to generate imported electricity}}$$

Electrical efficiency (fuel chargeable to power basis) =

$$\frac{\text{cogeneration gross electrical output + imported electricity - parasitic electricty consumed}}{\text{fuel to cogeneration unit + fuel used for imported electricity + fuel used for steam}}$$

where fuel used for steam is calculated at the efficiency of the auxiliary boiler on-site.

Overall thermal efficiency =

$$\frac{\text{cogeneration gross electrical output + imported electricity - parasitic electricty + site steam and/or hot water}}{\text{fuel to cogeneration + fuel used for imported electricity + fuel used for steam}}$$

Usually generator manufacturers quote the gross output capacity of their plant as measured at the terminals (since this is a bigger number!) rather than the net output, which is the power left for use after parasitic losses used to run the plant such as running engine auxiliaries and any transformer losses. Efficiencies and GHG emissions (gCO_2/kWh, for example) can also be measured and quoted on a gross or net basis, so care must be taken to ensure which one is being used and to allow for the inevitable parasitic losses if using gross output figures.

Typical efficiencies are:

- conventional imported electricity supplied from brown coal and separate on-site steam raising using natural gas (Figure 8.3): 50%

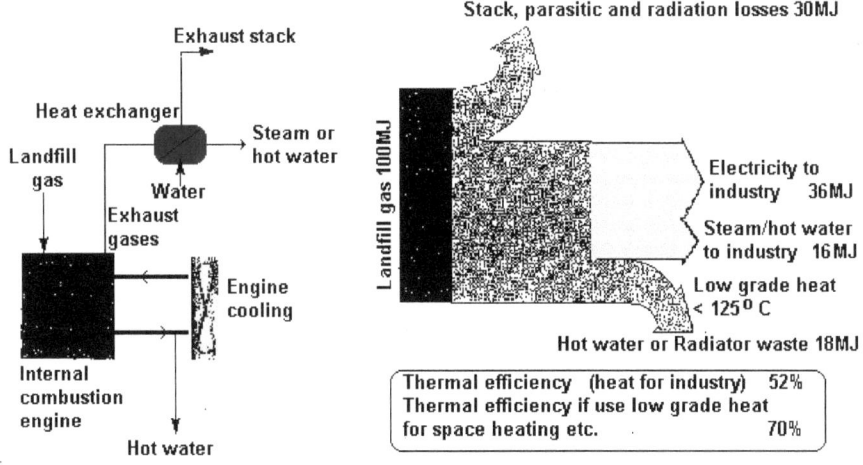

Figure 8.9 Energy flows using a reciprocating engine for cogeneration fired by natural or landfill gas with exhaust heat for steam raising (52% efficiency) and giving 70% efficiency when the low-grade heat can be used for hot water or radiators for space heating

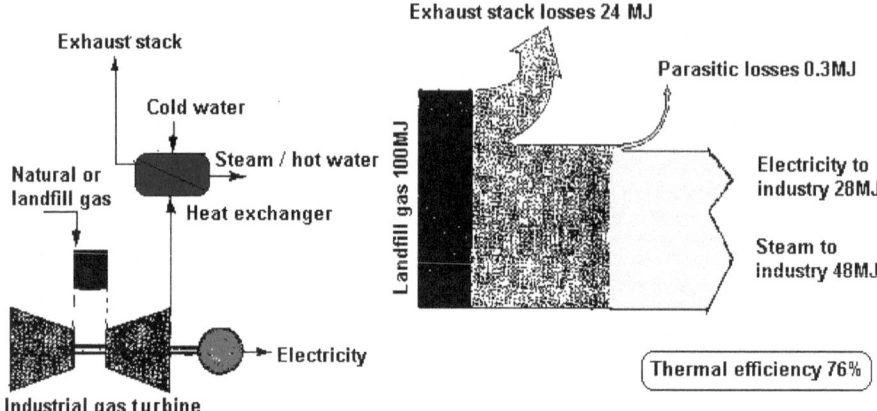

Figure 8.10 Energy flows from landfill gas, natural gas or producer gas using an industrial gas turbine for cogeneration

- topping cycle steam turbine cogeneration using biomass or coal (Figure 8.4): 83%

- reciprocating gas engine cogeneration (Figure 8.9): 52%

- as above but with residual low-grade heat used for space heating (Figure 8.9): 70%

- industrial gas turbine cogeneration (Figure 8.10): 76%

- aero-derivative gas turbine cogeneration: 72%

The aero-derivative gas turbine is similar to the industrial design but has higher stack emissions, and from 100 MJ of gaseous fuel produces equal amounts of heat (36 MJ) and power (36 MJ), giving slightly lower thermal efficiency.

These various configuration options provide different natural proportions of steam and electricity, which typically for every 100 MJ of energy input are:

- Conventional imported electricity supply from brown coal (59 MJ) and separate on-site steam raising using natural gas (41 MJ):
  Electricity 17 MJ   Heat 33 MJ

- Topping cycle steam turbine cogeneration on-site using coal:
  Electricity 13 MJ   Heat 70 MJ

- Reciprocating gas engine cogeneration on-site using landfill gas:
  Electricity 36 MJ   Heat 16 MJ

- As above but with the low-grade heat used for space or water heating:
  Electricity 36 MJ   Heat 34 MJ

- Industrial gas turbine cogeneration from landfill, producer or natural gas:
  Electricity 28 MJ   Heat 48 MJ

- Aero-derivative gas turbine cogeneration as for industrial design:
  Electricity 36 MJ   Heat 36 MJ

If a site project can be quickly assessed for the likely proportion of heat to power demands, then the ratios above could help to indicate which system to choose as a start for the analysis. For example, from the list the aero-derivative gas turbine and the reciprocating gas engine, when able to utilize the low-grade heat, both fit around the 50/50 split. Both options have similar thermal efficiencies, but the gas turbine is less suitable for smaller installations owing to the higher unit cost at this scale.

Case study 14

Wood-fired district heating scheme and power plant, Vaxjo, Sweden

Sweden is a world leader in bioenergy, and the original Vaxjo wood-fired power plant was one of the first built at this scale (Figure 8.11). Vaxjo is a town in southern Sweden with a population of around 70,000. It is one of 100 European communities that intend to become fully dependent on renewable energy by 2010. This aim is supported by the mayor, the council and the majority of residents, and simple changes are already occurring to help make it happen. For instance, when it snows every winter, salt is spread on to the footpaths and cycle tracks first before the roads, thereby giving priority to cyclists and pedestrians. It is this sort of change of mindset, supported by the whole community, that is needed for a major change in lifestyle to occur such as reducing fossil fuel dependence. A key part of achieving the 100% renewable energy community challenge is already in place, a large wood-fired cogeneration plant having been developed by the utility company VEAB.

The district heating network of underground pipes was built in the 1970s, and replaced numerous small furnaces and boilers around the town. The original power station to provide the heat for the scheme was built in 1974, fired by oil. In 1980, for economic reasons following the crude oil price rise after the Iran/Iraq war, it was retrofitted for $5.5 million to run on biomass (which was considered by some observers to be impossible). So it can now use wood process residues or peat or oil.

In 1997 a new cogeneration plant was constructed (Figure 8.12) to generate an additional 38 MW$_e$ and supply 30–40% of the town's power demand, together with 66 MW$_{th}$ of heat. Old and new plants together now provide all the district heating. The wood ash, containing nutrients and trace elements, is returned to the forest (though there are some concerns over a possible build-up of cadmium in the longer term).

In recent years the plant has been running on over 95% woody biomass, mainly wood chips and bark. It is designed to use high moisture content fuels to increase the flexibility of supply in order to obtain the lowest cost per GJ of fuel delivered. Use of more biomass was encouraged by the new Swedish

Figure 8.11 The original wood-fired power station at Vaxjo together with the new cogeneration plant on the right

company tax policy of 1997, which reduced taxes based on income but increased the carbon tax on fossil fuel use.

The new deregulated market for electricity led to the wholesale power price dropping considerably, so power generation was less viable. However, the flexible design of the plant enables less electricity and more heat to be produced by using the flue gas condenser, and hence the overall revenue from the plant was maintained.

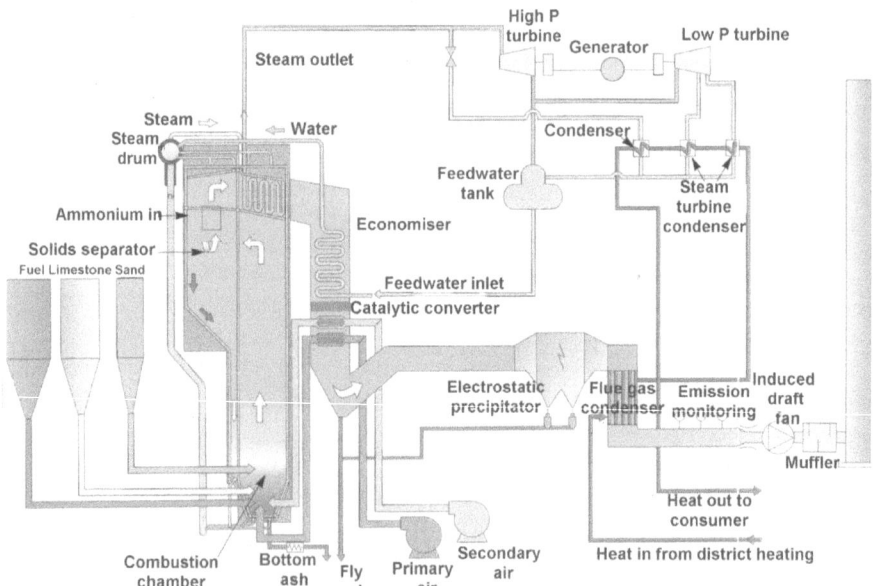

Figure 8.12 The new 104 MW cogeneration plant at Vaxjo
Source: Vaxjo Town Council

The more district heat used, then the more electricity that can be produced. It is less profitable to run the plant to produce electricity alone, so the several generating units at the power plant are brought on-stream to match the steam available, based on the heat demand in order to produce the environmentally and economically optimum combination of heat and power. Plans for the future are to convert the plant to a gasifier system to provide an additional 15–25 MW$_e$ should the electricity price rise. The steam turbines would then be used as a bottoming cycle to produce more heat.

The new cogeneration plant generates approximately 160 GWh of electricity and 1.26 PJ of heat annually. It cost $40 million, is fired only on biomass, has the flexibility to produce a reasonably high output of electricity in proportion to heat and has acceptable emissions closely monitored and well within environmental regulations since it uses a catalyst in the flue. For example, NO$_x$ emissions are only 20 mg/MJ output. It has saved the consumption of around 50 million litres of oil per year to produce the same amount of heat and power. The plant has a reasonably high thermal efficiency of around 38% when generating electricity alone. Since the total plant conversion efficiency includes the latent heat reclaimed by the condensers, then over 100% efficiency is claimed!

Biomass fuel contracts

About half the fuel comes from wood process residues and half from forest arisings. Tenders were sought from fuel suppliers, and several supply contracts were let for 1–5 years for various prices and quantities. It was thought that contracts of more than 5 years would be of no use as this would be too far ahead to predict changes in market prices or energy demand. Fuel is paid for by its energy content. Each truck is weighed on entering the plant, and samples of fuel are taken by an independent company to avoid any conflict between fuelwood suppliers and the plant owners. The moisture content is calculated and payment made according to the contract. Wetter fuel receives a lower price. Additional rules relate to fuel quality. For example, if two or more stones are found in one truck load, then payment is withheld for that load.

Fuel is delivered 6 days a week, 24 hours a day at two trucks per hour on average, each of around 100 m^3 load capacity. There are fewer deliveries in the summer (as the heat demand drops from 180 MW to just 10 MW for hot water only), so overall there are 6000–7000 truck loads per year. In summer some of the boilers are shut down. To anticipate demand the company follows the weather forecasts, and can predict heat and power use quite closely.

The boiler system

The Pyroflow Compact boiler and integrated biomass system was supplied by Foster Wheeler Energia Oy from Finland. It is a circulating fluidized bed, the sand being fed with the fuel into the combustion chamber to create a bed at the bottom of the boiler through which the primary combustion air is blown. (Limestone can also be added in with the mix when peat is used as fuel in order to reduce any sulphur emissions, the limestone and sulphur forming plaster, which becomes part of the bottom ash.) The primary air intake speed is increased until the bed is fluidized, when it acts like a boiling liquid. The combustion process takes place throughout the whole combustion chamber at an even and relatively low temperature of 800–900°C. The sand is recycled through a novel solids separator located above the combustion chamber, and is then returned to the boiler.

The boiler is constructed using panel tube walls made from many standing tubes that are part of the water circulating system. They are welded together with fins between so that the combustion chamber forms an enclosed box. In the lower parts of the boiler the panel walls are covered in a fireproof refractory layer, which protects the steel against corrosion at the high temperatures experienced. The biomass fuel is fed into the base of the combustion chamber by variable-speed auger screws. The sand returned from the solids separator is also supplied here, with any new sand added as needed. Secondary air is introduced above the bed, and the bottom ash is removed by a conveyor.

After passing through the solids separator the gases are returned to the upper part of the boiler and pass on to the superheater. This is a series of U-shaped pipes enclosed in the flue gas channel that remove much of the remaining heat to produce superheated steam for the high-pressure turbine. The warm gases continue to flow to the economizer, which uses the medium-temperature heat to preheat the feedwater returning to the boiler from the steam turbines and condenser. Any remaining heat is then used to preheat the primary and secondary combustion air in the air preheater.

The boiler is a single-dome design with natural circulation and steam temperature adjustments made by injecting water as necessary after the superheater. Owing to the high temperature and pressure (43 kg/s steam at 540°C and 14.2 MPa), stringent demands are placed on the quality of the feedwater treatment system, treatment taking place in a separate plant.

Steam turbines

The ABB (Alstom) dual-body steam turbine has high-pressure and low-pressure modules with double condensers. The high-pressure turbine module (Figure 8.13) is of barrel design, in which the rotor and diaphragms are assembled to form an integral unit that runs at 10,727 rev/min at the 140 bar and 540°C conditions of the inlet steam. The low-pressure turbine (Figure 8.14) runs at 6950 rev/min and has a split exhaust to generate more power and to make possible two-stage heating of the district heating water. The two turbines both drive the same ABB generator at 1500 rev/min via their own gearboxes.

The generating plant can be shut down or started up within 30 minutes when the boiler is operating (rather than when starting from cold), which enables the company to seek higher power prices at peak periods.

Emissions

Environmental emission controls were important. To lower the NO_x the boiler is equipped with a selected catalytic reduction system whereby ammonia is

Figure 8.13 The high-pressure steam turbine rotor cartridge being inserted into its barrel-type housing
Source: ABB/Alstom

Figure 8.14 Assembly of the low-pressure steam turbine
Source: ABB/Alstom

injected into the upper part of the solids separators (Figure 8.12). Any NO_x that were produced during the combustion process are reduced to nitrogen. A catalytic converter is also installed in the economizer (where the flue gas is at the optimum temperature) to lower the NO_x level even further by reducing any residual NH_3 present to N.

Separation of dust and recycling heat from the hot and moist flue gases is made possible by the particulate control system (Figure 8.15). An electrostatic precipitator separates the dust from the flue gas as it is transported through it by the induced draught fan. The dust is electrically charged and then sticks to the collecting electrodes, which are shaken regularly to clean them; 99.5% of the dust present in the gas is collected by this filter and transported to the dust hopper. The clean gas continues to the flue condenser, where it is cooled by the returning district heating cold water so that any moisture condenses. The energy contained as late nt heat in the flue gases can be recovered in this way to provide 10–20 MW_{th} of district heat without the need for any extra fuel. The condensate also helps to wash dust from the gas. Any fly ash collected in the precipitator is blown to a silo, where it is stored before treating and returning to the forest.

Controls

All equipment is computer controlled using a Siemens control system especially developed for power plants. All operating, monitoring and supervision are

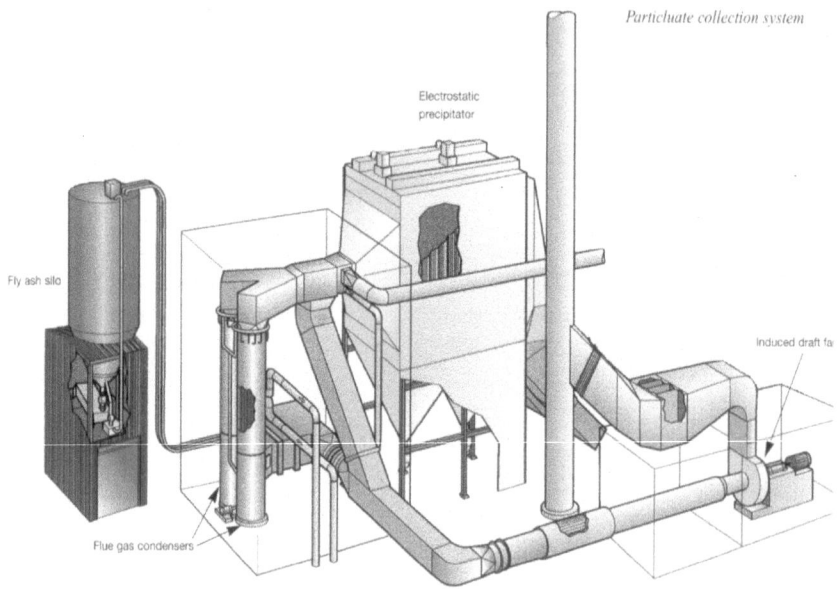

Figure 8.15   Particulate control and heat recovery system

Figure 8.16 Plant operators in the central control room

carried out via the terminals in the central control room (Figure 8.16). Communication between systems is by fibre optics. Over 4200 kinds of binary and analogue signal and 580 open and closed loops (as electrical motors and actuators) are monitored. The current plant condition is supervised by operators through graphic displays of the plant on screen, process displays, operating logs and alarms.

The plant management system is fail-safe if a fault develops. Personnel can use historical records to optimize plant efficiency, fuel costs, environmental emissions or power consumption on-site.

District heating

The temperature of the hot water as it leaves the plant is 80°C, and it returns at around 45°C. It is important that the flue gas condenser does not exceed this temperature, since each degree higher would give less 'free' latent heat recovered. Since the consumers pay less if the plant is working more efficiently they are willing to accept slightly lower heat temperatures. Customers pay for their heat use based on heat meters (calorimeters). Originally there was a fixed charge, but when the energy company changed to meters the consumption decreased by 30% as the householders stopped opening their windows to control the temperature! The incentive for householders to connect to the district heating was that it cost them less than would be needed to replace a domestic oil-fired hot water boiler. The air quality in the town improved as a result.

Notes

[1] Intergovernmental Panel on Climate Change 2001. *Third Assessment Report.*

Chapter 9

Biofuels for transport from the biochemical conversion of biomass

The production of vehicle fuels from biomass can technically be achieved in several ways, either as gaseous fuels originating from biogas or wood gasification plants, or as liquid fuels from a number of conversion routes suitable for a range of biomass resources including purpose-grown sugar and oil crops. In the future biomass may also have a role to play as a source of hydrogen for fuel-cell-powered vehicles (Chapter 10).

There are a number of important reasons for producing and use biofuels for transport:

- Petroleum is finite, but long before the reserves are depleted the atmospheric emissions of carbon dioxide will have caused potentially irreparable damage to the global environment.

- The feedstocks for biofuel manufacture are normally produced in the country of use, which helps to reduce trade deficits and create jobs.

- The agricultural sector, which is experiencing a recession in many developed countries, especially stands to benefit since biofuels are made from crops and agricultural residues, providing options for land use diversification and new revenue streams.

- The production of fuel domestically also leads to energy security, with less dependence on the strategic, political and economic whims of other governments.

When Henry Ford first designed his Model T automobile in the USA in 1908, he expected ethanol, made from renewable resources, to be the major fuel used. About that same time in Germany, Rudolf Diesel anticipated that his revolutionary new compression ignition engine design would be run on vegetable oils. Since then the global dominance of cheap crude oil has led to the consumption of over 900 billion barrels of it for vehicle propulsion. During the interceding decades, the designs of vehicle engines have been improved to use petroleum fuels more efficiently while, in parallel, the refining and quality of the petroleum fuels have been improved to meet better the tighter specifications of the engines and the more stringent environmental constraints.

This has led to a symbiotic relationship between the oil industry and the vehicle manufacturing industry.

In spite of huge investments of research funding, the internal combustion engine remains 'thermodynamically challenged'! The usual situation is for over two thirds of the chemical energy stored in the petrol or diesel to end up as waste heat, while only 25–35% of it is actually utilized to propel the vehicle.

Opinions vary as to how many oil reserves there are remaining underground, but the peaking of global production after 2010 is a popular view since, at present, it is being consumed faster than more oil is being discovered. In addition the climate scientists are adamant from their modelling work that we need to move towards a 'decarbonized world' (though of course this excludes bioenergy, which is carbon neutral and therefore acceptable). Regardless, there will still be a considerable volume of oil extracted for use as a transport fuel over the next 50 years or more. However, the Stone Age didn't finish because people ran out of stones! So perhaps the Oil Age will also finish before the oil reserves run out, for other reasons such as cheaper alternatives becoming available driven by growing environmental concerns, both locally, such as air pollution in city centres, and globally. One of the future mitigation options is to use biofuels produced sustainably from biomass sources.

The term **biofuels** is usually intended to imply fuels sourced from biomass that are used for transportation purposes (though at times the term is also used, perhaps misleadingly, for biomass when used for heat and electricity generation). Biofuels include methane, producer gas, alcohols, esters and other chemicals made from cellulosic biomass. Herbaceous and woody plants, sugar crops, oil crops, agricultural and forestry residues, and the 'green' fractions of municipal solid and industrial wastes are all suitable feedstocks used to produce a range of fuels.

- **Biodiesel** , is commercially available in several countries from the inter-esterification of vegetable oils including rapeseed, soybean, coconut and sunflower oils as well as from tallow and waste cooking oils.

- **Bioethanol** , from the fermentation of sugar, has been used widely in Brazil from sugar cane and in the USA from maize (or corn). The ethanol fuel has been sold either neat or blended with gasoline[1] for nearly two decades. There is also considerable worldwide interest in producing ethanol from small cereal crops, from other sugar crops such as beet, and from ligno-cellulose feedstocks using acid or enzyme hydrolysis to first produce the sugars.

- **Biomethanol** , produced from synthesis gas or biogas (or more commonly perhaps from natural gas), has also been evaluated as a fuel for vehicle internal combustion engines. It can be particularly useful where liquid fuels are preferred to gaseous fuels for their relative ease of transport and storage.

- **Pyr olysis oils** and **biogas** can be cleaned, refined and compressed and then also used for transport fuels, as discussed in Chapters 6 and 7.

Biofuels offer many benefits. They are beneficial for the environment because they add fewer emissions to the atmosphere than petroleum fuels on a per kilometre travelled basis, and they often utilize waste biomass resources that currently have no value and require disposal. Unlike petroleum, which is a non-renewable natural resource, biofuels are renewable and inexhaustible sources of fuel, assuming the feedstock is produced in a sustainable fashion. Where energy crops are grown domestically, or other biomass sources are readily available for conversion, biofuels can reduce a country's dependence on the vagaries of imported oil price fluctuations and uncertain supplies. The use of biofuels could therefore help to strengthen the energy security and boost a nation's economy should crude oil prices reach and maintain levels above around $30/barrel.

All petroleum-derived fuels suitable for transport vehicles are compounds containing predominantly carbon and hydrogen atoms. Other constituent elements generally regarded as undesirable contaminants include nitrogen, sulphur and phosphorus. Compounds such as tetraethyl lead have been added in the past to modify fuel properties so as to reduce the tendency of the fuel to 'knock' during combustion in an internal combustion engine. Other chemical additives such as nitromethane have been added to speciality fuels to improve power output.

Liquid biofuels differ chemically from fossil fuels in that they contain oxygen in addition to the carbon and hydrogen atoms. As with fossil fuels, they may also contain other elements, notably nitrogen, and once again this is generally regarded as an undesirable impurity.

In most cases a specific fuel consists of a mix of compounds, which varies not only with the source but also with the time of year. For example, to assist in easier vehicle starting in cold conditions, winter formulations of gasoline tend to have a higher proportion of the more volatile components than do summer formulations. Likewise summer formulations are less volatile than winter formulations to avoid evaporative losses and vapour lock.

Owing to the variable nature of petrol and diesel, depending on the crude oil source and the refinery process, a sample may contain as many as 500 compounds, of which perhaps 10–20 will dominate and constitute 60–80% of the total volume. Hence petrol and diesel fuels tend to be characterized not by a specific chemical formula but by their physical and bulk chemical properties. The major parameters are their octane and cetane numbers respectively.

- The **octane number** is the resistance of the unburnt end gases to spontaneous ignition under specified test conditions, and is a measure of the 'combustibility' of a fuel when used in spark ignition engines: the higher the number the better the fuel quality – within limits. The two reference points are straight-chain heptane (C_7H_{16}), which is given a rating of 0, and iso-octane (C_8H_{18}), which is given a reference rating of 100. A fuel that behaves similarly to, for example, a test mixture of 8% heptane and 92% iso-octane is thus given a research octane number of 92.

- The **cetane number**   is a measure of the auto-ignition quality of fuels used in compression ignition engines. A fuel with a cetane rating of 50 or more is considered to be 'fast' burning, implying more thorough combustion, higher efficiency and lower particulate emissions. A cetane rating below 40 denotes a 'slow' fuel. It relates to the delay between the fuel's being injected into the cylinder and its ignition. The cetane rating of a fuel can be an indication of cold startability, exhaust emissions and combustion noise.

For biofuels there is some discrepancy between their specifications and the measured octane or cetane number since the tests were originally developed for hydrocarbon fuels. A direct comparison between the octane and cetane numbers for fossil fuels and biofuels may therefore be misleading when compared in terms of engine performance. However, the measures can be used indicatively to show the suitability of the fuels for various applications.

In the mid 1970s crude oil price rises and fuel shortages led to renewed interest in many countries in diversifying their transport fuel resources by developing biofuels as an alternative to petroleum-based diesel or petrol. More recently there have been increasing concerns about the potential problems from emissions and their effects on global climate change. Since biofuels are produced from biomass feedstocks that originally consumed carbon dioxide from the atmosphere as the crop plants or trees grew, combusting biofuel adds no net carbon emissions to the atmosphere even though the carbon dioxide is once again released to the atmosphere. Understanding this carbon cycling is an important concept for all sources of biomass, especially when comparing methods for mitigating the build-up of greenhouse gases in terms of $ investment per tonne of carbon equivalent avoided.

An interest in the potential for biofuels exists even in the USA, where President George W Bush, soon after coming into office, announced a major expansion of domestic oil and gas exploration activities in order to maintain energy security. Continuing in parallel, however (although no doubt with far lower resources), is an active National Biofuels Program administered by the US Department of Energy's Office of Fuels Development (OFD). It aims to collaborate with private industry and other stakeholder groups to commercialize biofuel technologies. It sponsors biofuels research, particularly at the National Renewable Energy Laboratory (NREL) and Oak Ridge National Laboratory, and is working to bring bioethanol and biodiesel to the marketplace at a competitive price. The first public biodiesel retail bowser in the USA was opened in San Francisco in May 2001.

At present the cost of production and processing of most biofuels tends to exceed the prices per litre of petroleum fuels ex-refinery by a factor of 2 or 3. The commercial production and sale of biofuels have therefore been implemented only in countries where some form of government support exists, possibly partly for strategic reasons to give some security of supply.

Bioethanol and biodiesel are the two most common types of biofuel currently used around the world. This is due to good availability of suitable feedstocks, a relatively good understanding of conversion technologies and the opportunity for practical implementation by being able to blend the biofuels with petroleum-based gasoline or diesel. Within certain limits, blending avoids the need either for any expensive engine modifications to be undertaken or for a new infrastructure to be developed to distribute and store yet another fuel type at retail service stations.

The chemical composition, C:H ratio, energy density and amount of carbon dioxide emitted during combustion on a stoichiometric basis have been compared for a number of fossil and biomass fuels (Table 9.1). Most of these fuels are a complex mixture of chemicals, and consequently only approximate values can be given.

Whereas fossil fuels are normally characterized as hydrocarbons since they have zero or low oxygen content, biofuels can be used as 'oxygenates' when blended with petroleum fuels as they have oxygen present in their molecular structure. This results in improved combustion and cleaner vehicle exhaust emissions.

Biodiesel

When Rudolf Diesel designed his prototype diesel engine nearly a century ago, he ran it on peanut oil. He envisaged that his diesel engine design would operate on a variety of vegetable oils, but when petroleum-based diesel fuel arrived on the market vegetable oils could not compete as mineral diesel was cheap, reasonably efficient, and readily available. So it quickly became the most popular fuel of choice for these new and efficient engines.

Although raw vegetable oil can be used directly, or blended with diesel as a substitute to power compression ignition engines, owing to its higher viscosity – particularly at lower ambient temperatures – it is normally first converted into

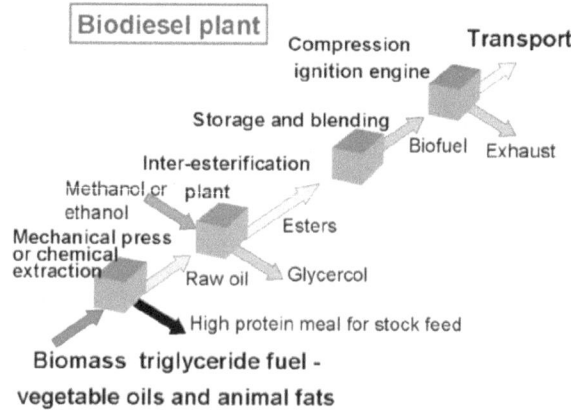

Figure 9.1 Biodiesel is based on esters produced from oils/fats

Table 9.1 Approximate chemical compositions and properties for selected fossil fuels and biofuels used for transport

Fuel	Approximate average formula	Average molecular weight	Approximate C:H ratio	Energy density (MJ/l)	Energy density (MJ/m³)	CO$_2$ emissions (g/MJ)
Natural gas (NG)	CH$_4$	18	1:4.0		38.2	51.3
Liquid NG	CH$_4$	18	1:4.0	25.0		51.3
Compressed NG	CH$_4$	18	1:4.0		38.2	51.3
LPG	C$_3$H$_8$	44	1:2.6	25.7		60.2
Petrol	C$_6$H$_{12}$	84	1:2.0	35.2		65.8
Automotive diesel	C$_{15}$H$_{22}$	202	1:1.9	38.6		65.8
Biomethanol	CH$_3$OH	32	1:4.0	15.8		60.8[a]
Bioethanol	C$_2$H$_5$OH	46	1:3.0	23.4		64.3[a]
Biodiesel (RME[b])	C$_{13}$H$_{29}$O	201	1:2.3	33.3		85.0[a]

[a] Carbon dioxide is recycled
[b] Rapeseed methyl esters

esters, which are the usual form of biodiesel. Esters can be made from several types of vegetable oil such as soybean, coconut, palm or rapeseed, from micro-algal oils, and also from animal fats since they are all triglycerides. Through a process called **transesterification**, organically derived oils are combined with alcohol (ethanol or methanol) and a catalyst and are chemically altered to form fatty esters. Most natural triglyceride oils are a mixture of 2–10 fatty acids. Chemical break-up of the triglyceride molecule results in a variable mixture of esters depending on which oil was used as the original feedstock and what fatty acids it contained.

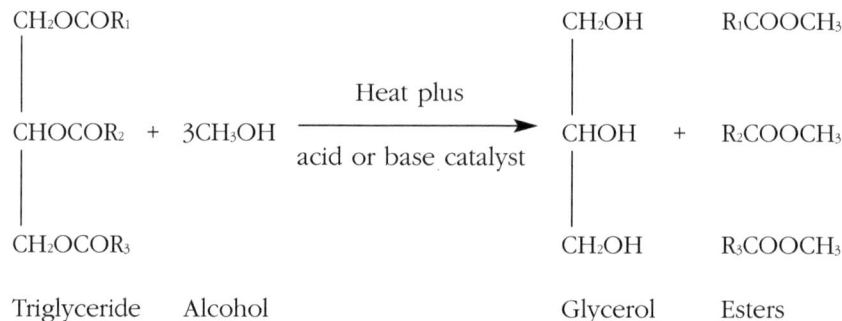

$$
\begin{array}{c|}
CH_2OCOR_1 \\
| \\
CHOCOR_2 \\
| \\
CH_2OCOR_3
\end{array}
\;+\; 3CH_3OH
\;\xrightarrow[\text{acid or base catalyst}]{\text{Heat plus}}\;
\begin{array}{c|}
CH_2OH \\
| \\
CHOH \\
| \\
CH_2OH
\end{array}
\;+\;
\begin{array}{l}
R_1COOCH_3 \\
\\
R_2COOCH_3 \\
\\
R_3COOCH_3
\end{array}
$$

Triglyceride   Alcohol   Glycerol   Esters

where R_1, R_2 and R_3 are specific fatty acids depending on the triglyceride used.

Product recovery is separated into phases, which provides for easy removal of glycerol, a valuable industrial by-product, in the first phase. The remaining alcohol/ester mixture is then separated and the excess alcohol is recycled. Then the esters are sent to the purification process, which consists of water washing, vacuum drying and filtration.

The ethyl or methyl esters have viscosities and energy values similar to those of conventional diesel fuel, with which they can be easily blended in any proportion without fear of phase separation; alternatively they can be used as a neat fuel (100% biodiesel).

In general, no engine modifications to the ignition system or fuel injectors are required to enable standard diesel engines to operate satisfactorily on biodiesel over the long term. However, this may not be the case when using raw vegetable oils, whose higher viscosities can lead to fuel pump damage and carbon build-up on the injectors. The solvent characteristics of the esters may in a few engine designs require substitution of certain hose and fuel line materials, and in some cases fine-tuning of the engine with regard to injection timing may be necessary in order to obtain optimal performance.

When used at ambient temperatures below 0°C there may be problems arising from the phase separation of some biodiesel types such as tallow esters. Biodiesel made from rapeseed oil and methanol (RME) has a cetane number of ~48, whereas soy-oil-derived biodiesel has a cetane number of ~56, coconut and palm kernel oil ~59 and tallow esters >70. Fuel consumption of biodiesel on a volume basis is about 10% higher as the energy value per litre of the biodiesel is lower than that of a typical diesel by a similar proportion.

Usually when comparing engine performance between that of biodiesel and conventional diesel fuels the power output is not significantly different, and vehicle operators can detect little if any difference between the fuels. Tests by Mercedes-Benz indicated that fuel consumption when driving on highways was essentially the same. In addition, comparative engine performance tests at the Austrian Institute of Agricultural Engineering showed lubricant consumption and engine wear to be similar for either fuel.

In addition biodiesel offers enhanced safety characteristics when compared with other diesel alternatives including petroleum, methanol and natural gas. It has a higher flash point, does not produce explosive air/fuel vapours, has very low mammalian toxicity if ingested and is biodegradable. The emissions are also expected to be less toxic, though this has yet to be fully evaluated.

Commercial processing plants for the medium-scale production of biodiesel from triglycerides have been developed in France, Germany, Italy, Austria, Czechoslovakia and the USA. Around 1.5 million tonnes of biodiesel is produced each year; the largest plant has a capacity of 120,000 t/yr. Commercial production of biodiesel in the USA, based on recycled cooking oils and soybean oil produced mainly by Proctor & Gamble, is over 100 million litres per year and growing. The Southern States Power Company (www.sspowerco.com) has linked with NOPEC, a biodiesel manufacturer in Florida using vegetable waste oils and greases, to produce and distribute around 20 million litres per year of 'OxyG B-60'.

Most of these countries have already adopted a national fuel standard for biodiesel, and many automobile manufacturers such as Volkswagen will maintain engine warranties after using such fuels. In Germany the Elsbett company designed an engine specifically for use with vegetable oils, with a swirl chamber incorporated into the piston head to encourage better combustion. It was tested in a wide range of vehicles and other applications in the 1980s (Figure 9.2) but made little inroad into the market.

An international conference was held in Vienna in November 1995 on the standardization and analysis of biodiesel. It included analytical procedures, material compatibility, emissions and storage properties. The intent is that these will soon be developed to a more common international standard, so that engine manufacturers and future users of biodiesel will have added confidence in its potential as a fuel.

Environmental benefits include low sulphur and particulates, resulting in a 99% reduction of sulphur oxide emissions and a 39% reduction in particulate matter compared with diesel fuel use. Other benefits include a reduction in greenhouse gases of at least 3.2 kg of CO_2 per kg of biodiesel, higher biodegradability and increased energy supply security.

In addition, the odour after combustion is less obnoxious than that from diesel. A positive energy ratio is also claimed, though other studies dispute this and show a poorer ratio, at times as low as 1:1 when the crop was irrigated. More typically a 1:3 ratio is quoted, 1 MJ from fossil fuel inputs (used to produce and transport the feedstock crop then process it) giving around 3 MJ in the biodiesel produced. For example, the production of a volume of soybean

Figur e 9.2 A specifically designed three-cylinder Elsbett engine powering a small generator on a mobile mechanical oil extraction plant suitable for small-scale vegetable oil production at the farm scale

methyl esters (SME) containing 1 MJ requires only around 0.3 MJ of primary fossil fuel input (Table 9.2). In other words, during its full life cycle, biodiesel contains more than three times the amount of energy that it consumes in its manufacture. It can be assumed that the methanol or ethanol used in biodiesel production is usually derived from natural gas. Since the alcohol accounts for almost 50% of the fossil fuel input into the process, the use of biomass-derived

Table 9.2 Primary energy inputs in terms of litres of diesel equivalent for soybean methyl ester biodiesel production at the commercial scale

Processing stage	Energy ratio (MJ in fossil fuel/ MJ out in the biodiesel)	Proportion of total energy input into the process (%)
Soybean production	0.0656	21.08
Soybean transport	0.0034	1.09
Soybean crushing	0.0796	25.61
Soy oil transport	0.0072	2.31
Soy oil conversion	0.1508	48.49
Biodiesel transport	0.0044	1.41
Total	0.3110	100.00

methanol or ethanol would significantly reduce the fossil fuel input further, giving an even higher energy ratio of fossil fuel input to biodiesel output.

It can be seen that, while biodiesel still has significant fossil fuel inputs, at least under current farming and transport practices, the positive net energy output is significantly higher than the negative net energy output of petroleum diesel production.

For biodiesel, oilseeds can be crushed and the oils extracted and used, either directly or after inter-esterification, to replace diesel or as a heating oil. Oil energy content is around 40 GJ/t, which is similar to that of diesel at 38–45 GJ/t. In the Philippines diesel is mixed with coconut oil and the blend is used in tractors, buses and trucks, though this would not be feasible in cooler countries as the viscosity of the oil increases and can cause damage to the fuel pumps. There are currently 85 biodiesel plants around the world (including one in Malaysia using palm oil) with a combined capacity of over 1.28 Mt/yr. The cost of the raw material is the most important factor affecting the overall cost of production.

Economically the cost of producing biodiesel far exceeds the current retail price of diesel in most countries, owing mainly to the high cost of producing the vegetable oil feedstocks, even when grown on land not used for food and fibre production and therefore with low opportunity cost. The costs of production are around double the retail price of fossil diesel, and it is unlikely to become more cost-effective before 2010. The production and sale of biodiesel have therefore only been implemented in countries where government subsidies of various forms currently exist. In some European countries, such as Germany and Austria, biodiesel is commonly available from filling stations as a neat fuel, and is competitive with diesel as there is no excise tax added. In Germany the mineral oil tax of DM0.62/l has been waived, and biodiesel retails at over 400 service stations for slightly less (around DM1.35/l) than standard diesel fuel (around DM1.40/l) (Figure 9.3). In France rapeseed oil esters are tax regulated and blended with fossil diesel fuel nationwide at the 5% level to 95% diesel. Some municipal authorities were the first to make the switch to biodiesel by fuelling their urban bus fleets with it. Others making the switch include owners of heavy-duty truck fleets, airport shuttles, marine and national park boats and vehicles, and military and mining operations

The cost of biodiesel is largely dependent on the choice of feedstock. According to a USA market analysis by NREL in 1995, if soybeans are used the fuel will cost approximately $0.66/l on a small-market scale. However, large-scale commercial use of biodiesel produced using the latest technology could reduce this cost to $0.40–0.45/l. Additional research advances using existing feedstock technologies or innovative feedstocks such as micro-algae could further reduce costs. The goal of the DOE/NREL programme is to produce biodiesel from micro-algae at a cost of $0.26/l.

Financial incentives by way of carbon emission levies, taxation benefits or capital grants may be required if a government wishes seriously to encourage the use of biodiesel. As the environmental concerns from using fossil fuels increase, their price rises, long-term security of supply becomes questionable

Figure 9.3 One of many German service stations retailing diesel for DM1.40 and biodiesel for DM1.35

and land use diversification from food and fibre crops to energy crops becomes feasible, then such incentives could well be warranted.

Since vegetable oils currently cost considerably more to produce per litre than diesel fuel ex-refinery, new techniques such as distillation under vacuum with microwave heating are needed to help bring down the price. However, low oil yields of around 1–3 t/ha of oil for most annual oil crops are a major limitation to being able to reduce the price significantly. Even at a higher oil yield of 3 t/ha, this equates to only around 120 GJ/ha.yr of available energy in the form of oil. Any straw from the crop also has an energy content of around 15 GJ/t, which can be included in the total biomass energy available if it can be combusted and used for useful heat purposes. However, if the land was used instead for growing say high-yielding short-rotation forest crops producing 20 odt/ha.yr, the total amount of solar energy captured and stored would then be closer to 400 GJ/yr from that same hectare of land.

The potential for biomass-derived products as lubricants is also gaining interest. High erucic acid varieties of oilseed rape have been bred for the good lubricity characteristics of the oils and have long been used in aircraft. They are also biodegradable, have low toxicity, and therefore tend to be more environmentally acceptable than mineral oil, which can cause adverse effects on soils and plants when poorly disposed of. Conversely the natural biodegradability of vegetable oils over time means that long-term storage of esters can lead to some degradation.

The European Commission Directorate-General XII on Science, Research and Development in their 1994 publication *Biofuels* concluded:

In comparison to other processes for production of fuel from renewable agricultural resources, rape oil and also sunflower oil production give a good output/input relation of energy and a considerable net output of fuel energy. Results of engine tests have shown that rape and sunflower oils and their methyl esters are energetically satisfactory as fuels in diesel engines. The behaviour of diesel engines in long term operation with esters alone or in blends under varying and hard field conditions must be the object of further investigation.

Case study 15

The potential for biodiesel in New Zealand

New Zealand's agriculture and forest industries account for over 60% of the country's annual export earnings, and depend on diesel fuel for operating tractors, trucks and heavy machinery. Since the 1970 world oil shocks a major proportion of imported gasoline has been substituted by locally refined synthetic petrol manufactured from natural gas via the Mobil methanol process. In addition compressed natural gas (CNG) and liquid petroleum gas (LPG) infrastructures were established and vehicle engine conversions encouraged. A research programme was also instigated in the mid 1970s to evaluate diesel fuel alternatives. At the time this was deemed to be strategic research in order to maintain the balance of products from the oil refinery and to combat the anticipated continued rise in the price of crude oil. In contrast to petrol, substitution of diesel fuel has not been easy, and no economic alternatives were identified. In the 1980s, as a consequence of the lower demand for imported petrol but a steadily increasing demand for diesel, a hydrocracker was upgraded at the sole oil refinery at Marsden Point to provide a wider cut of middle distillate fuel. More diesel and less petrol could then be produced per barrel of crude oil. The diesel fuel demand was then met either by importing crude oil and refining it or by importing the refined product directly.

Much of the R&D work on diesel alternatives was contracted out by the Liquid Fuels Trust Board, funded by a $0.001 levy on every litre of gasoline or diesel sold. The Board was disbanded in 1987 following the continued low price of crude oil, but during its life much useful research was accomplished. The results are still relevant today, including the research programme for biodiesel as outlined here.

A series of oil crop production trials were undertaken in the 1970s to assess the yields and costs for various regions of New Zealand, since vegetable oil crops were not commonly grown. This work was followed in 1982 by an overview study to ascertain the availability, yields and costs of supply of a range of natural oils and fats. Oilseed rape was the preferred crop, for the following reasons:

• Its oil yield potential was relatively good, and it grows throughout New Zealand (from latitudes 34°S to 47°S) in a temperate climate.

- The ability to use conventional cultivation and harvesting equipment reduced the investment costs for new machinery that other new crops such as sunflower would require.

- Its high protein meal by-product was suitable for pig and poultry feed.

- The chemical and physical properties of the oil were well suited for use as a fuel for compression ignition engines.

In addition to oilseed rape, inedible tallow, a by-product from the meat industry, was also selected for further evaluation as a diesel substitute. Tallow is exported mainly for soap and candle making, and volumes exceeded 10% of the national diesel demand when measured on an energy content basis. Research on the use of CNG, LPG and alcohol fuels as diesel fuel substitutes was also being conducted at this time, but the problems of low energy density, high production costs and expensive engine modifications were limitations to their uptake.

The overall aim of the New Zealand research programme on triglyceride fuels was to develop a fuel that would be technically feasible to produce, could be distributed nationwide, would have a high strategic value, would produce minimal environmental pollution, would be competitively priced, and would be suitable for use in the existing engine fleet without modifications being required. Also considered were the environmental benefits resulting from using biofuels as substitutes for fossil fuels, and that the feedstock had to be produced on a sustainable basis.

Oil and fat resource assessment

An important early component of the programme was learning how much resource was available. The availability of suitable land for growing oilseed rape crops was evaluated, and the returns to the farmer – based on assumed seed yields of up to 3 t/ha (but typically 1.6–1.8 t/ha) and 45% oil content – were compared with other cropping enterprises. The volumes and quality of tallow produced at the many meat works, and already collected and transported to several ports for export, were also evaluated.

The oil/fat qualities in terms of engine fuel specifications were measured. These included properties such as free fatty acid content, iodine value, viscosity, energy value and degree of chemical saturation, as well as variations between batches and from different extraction plants. The existing infrastructure already in place for processing and transporting the oils/fats was also examined.

The comparative values of the edible and inedible oil/fat supplies both for export and for the domestic market were determined from market research. The markets for the by-products glycerol and protein meal were also examined, as it was realized their future value could in part offset production and processing costs, which could affect the economic viability of the enterprise. Future trends were considered but, as with most commodities, it proved difficult to assess them with any reliable degree of accuracy. For example, in the decade following this resource assessment the sheep meat and wool prices fluctuated widely such that the New Zealand flock number reduced to 50 million from its peak of around 65 million.

Raw vegetable oil fuels

The use of untreated, home-grown rapeseed oil as a tractor fuel by a number of over-enthusiastic New Zealand farmers in 1982 created national interest but, as predicted, eventually led to major engine operating problems. Some power loss became evident, fuel pumps seized, carbon build-up appeared on injectors resulting in poor fuel atomization, and dilution of the lubricating oil occurred: this led first to polymerization and then suddenly, under specific conditions, the oil turned into a gel, leading to disastrous results. Problems also appeared during storage of the oil after only a few weeks following extraction. These factors served to confirm that although it was relatively easy to grow the rapeseed crop, extract the oil and put it in vehicle fuel tanks, continuation of a proper scientific evaluation was justified. Blending the oil with diesel could serve to reduce the problems listed, particularly those resulting from high viscosity, but did not eliminate them.

It was therefore strongly recommended that farmers should not use raw vegetable oil as fuel in their compression ignition engines, particularly those with low compression ratios, rotary fuel pumps and direct injection. Those with worn injectors or likely to be operated under light duty cycles or cold conditions were also at higher risk. Furthermore, for any engine using biodiesel, daily checking of the lubricating oil for thickening, together with regular cleaning of the injector nozzles, was strongly advised. The solution to many of the problems experienced by those early rapeseed oil users appeared to be to convert the triglyceride oils to esters by trans-esterification.

Ester production and blending

Selection of the most appropriate process and design of plant was undertaken by chemical engineers. A base catalysis process using methanol and sodium methoxide as catalyst was selected, the methanol already being produced from the natural gas resource. The choice was also partly dependent on the availability of chemicals and the scale of operation. Assessments were undertaken for rapeseed oil when extracted by a single grower on-farm (4000 l/yr), by small co-operatives of eight or nine farmers (36,000 l/yr), by larger farmer co-operatives (190,000 l/yr), and at the regional plant scale with production of 2,750,000 l/yr. Tallow ester plants of 4,200,000 and 25,000,000 l/yr capacities were also evaluated.

The smaller plants were more expensive to operate, and the by-product values were more difficult to capture. Commercial costs of oil production, extraction and trans-esterification were all significantly less at the larger scale.

The goal of the research programme was to produce a fuel blend that would be totally within the current diesel fuel specifications in all respects. Hence no engine modifications would be necessary and no changes in fuel storage or handling regulations would be required. The intention was then later to relax these standards to make ester production more profitable by employing less stringent processing conditions, but only after considerable experience of using the fuel had been gained. This cautious approach was to provide early confidence in the fuel by the consumers who would have a choice between diesel and biodiesel. A summary of the stringent fuel specifications set for the tallow esters is given in Table 9.3.

Table 9.3 Summary of tallow ester fuel specifications used during production of the fuels for fleet testing

Fuel component	Recommended level	Reason
Free fatty acids	<0.1%	Attack metal
Water	<0.5%	Storage deterioration
Methanol residue	<0.05%	Fire/explosion
Catalyst residues	<0.5 ppm	Excess wear in engine
Monoglycerides	<0.05%	Filter blockages
Iodine value point	42–52	Solubility/oil degradation/ melting point
Insolubles	Pass through a 4 μm filter	Fuel system blockage

To meet these initial conditions a test blend of tallow esters was restricted to 10% methyl tallow ester and 90% diesel, although engine performance testing was also carried out on 20% ester blends and on 100% esters. If a 10% blend was to be introduced nationally to replace diesel, the total volume of tallow available would be utilized. At higher blends not only would there be a scarcity of tallow but cold temperature limitations were expected from separation of the esters at temperatures below 3°C. Blends using greater volumes of MRE remained within existing fuel specifications since rapeseed oil esters have higher degrees of saturation and higher melting points than tallow esters. There was also the potential to produce sufficient volumes of vegetable oil esters to supplement the finite supply of tallow, assuming land was available.

Fuel property evaluation

A wide range of standard fuel tests were undertaken on a range of esters and blends. The major concern was the phase separation under cold conditions when higher blend levels of methyl tallow esters were produced. This was thought to result from supercooling of the esters as ambient temperatures dropped. At warmer temperatures homogeneous mixes occurred at all blend levels.

A high cetane value of over 70 was measured by Perkins Engines (UK) for methyl tallow esters when evaluating these fuels under contract to the Liquid Fuels Trust Board. It is not clear what produced this extremely good combustion property. As a result of this characteristic, tallow esters could be used as an additive to upgrade poorer-quality diesel produced more cheaply from a wider-cut middle distillate. This would help to improve the product balance at the oil refinery.

The stability of vegetable oils and tallow after prolonged storage is questionable, mainly as a result of biological degradation by bacteria, yeasts and other organisms, particularly under warm conditions. A field trial of tallow esters stored in 200 l drums for 14 months showed no thermal, oxidative or biological degradation. A more thorough evaluation of storage potential conducted in Germany confirmed that oils/fats in the form of esters can be stored successfully for at least 2 years.

Engine performance tests and fleet demonstrations

A series of screening trials were initially conducted on a wide range of oils, esters and blends. No major limitations became apparent in these short-term tests other than when using raw tallow, which, as might have been expected, solidified in the fuel system on cooling of the engine after shutdown! Comparisons of esters and blends with diesel when fuelling the same engine showed that the esters tended to give slightly lower power outputs, lower exhaust temperatures and reduced exhaust particulate emissions (Figure 9.4).

A series of engine bench tests using tallow ester fuels and blends was undertaken by Perkins Engines (UK). Compared with diesel it was noted that

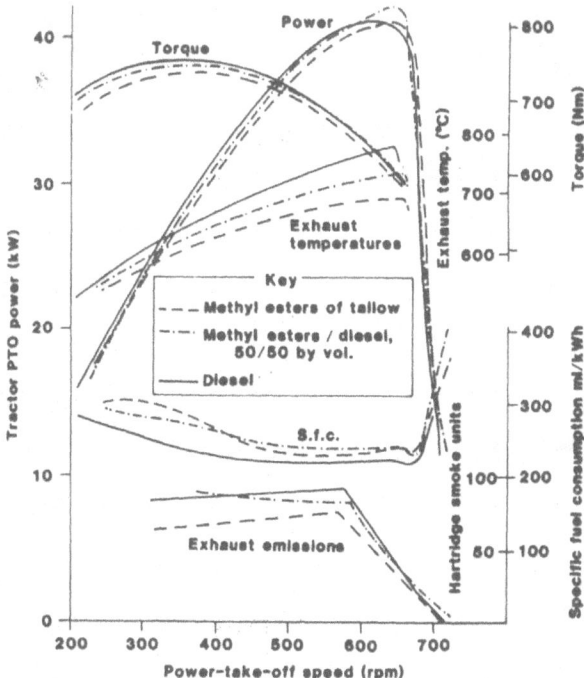

Figure 9.4 Engine performance test results comparing rapeseed oil, methyl esters of tallow and rapeseed oil and 50% blends with diesel

combustion improved, emissions were reduced and that even under worst-case operating conditions lubricating oil dilution was not a limiting factor.

A fleet of 42 vehicles was then selected to represent open road use (long- and short-haul distances), urban transport, agricultural vehicles and marine engines. Over 300,000 l of 10% ester blend diesel was consumed during the trials. No technical problems became apparent even under extreme operating conditions. Operators reported visible reductions in smoke emissions and reduced engine noise. This served to confirm the laboratory bench tests, and showed that operators would probably notice little difference in performance between using diesel or ester/diesel blends should the latter be introduced as a commercial fuel.

More detailed evaluations were conducted on four vehicles:

- A John Deere 3140 tractor was run on a tallow ester/diesel blend for over 500 h and closely monitored before a turbocharger was retrofitted. Temperatures were recorded throughout the engine lubricating oil circuit, and oil samples were taken at frequent intervals. The tests included immediate engine shutdown from full power to gauge maximum oil gallery soak temperatures in the turbocharger housing. The objective was to investigate the baking propensity of the lubricating oil if diluted with non-combusted ester contaminants. No problems became evident.

- A 15 t Ford N 1317 truck with Hino EH700 engine (Figure 9.5) was run under normal road conditions for double the recommended period between lubricating oil changes on the 10% ester/diesel blend. After 12,000 km no significant lubricating oil dilution was observed. The viscosity and total base

Figure 9.5 One of several vehicles tested using tallow ester diesel blended fuel

numbers reduced with travel distance and the total acid number rose slightly before stabilizing after 2000 km, but no significant oil deterioration occurred.

- A Ford 6610 tractor and a Ford Courier utility with a 2 litre Mazda engine, both fuelled by 10% esters/diesel, were compared with diesel-fuelled companion vehicles of similar workloads and ages after over 300 hours of operating. Monitoring of the lubricating oils was also undertaken. A slight ester dilution was noted in the tractor oil but dispersancy tests remained satisfactory for both vehicles. The diesel-powered utility gave a rise in lubricating oil viscosity and acid number coupled with a drop in base number towards the end of one oil change period, showing that oil failure was imminent. This did not occur with the ester/diesel vehicle. On completion of the trials, engine strip-downs showed the wear and cleanliness of the test fuel engines to be superior to those running on diesel fuel alone (Figure 9.6).

Figure 9.6 Comparisons of engine blocks, injector nozzles and valves from the number 4 cylinders of the sister 2 litre Mazda engines in Ford Courier vehicles running on diesel (left) and on 10% tallow ester/90% diesel blend (right)

Economic analysis

A national economic analysis was undertaken for rapeseed methyl ester production, which included feedstock costs, transport costs and opportunity costs together with by-product credits in addition to the normal commercial costs. The cost of ester fuel production from a large central processing plant was less than half the cost of a small-scale on-farm process but was still well over double the ex-refinery diesel fuel price. This has since been confirmed by several analyses undertaken for US and European conditions. A further disadvantage of smaller-scale installations would be the variation in fuel quality that would result from variations of feedstock and processing. This could lead to a miscellany of engine operating problems.

Production costs of the esters depended mainly on oil yield per hectare, opportunity cost of land for other enterprises, the assumed relative export value of the vegetable oils, and world prices for protein meal, glycerol and methanol. Socio-economic factors such as rural employment and health effects from vehicle emissions in towns should also be considered. In addition the environmental benefits from avoidance of burning fossil fuels should be included in any comparative calculations.

A similar national economic analysis was also undertaken for tallow esters. In both, transfer payments such as interest were ignored since they do not represent real resource costs. More favourable economics resulted for tallow, as it is a by-product of the meat industry rather than having to be especially grown for fuel as is oilseed rape. A concept study determined that a single processing plant adjacent to the existing oil refinery was the best option, the tallow being back-loaded by road from meat-processing plants or by boat from ports after delivery of fuel to the regions by the existing fleet of coastal and road tankers. The concept study involving feedstock collection, oil processing, product distribution and quality control was an essential part of the national study.

Under the base cost assumptions of \$200/t to purchase the tallow, \$0.08/t.km to transport it, \$70/t to process and blend the esters and \$800/t for the glycerol by-product credits, the crude oil price needed to reach \$27/barrel to provide a 10% real rate of return after tax. The by-product value is critical, such that a lower glycerol price resulting from saturation of the world market would significantly increase the ester costs.

Several constraints were identified, which will require further careful assessment before implementation of biodiesel fuels could successfully occur:

• Engine manufacturers will need to meet their usual warranties.

• Distributors and consumers will need to be convinced that the ester fuel supplies will remain secure and readily available. This is another justification for using modest blends with diesel at first and then, once proven, increasing the ester proportion if appropriate.

• If the ester is to be blended with diesel, the refinery company must be agreeable or regulations must be put in place to enforce it. The petroleum suppliers will also need to be in agreement.

In summary, based on the New Zealand experience, technically the use of triglyceride fuels in the form of neat esters or blended with diesel for use in compression ignition engines is well understood. Economically it is not a commercially viable proposition compared with current diesel prices in most countries. To become commercial one or more factors need to occur:

• Some form of government incentive is needed.

- The environmental external costs of using fossil fuels must be recognized.

- The crude oil price must remain above $27–30/t.

- The costs of producing and processing the biodiesel need to decline by increasing crop yields or developing cheaper process systems.

Bioethanol

Ethanol is the most widely used biofuel today, with billions of litres added to gasoline each year to improve vehicle performance and reduce air pollution. USA and Brazil are large-scale producers, as is France. Canada has more than doubled its production since 1996, with six plants producing 200 million litres per year. Mexico is to begin producing ethanol to add value to its sugar cane industry and to reduce air pollution in Mexico City by using blended fuels. Australia has had an industrial ethanol plant operating for several years, and recently BP Australia announced that it is currently evaluating the potential for developing a new ethanol plant alongside its Bulwer Island oil refinery in Queensland to enable blending to be easily achieved. The company anticipates producing a 10% ethanol blend at up to 80,000 barrels per day, and the Australian Government is providing a $5 million grant to undertake a feasibility study of the $40–50 million project to upgrade the refinery in order to store, blend and deliver the fuel. Sugar cane and wheat sources will be studied, as well as the changes needed at the refinery to modify plant to process and handle the new fuel. A similar-scale project to produce 3 million litres per day is planned for Jilin, China, also to be used in a 10% blend and designed by Austrian biotechnology engineers. It is in the main corn-growing area, and the target is 99.5% pure alcohol.

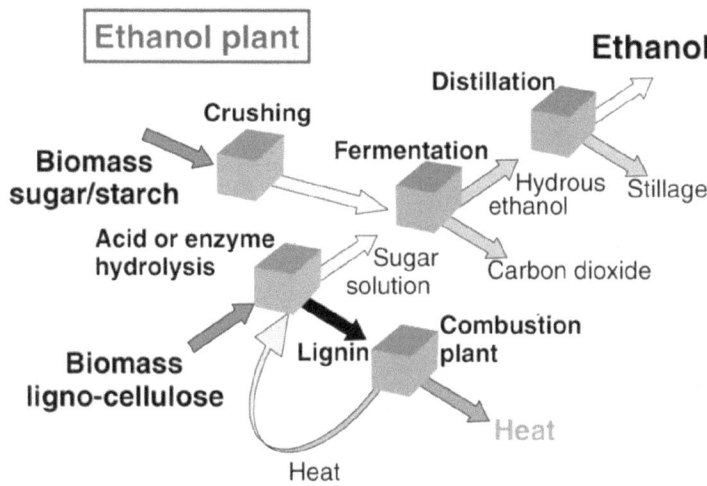

Figure 9.7 Ethanol can be produced from a variety of biomass sources

Ethanol is an alcohol, mostly produced using fermentation, in which natural hexose (6-carbon) sugars are fermented by yeasts. The ethanol is then distilled into its final pure form of 95% or above. Sugar cane has easily fermentable free sugars present, but crops such as beet or cereals growing in temperate climates have few if any. Extra processing is therefore needed to release yeast-digestible sugars. These various fermentation conversion routes have been well known for many centuries to wine and beer makers. Ethanol can also be made from woody and fibrous biomass materials by the hydrolysis of the cellulosic material into sugars, followed by fermentation.

Typically bioethanol contains 5% water and other trace impurities, and has a reasonably high octane number of 105–120. Spark ignition engines, based on the Otto cycle, will operate on pure ethanol after retuning and minor modifications including the replacement of certain seals and other non-compatible components. However, owing to the additional cost, blending with gasoline at up to 15% to form **gasohol** as in the USA is often preferred to enable existing engines to run unmodified. It can also be emulsified with diesel to form **diesohol** and used in unmodified compression ignition engines, though this is less well proven, and many engine manufacturers will not maintain warranties.

Just as for biodiesel, the use and sale of bioethanol for engine fuel is not a new concept. From 1920 to 1924 the Standard Oil Company of the USA marketed a 25% (by volume) mixture of absolute ethanol blended with gasoline in the Baltimore area, but high corn prices combined with storage and transportation difficulties terminated the project. Subsequent efforts to revive an ethanol fuel programme in the late 1920s and 1930s through federal and state legislation, particularly in the Midwest corn belt, failed. Then Henry Ford and several experts joined forces to promote the use of ethanol, and a small fermentation plant was built in Aitchison, Kansas, to manufacture 38,000 l/day specifically for motor fuels (once the problems arising from the Prohibition Law for alcohol had also been resolved!). During the 1930s more than 2000 service stations in the Midwest sold this corn-based ethanol. However, the continuing trend towards lower petroleum prices resulted in the closure of the ethanol plant in the 1940s.

More recently in 1979, ethanol–gasoline blends between 5% and 20% ethanol by volume were reintroduced to the American market following crude oil supply disruptions in the Middle East, which then became a national security issue once gasoline became scarce in some states. Alternative fuels were seen as a solution to the problem, although those arising from corn and other cereals are subject to world price fluctuations of these commodities. The American Oil Company and several other major oil companies began to market ethanol–gasoline blends as a 'gasoline extender' and an octane enhancer.

Most ethanol production in the USA today is still based in the large grain-growing states of the Midwest, where about 13 million m^3 of corn and other cereal starch crops are used to produce over 6 billion litres of ethanol each year. In 1998 there were 51 production plants in 19 states, which created farm income of $4.5 billion and resulted in almost 200,000 jobs. Since any further increase in demand for corn will eventually push up prices, increased demand for the fuel will ultimately require the use of other low-cost feedstocks such as agricultural and forestry residues, municipal solid wastes, industrial wastes, and crops grown specifically for energy purposes. The Department of Energy (DOE) is therefore concentrating its efforts on developing ethanol feedstocks from a range of cellulosic biomass resources, which are abundant and inexpensive. For example in Middletown, New York State, Masada Oxynol is planning an ethanol plant using municipal solid waste. To ensure that a low-cost energy feedstock is available in sufficient quantities, researchers are also examining dedicated energy crops including short-rotation forest and vegetative grass species that have been selected and bred to produce high biomass yields. With proper management,. researchers estimate that an average of 2.45 billion tonnes of cellulosic biomass could be available in the US each year for fuel conversion.

The Clean Air Act Amendments of 1990 mandated the sale of oxygenated fuels in areas of the country with unhealthy high levels of carbon monoxide, most notably in California. However, this measure was later changed in favour of oil companies' adding the cheaper MTBE (methyl tertiary butyl ether) instead. The environmental authorities then became alarmed at finding this carcinogenic chemical entering the groundwater, having been washed off road surfaces by rainwater. This was predictable, and should have come as no surprise! In some states ETBE (ethyl tertiary butyl ether) was therefore substituted instead, being less toxic. In other states ethanol was advocated. In Illinois, for example, concerns at the use of MTBE resulted in a strong demand for bioethanol to be blended with gasoline as an oxygenate. This is due to the relatively high oxygen content in the C_2H_5OH molecule (Table 9.1) compared with hydrocarbon fuels, which as their name implies are composed of complex strings of mainly carbon and hydrogen atoms. However, the petroleum-refining companies would prefer to exclude all such mandatory additives by refining the gasoline to give it a higher oxygen content in the first instance.

Both MTBE and ETBE are claimed to have a more favourable set of fuel additive properties as a gasoline additive than those of biomethanol and bioethanol, from which they can be derived. However, the production costs of hydrolysis of the feedstock using iso-butylene as a reagent for the catalytic etherification of the alcohols, and then fermentation followed by additional distillation, remain relatively high. Alcohol biofuels also have the potential to power fuel cells through reformers to produce hydrogen (Chapter 10).

All automobile manufacturers in the United States approve the use of specific ethanol–gasoline blends. They are successfully used in all types of vehicle and engine that normally are fuelled by gasoline. E10 (a blend containing 10% ethanol) has been the most commonly distributed, but E85 and E95 blends have also been successfully tested in modified or dedicated engines installed in government fleet vehicles, urban transit buses and flexible-fuel passenger vehicles (FFVs), which can run on either 100% gasoline or E85. The number of E85 and FFVs has increased since the mid 1990s. In the summer of 1997 Ford and Chrysler separately announced that they would each manufacture as many as 250,000 FFVs per year and sell them for the same price as gasoline models, a move prompted partly by the Californian drive towards zero-emission vehicles. Approval of ethanol blends is usually found in vehicle owner manuals.

Hydrolysis of ligno-cellulose

Cellulosic biomass is a complex mixture of carbohydrate polymers from plant cell walls known as cellulose and hemicellulose, plus lignin and a smaller amount of other compounds generally known as extractives. To produce ethanol is more complex than from simple hexoses using yeast fermentation since a pretreatment process is used to reduce the feedstock size, break down the hemicellulose to pentoses (5-carbon sugars) such as xylose and open up the structure of the cellulose component. It involves the decomposition of biomass at lower temperatures by the use of steam explosion or acid hydrolysis. Pressure and heat are used in a controlled environment under acidic conditions. Considerable research was undertaken at the New Zealand Forest Research Institute in the 1980s and is now being continued in the USA at the National Renewable Energy Laboratory (NREL) and elsewhere. Research was also undertaken by Convertech Ltd in Christchurch, New Zealand, and a 'biorefinery' pilot plant was built. The focus of that project was on the production of ethanol together with high-value chemical feedstocks used for other industrial processes. In Canada, Iogen has initiated construction of a $25 million cellulose-to-ethanol demonstration plant.

Normally the cellulose is hydrolysed by acids or enzymes and hence is broken down into glucose sugar, which is then fermented to ethanol (Figure 9.8). The pentoses from the hemicellulose can also be fermented to ethanol but not by natural yeasts, which are unable to react. So several new technology

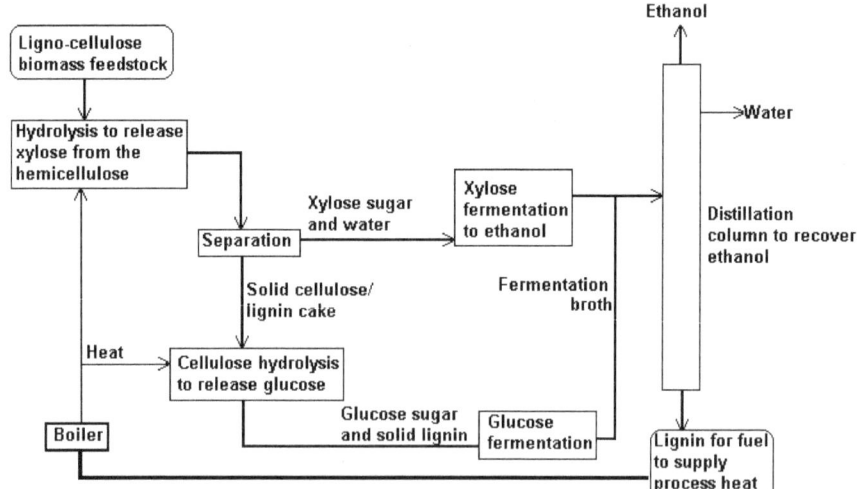

Figure 9.8 An example of a process to produce bioethanol from ligno-cellulose biomass
Source: Based on BC International Corporation

developments have evolved, some of which are only just becoming commercial. For example, genetically modified *E. coli* bacteria have been developed by BC International Corporation and used successfully for this purpose. The lignin is burned as fuel to power the process.

The use of thermophilic production processes using bacterial organisms that can stand higher temperatures (70–80°C) than yeast during the fermentation process was first evaluated for bioethanol production in the 1970s. More recently Agrol, an American company, also realized that for the process to have any real advantage over yeast fermentation it would be necessary to ferment the pentose sugars present as well as the hexoses to give a higher conversion efficiency. This would also allow greater tolerance to the ethanol concentration in the fermentation broth.

One company has developed a new process including the technologies required to develop a new strain of bacteria, the strain itself and the fermentation process for the production of chemicals. A worldwide patent was first granted in 1988 for a process in which thermophilic anaerobic bacteria can be selected for their characteristics of fermenting sugars above 70°C. The fermentation is performed in a continuous two-stage reactor. The activity is maintained by continuously withdrawing ethanol as well as cells from the anaerobic vessel. The cells are then encouraged to grow aerobically on residual sugars present in the medium, before returning them to the anaerobic vessel. In 2000 a patent was lodged to introduce the pyruvate decarboxylase enzyme into a thermophilic strain of bacillus by gene interaction and to use other genetic techniques to transform foreign DNA into a strain of bacterium. Later that same year a further patent was sought to use the bacterium strain in a novel fermentation process to provide an extremely fast rate of conversion. The target was for 20 ml of ethanol to be produced per litre of fermentation volume per hour, which is 10 times faster than traditional yeast fermentation. The yields of ethanol (grams per gram of sugar) are similar. Since it is a continuous process, vessel size is reduced, plant downtime is less, and capital costs are lower to give the same output of bioethanol.

Simultaneous saccharification and fermentation (SSF) is yet another novel process for converting cellulose to ethanol that is being investigated by NREL. It combines the cellulose hydrolysis and fermentation steps in one vessel, and produces higher yields of ethanol. Conversion of these cellulosic materials using enzymatic hydrolysis is being evaluated in a 1 t/day pilot plant, and is claimed to be nearing the production phase. As the cost-effectiveness of this technology increases, industrial companies are becoming increasingly interested in commercializing the technology. The first commercial ligno-cellulose biomass ethanol plants will possibly begin operation within the next few years. Continued improvements in such key technical areas will make biochemical conversion of biomass to ethanol a more efficient and economical process.

Research activities such as those at NREL over the past 20 years have developed new technologies to convert ligno-cellulosic materials such as bagasse, rice husks, municipal green waste and straw to ethanol. However, as a consequence of the additional acid or enzymatic hydrolysis process needed to first convert the cellulose to free sugars, the process is still far from being competitive with the production of bioethanol from starch or sugar crops. One further barrier is the high cost of manufacturing the cellulase enzymes. New production processes are being developed and evaluated. Various fungal enzymes have been isolated by commercial companies in the USA, such as Genencor International and Dyadic International. The objective is to reduce the current costs of ethanol of around 7–8 c/l down to almost 1 c/l. Advances in the genetic engineering of enzymes and bacteria are making the conversion process more productive, so that the target is achievable.

The cost of bioethanol is related directly to the cost of biomass material available and the conversion efficiency of the process. As an example, if say bagasse feedstock costs $12/t to acquire and deliver, and the process used is 35% efficient, then the cost of producing the bioethanol will be around $330/t. If, however, the bagasse can be delivered for $2/t and an improved process is used with a conversion efficiency of 45%, then the ethanol could be produced for closer to $230/t.

Other research is investigating using the stillage from the fermentation as an animal feed or another chemical feedstock rather than disposing of it. A systems approach to biofuel production, so that all the feedstock is consumed in some useful way in the process, will increase the economic value of the biomass processing operation.

Another American company, Bioengineering Resources Inc. (BRI), has innovatively linked the biomass gasification process with bioethanol production. At present gasification for heat and power remains uneconomic in many countries, in spite of the intensive development of several research and demonstration plants. The concept of BRI was to couple gasification with producing higher-value liquid fuels in order for it to become a viable operation. This has been achieved by using the anaerobic bacterium *Clostridium ljungdahlii* to convert the H$_2$ and CO to ethanol. The claim is that reactor fermentation vessels have been developed that result in very short retention times, even when at atmospheric pressure, thereby keeping the equipment costs low. Slow growth of the bacteria in the reactor apparently regenerates the biocatalyst, but commercial feasibility of the project has not yet been demonstrated in a pilot plant. Many seemingly good concepts for producing biofuels have been initiated in the laboratory but unfortunately have failed at the commercial scale for one practical reason or another.

Case study 16

The Brazilian ethanol fuel programme

The Brazilian Alcohol Programme, the largest in the world, was initiated in the 1970s when the world oil price was high and the world raw sugar price low. It faced several difficulties in the 1990s. For example, a serious ethanol fuel supply crisis occurred in 1989 because an increase in the world sugar price reduced the feedstock availability. This led to a lack of public confidence in the security of supply of the biofuel, which consequently reduced the demand for dedicated ethanol fuel vehicles. The share of production of these cars achieved 96% of all those manufactured in Brazil in 1985, but declined to 3.1% in 1995 and to only 0.1% in 1998. To counteract the trend the Brazilian government approved a higher blend level of ethanol in gasoline, up to 24%. With the growth observed in annual sales of automobiles, and the higher fraction of blended ethanol now possible, absolute production of ethanol continued to increase, achieving a peak in the 1997/98 harvesting season of 15,307 m^3 against 11,900 m^3 in the 1990/91 season. In 1997 gasoline consumption had reduced to 16,032 m^3 since ethanol represented 42.7% (when measured as gasoline equivalent) of the total fuel consumption by all Otto cycle motors. As a result the annual net carbon emission levels were abated by 18% of the total national carbon emissions from the use of fossil fuels. Since 1975 over 4 million vehicles have been run on neat ethanol, with over 100 billion litres produced from sugar cane.

Brazil's sugar cane growers remain confident, total yields having increased to 312 million tonnes of saccharose as produced in the 1997/98 season. At this time, however, only 60% of the total saccharose produced was converted to meet the bioethanol demand as against 74% in 1989/90. Technology for sugar cane harvesting in Brazil and elsewhere has improved, driven partly by environmental legislation that limits the use of pre-harvesting fires in the field to burn off the trash and make access easier for harvesting equipment. A small fraction of the harvest is undertaken by machinery, the rest by hand cutting.

Evaluations are under way to determine whether the biomass residues, or trash, which used to be burnt off in the field, should now be separated from the sugar cane and left in the field or transported to the sugar mills. The major driver for the last option in many countries is the existence of a significant new market for electricity generation from independent power producers (IPP). Appropriate legislation to encourage IPPs is in effect in Brazil, but the effects have been insignificant compared with power generated by the traditional state-owned generators.

Breeding more productive sugar cane plants is continuously advancing, with average saccharose yields increasing since 1990 from 129 to 143 kg/t of green sugarcane owing to the growing of new selected varieties. In transforming saccharose to ethanol no significant technological advances have been observed, but the use of 'best practices' in a large number of factories has increased the average conversion efficiency to almost 50%. Adoption of best practices using known technologies is anticipated to bring down costs by around a further 25%.

Even with the slowdown in demand for ethanol as a fuel, the Brazilian programme provided a significant economic gain for the country. As much as $33 billion in hard currency was saved by avoiding the purchase of imported oil from 1976 to 1996. With the addition of interest that would have been paid because of the increase of the country's external debt if the ethanol was not available, the total national saving is around $50 billion.

Expansion of the 25-year-old Brazilian yeast–bioethanol industry can be achieved by increasing the demand for locally made dedicated ethanol-powered vehicles once again, by increasing the proportion of ethanol to be blended with gasoline, or by seeking exports to the USA and other countries. The industry has a good future.

Other biofuel production options

Biomethanol

Methanol is the alcohol of methane, formerly called wood alcohol as it was historically made from the pyrolysis of wood. It is a liquid transport fuel that can be produced from fossil or renewable domestic resources. It is technically feasible to produce it from synthesis gas (CO and H_2) formed from the thermochemical gasifying of biomass. The gas mix is then reformed into methanol.

Nowadays most methanol is made from natural gas feedstock when there is gas available surplus to other demands, but it can also be produced from other feedstocks including coal and residual oil. When produced from these sources it should really be classed as a fossil fuel. It is most commonly used as a chemical feedstock, extractant, solvent or as a feedstock for producing methyl tertiary butyl ether (MTBE), an octane-enhancing gasoline additive. The US methanol industry, as an example, produces approximately 4.7 billion litres annually, mostly from natural gas. About 38% is used in the transport sector, mostly in the production of MTBE. It can also be used in neat (100% pure) form as a gasoline substitute or in gasoline blends such as M85 (85% methanol and 15% gasoline).

Cost-effective, efficient and environmentally sound processes for producing methanol from biomass are being pursued by both government and industry researchers. The efficiency with which syngas can be converted to methanol is between 40% and 50% depending on the gas composition. One tonne of biomass feedstock can be converted to around 720 l of methanol. As a renewable resource, biomass represents a potentially inexhaustible feedstock supply for methanol production. The biomass is subjected to elevated temperatures (and higher pressures too, with some systems) during the gasification process to form the synthesis gas. This mix of CO and H_2 is conditioned to remove impurities such as tars and methane, and the hydrogen-to-carbon monoxide ratio is adjusted to 2:1. The syngas is then reacted over a catalyst at elevated temperatures and pressures to form methanol.

Methanol has a lower boiling point (64.5°C) than ethanol (78.3°C), has a lower heat value on a volume basis and is toxic. At present it is unclear whether the owner of a ligno-cellulose biomass resource wishing it to be used for transport fuels would do better to convert it to ethanol or to methanol. Research into methanol production from woody biomass continues, with the successful conversion of around 50% of the energy content of the biomass at a cost estimate of currently around $0.45/l. The goal of the NREL research programme on selecting catalysts and improving the gasification process is to reduce the cost of methanol to $0.13/l ($10/GJ) when using biomass fuel delivered at $2–3/GJ. It will then be competitive with the wholesale price of gasoline when refined from crude oil at around $25/barrel, bearing in mind that the energy content per unit volume is only around 40% that of gasoline. This means that extra on-board storage will be required to provide the same range as when using petrol or diesel.

Biogas

The anaerobic digestion of organic material produces biogas, as described in Chapter 7. If the gas is 'scrubbed' to remove the carbon dioxide and hydrogen sulphide present then it consists mainly of methane and can be used as a vehicle fuel similarly to natural gas (Figure 9.9). It is usually compressed and carried on board in cylinders. Most spark ignition engines can be converted to run on dual fuels, so that either methane or petrol can be used at any time; alternatively dedicated gas engines can be used.

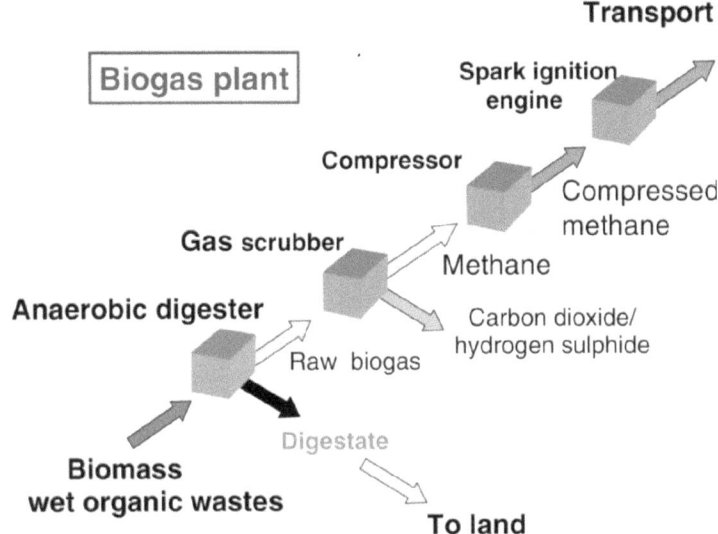

Figure 9.9 Biogas plant to produce a gaseous transport fuel

Bio-oil

Bio-oil produced from pyrolysis as discussed in Chapter 6 can also be used as fuel for diesel engines but mainly for stationary applications. The fast pyrolysis process 'Biotherm', commercialized by Dynamotive Technologies in Vancouver, Canada, was first evaluated in a 2 t/day pilot plant. Partners were then sought for developing several demonstration plants of around 10–25 t/day in Europe, North America and Brazil, with the aim of eventually moving towards full-scale commercial plants of over 100 t/day. Refining the bio-oil and then blending the product is being researched and showing some promise, according to various independent due diligence reports.

Syngas conversion to Fischer-Tropsch liquids

As discussed in Chapter 6, a disadvantage of synthesis gas is that essentially it must be utilized on-site. If converted to a liquid fuel then it would have wider application. The Fischer-Tropsch technology, which is well established, takes CO and H_2 (the basic products of biomass gasification) and converts them over iron or cobalt catalysts into a range of liquid fuels. Of particular interest is the production of a middle distillate (diesel) fuel equivalent with a high cetane number and little or no sulphur or aromatics. This product can be blended with conventional diesel fuels to assist with the reduction of exhaust pipe emissions. So there is considerable interest in its development by major vehicle manufacturers of modified compression ignition engines. It has been estimated that Fischer-Tropsch liquids can be produced at similar prices to petroleum-based diesel at the refinery, but these claims are yet to be verified.

Some simplified smaller-scale systems linking gasification processes, a Fischer-Tropsch plant and a combined-cycle gas turbine plant have been suggested but have yet to be built. These would produce electricity together with liquid fuels, and could be ideally suited for rural areas in developing countries such as China.

Pollution reduction potential

Air pollution

Air pollution is a growing concern in most urban areas. The transport sector is responsible for a large majority of it, including the production of carbon monoxide and nitrogen oxides, which together can form ground-level ozone. Millions of people live in areas that fail to meet at least one air quality standard.

In 1990 the US Congress passed the Clean Air Act Amendment to combat high emission levels of carbon monoxide, nitrogen oxides and the creation of ground-level ozone by petroleum-based transportation fuels. This Act specifically required the production and distribution of cleaner-burning gasoline in America's most polluted cities. Additives containing oxygenates such as ethanol were therefore used. Tougher emissions standards are also causing diesel engine users to find fuel additive or substitution options, resulting in cleaner air. Both ethanol and biodiesel have been proven to reduce emissions that are contributing to urban air pollution.

Particulate matter (PM), various oxides of nitrogen (NO_x) and volatile organic compounds (VOC) should be avoided, as many of them have health effects, are precursors to photochemical smog, contribute to acid rain and can cause other environmental problems. A comparison of emissions from the use of biodiesel and 95% ethanol blends in buses compared with various fossil-fuel-based fuels is given in Figure 9.10.

The values for biodiesel and diesel are similar, with specific emissions being only slightly higher or lower. The major exception (not shown) is that biodiesel, and also E95, are naturally very low in sulphur and thus have sulphur emissions several orders of magnitude lower than those of the equivalent fossil fuels.

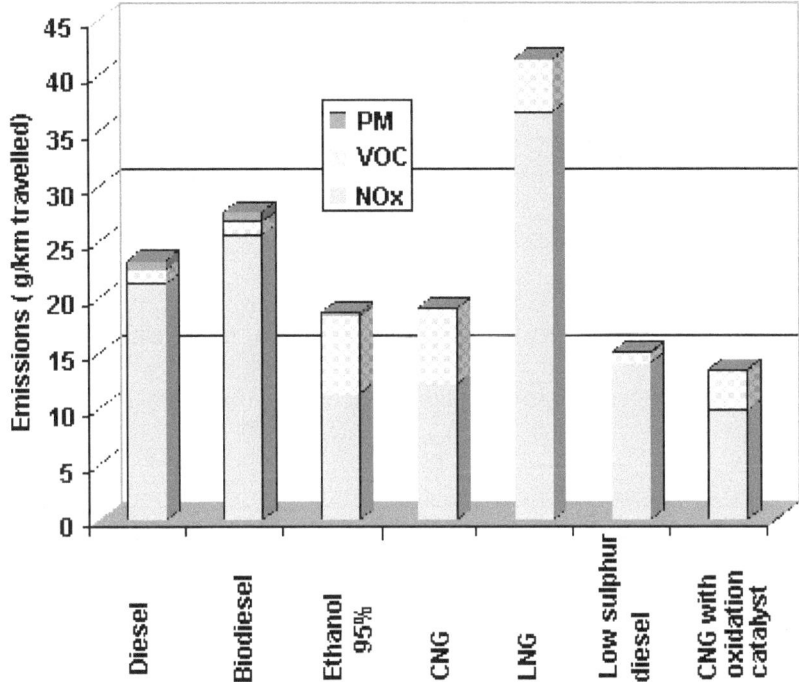

Figure 9.10 Emissions of PM, VOC and NOₓ from biodiesel and 95% ethanol compared with fossil fuels

The pollutant of perhaps the most concern is PM, as it is a carcinogen and is known to cause respiratory disease. Particulates form as a result of incomplete combustion of the fuel. This is caused by a number of factors, including quenching of the flame front by the colder cylinder walls, poor injector maintenance and design and wrong timing resulting in fuel entering the combustion chamber at inappropriate parts of the cycle.

Particulates and other pollutants may not be such a problem in the future if countries enforce emission standards. The levels of PM, NOₓ, CO and VOC from ethanol and biodiesel vehicles may be reduced to levels lower than those for vehicles currently using CNG or LPG. When ethanol is added to gasoline, it displaces some gasoline, which generally also reduces all of the pollutants. In addition, since ethanol causes fuels to burn more completely, it further reduces emissions of CO and VOC and toxic air emissions. Biodiesel, in a 20% blend with diesel, reduces visible smoke and odour, PM, CO, total hydrocarbons, sulphur dioxide and lead. When the blend is used with an oxidation catalyst, particulate matter is reduced even further.

Water pollution

Water pollution is associated with petroleum including marine oil spills, groundwater contamination from underground gasoline storage tanks and the runoff from vehicle engine oil and fuel from service station forecourts and road surfaces. Oil and fuel leakages from vehicles are carried to streams and lakes whenever the surface becomes wet and water drains across it, causing environmental damage to aquatic plant and animal life.

Marine oil spills such as the Exxon Valdez spill in Alaska in 1989 continue to cause considerable environmental damage. Acute oil spills can damage individual organisms and wipe out entire populations of marine and coastal species. They also require large-scale, costly clean-up operations. However, catastrophic marine oil spills are actually less damaging to the environment overall than the thousands of smaller spills that are reported annually. For example, pipeline spills in the USA alone, as reported to the Department of Transportation, average 50 million litres of petroleum products per year. According to the US General Accounting Office an average of 16,000 small oil spills account for more than 200 million litres seeping into waterways each year. By comparison the Valdez disaster released some 40 million litres. In addition, according to the US Environmental Protection Agency, more than a quarter of the nation's one million underground gasoline and oil tanks leak, causing considerable groundwater contamination.

Bioethanol and biodiesel can replace the more toxic gasoline and diesel with fuels that biodegrade in water relatively quickly, thereby reducing the threat that petroleum products pose to waterways and groundwater. Bioethanol and biodiesel spills or leaks do not usually result in an environmental hazard. Such fuels are becoming an attractive alternative to using petroleum-based fuels for boating activities in order to protect and improve water quality.

Greenhouse gas emissions

The transportation sector is responsible for one third of global carbon dioxide emissions. The production and use of biofuels can help to reduce atmospheric CO_2 build-up by displacing the use of fossil fuels, and by recycling the CO_2 that is released when it is combusted as fuel. Through the use of biofuels instead of fossil fuels the emissions resulting from fossil fuel use are avoided, and the carbon content of the fossil fuels remains in storage underground. Further CO_2 reductions occur because the plants and trees that serve as feedstocks for biofuels require CO_2 to grow, so they absorb what they need from the atmosphere. Thus much or all of the CO_2 released when biomass is converted into a biofuel and burned in automobile engines is recaptured when new biomass is grown to produce more biofuels.

Tropospheric ozone formation

Tropospheric ozone formation can create smog, which consists mainly of ozone gas. When most people think of ozone, they think of the protective layer of ozone in the Earth's upper atmosphere that protects the surface from harmful ultraviolet rays. At ground level, however, ozone is a toxic gas and a powerful oxidizing agent. While 'ground-level ozone' does occur naturally, it only does so in small quantities.

When fossil fuels are burned, a variety of pollutants are emitted into the troposphere, the lower region of the atmosphere where plants and animals exist. Among these pollutants, hydrocarbons, nitrogen oxides and carbon monoxide are all ozone precursors. A variety of complex chemical reactions occur between these chemicals, some resulting in the formation of nitrogen dioxide, which gives the distinctive brown haze to urban air that is so obvious when flying into city airports such as Los Angeles or Bangkok. Nitrogen dioxide also can react photochemically to form ozone, which is a known health hazard. Biofuels help to combat ground-level ozone formation because they emit fewer of the ozone-forming pollutants than petroleum fuels do.

Acid rain

Acid rain stems mainly from fossil fuels containing sulphur. Depending on the refining and blending process used, gasoline and diesel contain varying amounts of sulphur, which after combustion is emitted into the atmosphere as SO_2. This gas is oxidized into an aerosol of sulphuric acid that is then deposited in tiny droplets on the Earth's surface when it rains. When the SO_2 concentrations are very high in rainfall, being acidic they can cause substantial damage to buildings, agricultural crops, forests and lakes, and severe respiratory damage to humans. Acid rain damages an estimated $2–3 billion of agricultural crops in the US each year. Natural forests are claimed to be dying from the acid rain, and their biological diversity of species is under threat. Metal deposits in the soil resulting from the acid rain are later released back into lakes and streams, which then become toxic to fish. Thousands of lakes in the USA and Canada may have suffered serious losses of aquatic life as a result. Replacing petroleum fuels with biofuels would dramatically reduce the amount of sulphur dioxide and other emissions from the transport sector.

Notes

[1] Gasoline, petrol and motor spirit are used synonymously here to describe the fraction of crude oil distilled off at the oil refinery that is most commonly used in spark ignition engines.

Chapter 10

Small-scale bioenergy systems: present and future

Bioenergy systems from 10 kW to 10 MW have the potential to provide heat, electricity and cogeneration at the village or community scale or for small industrial use. They include primary conversion technologies that convert biomass into heat, gaseous or liquid products, together with secondary conversion technologies that convert these products into useful forms of energy. An overview of the relevant technologies is given in Figure 10.1.

The utilization of biomass-derived producer gas and biogas has particular potential at this scale because fuel cells, Stirling engines and micro-turbines, which can all be run on these gases, are promising technologies for distributed energy systems. Only the thermochemical conversion technologies of combustion, gasification and pyrolysis are covered here; biochemical

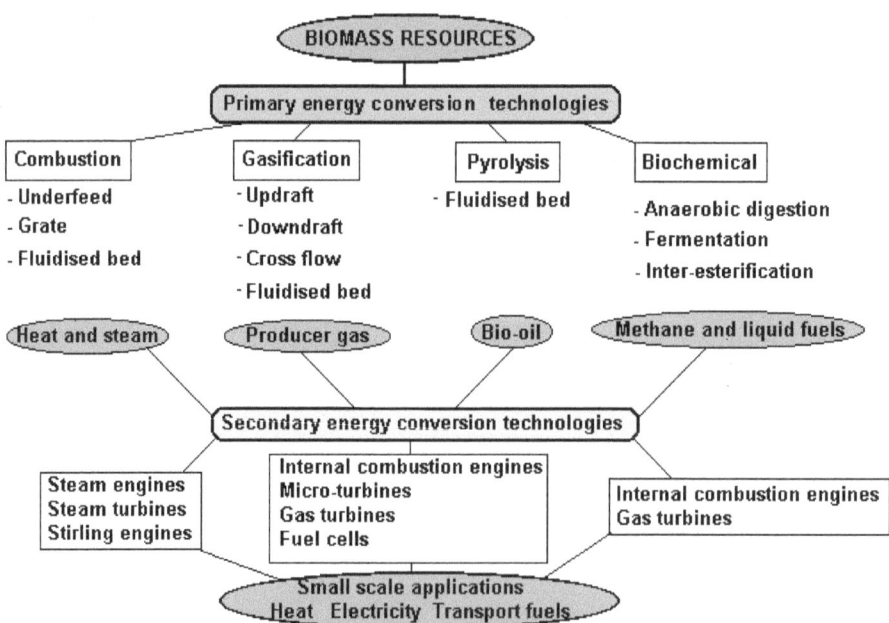

Figure 10.1 Primary and secondary conversion technologies suited to small-scale biomass projects

conversion technologies, including small-scale conversion, have been covered in Chapter 7.

The wide range of technologies that exist to convert the energy stored in biomass to more useful forms of bioenergy can be classified according to the principal energy carrier produced in the conversion process. Carriers in the form of heat, gas, liquid or solid products depend on the extent to which oxygen, usually as air, is admitted to the conversion process. As discussed in Chapters 5 and 6, the three principal methods of thermochemical conversion corresponding to each of these energy carriers are combustion in excess air, gasification in reduced air, and pyrolysis in the relative absence of air. They are briefly reconsidered here, but from the small-scale perspective.

Primary energy conversion technologies

Combustion

Biomass fuel is burnt in a range of burner and boiler designs with an excess of air to produce heat. The method of feeding in the fuel characterizes the technology at both the large and small scales.

Underfeed stokers

Underfeed stokers, which feed the biomass into the combustion zone from underneath the firing grate, are suitable only for small-scale systems up to a nominal boiler capacity of 6 MW_{th}. They are better suited to biomass fuels with a low ash content such as wood chips and sawdust, since high-ash fuels such as bark and cereal straw need more efficient ash-removal systems. Sintered (melted) ash particles covering the upper surface of the fuel bed can cause problems in underfeed stokers, because unstable combustion conditions can occurring if the fuel and the air break through the ash-covered surface.

Grate stokers

Grate stokers are available as fixed grates for small-scale combustion systems less than 1 MW_{th}, reciprocating grates at larger scales, and newly developed designs that allow a horizontal and vertical movement of the grate. When the supply of primary air is controllable, it is possible to operate grate firings efficiently, even at partial loads down to 25% of the nominal maximum furnace load. Water-cooled grate designs avoid slagging and extend the lifetime of the materials.

Grates are well proven and reliable, and can tolerate wide variations in fuel quality including moisture content and particle size, as well as a high ash content. The advantages of grate furnaces for plants smaller than 10 MW_{th} include low investment costs, low operating costs and good operation at partial load. Disadvantages include NO_x production, excess oxygen, which decreases the thermal efficiency, and combustion conditions that are less stable than in other types of furnace. Recent design developments have been driven by the desire to reduce emissions, which also led to the development of the main alternative to grate-based systems, the fluidized bed.

Fluidized-bed combustors

Fluidized-bed combustors burn the biomass fuel on a bed of inert material (usually sand particles) suspended by blowing primary air through it. For plants with a nominal boiler capacity greater than 10 MW_{th}, bubbling fluidized-bed (BFB) combustors start to be of interest, and if larger than 30 MW_{th}, circulating fluidized-bed (CFB). The plant size for which CFB and BFB technologies are economically viable is normally larger than 5 MW_e though several cogeneration plants are operating successfully in Scandinavia around 3–4 MW_e but with 16–17 MW_{th}.

The conventional technology for generating electricity from biomass is to use a combustion system to raise steam, and then to expand this steam in an engine or turbine to produce mechanical power or drive a generator set. The production of steam is efficient, but the conversion of steam to electricity is much less so. Biomass combustion systems of all designs can be very efficient at producing hot gases, hot air, hot water and steam, typically producing 65–90% of the energy contained in the fuel as useful heat. The lower efficiencies are associated with wetter fuels (50–60% m.c.w.b.). Where the production of electricity is the main goal, the steam engine or turbine will exhaust into a vacuum condenser. Conversion efficiencies are then likely to be in the range 5–10% for plants smaller than 1 MW_e, 10–20% for plants of 1–5 MW_e, and 15–30% for plants of 5–25 MW_e. Heat available from the condenser will be at low temperatures, usually below 50°C, which is insufficient for most heating applications, even for low-grade heat, so it is normally wasted.

There is little scope for achieving significant efficiency improvements with steam technology because it is mature. However, where there is demand for process heat as well as electricity, the plant can be arranged to take some high-temperature steam directly from the boiler, or partially extract expanded steam from a turbine designed for the purpose, or provide the exhaust steam at the required temperature. All three options will reduce the amount of electricity available, although the overall energy efficiency may be much higher, 50–80% being common for such a cogeneration system.

Steam technology from biomass can be regarded as robust and well proven, but relatively expensive compared with fossil fuel systems unless waste fuels and residues are available at low or zero cost. Where the biomass has to be purchased at market prices, electricity prices will probably not be competitive at the small scale. In these circumstances biomass-to-electricity schemes need other reasons for their existence. The main problems with combustion technology are the reactions that take place that cause corrosion and fouling, especially when potassium- and chlorine-rich biomass fuels such as straw, cereals and grass are used. Ash melting behaviour and its influencing variables also need further evaluation.

Gasification

The gasification of biomass is based on partial combustion in a restricted supply of air or oxygen at elevated temperatures up to 1200°C. The variability of biomass fuels with respect to moisture content and particle size affects gas composition. This makes gasifier operation significantly more demanding than the operation of combustion systems.

Gasifiers that use air as a gasification medium (directly heated gasifiers) use the exothermic reaction between oxygen and organics in the fuel to provide the heat necessary to devolatilize biomass and to convert residual carbon-rich chars. The heat to drive the process is generated internally within the gasifier. The high temperatures used during the gasification process make it unsuitable for fuels with low ash softening and melting temperatures, such as many annual crops and their residues.

The final fuel gas consists principally of carbon monoxide (CO), hydrogen (H_2) and methane (CH_4) with small amounts of higher hydrocarbons such as ethane (C_2H_6) and ethylene (C_2H_4). Because the combustible gases are diluted with carbon dioxide and nitrogen, which have no calorific value, the calorific value of the final fuel gas mixture is low (4–6 MJ/Nm^3). This is only 10–14% of the calorific value of natural gas, for which commercial gas engines and gas turbines have been designed. The low calorific value also makes the gas unsuitable for transportation, as high on-board storage costs are involved.

The calorific value of the gas is increased to 10–15 MJ/Nm^3 if oxygen is used instead of air. This would allow unmodified engines and turbines to be used to generate electricity. However, the cost of production of oxygen and potential hazards associated with its use have made the use of oxygen blown gasifiers unattractive to date, especially at the small scale.

A pressurized gasifier will produce gas suitable for direct gas engine or turbine application and provide the highest overall process efficiency. To take full advantage of operating under pressure however, a number of ancillary systems must be developed such as reliable feed systems and hot gas clean-up systems. Alternatively, gasifiers can be operated at low pressure and the product gas then cleaned and compressed to the pressure required for gas-turbine application. Pressurized gasifiers will be physically smaller than atmospheric gasifiers for the same output, but they will also be more expensive to manufacture. The pressurized gasifier would possibly be more efficient overall due to more complete conversion of tars, the retention of sensible heat and the absence of the need for gas compression. However, the fuel feeding system would be much more complicated as it would involve pressure seals, and would need purging with inert gas. These factors increase both the capital and operating costs of such a system and make it less suitable at the smaller scale. Though some research on small pressurized downdraught gasifiers has been undertaken, the comparative cost/kW installed is unknown.

Figure 10.2 Waterwide close-coupled gasifier achieves good mixing of the gases with air in a vortex action to reduce the tar content, here shown in a demonstration facility coupled to a boiler
Source: Renewable Energy Corporation, Sydney

For heating applications only, it is possible to burn hot fuel gas as produced by the gasifier directly in a boiler or furnace as used by the Waterwide close-coupled gasifier. Maintaining high temperatures prevents the condensation of tars which are cracked and burned at high temperature in the combustor (Figure 10.2).

It is difficult to clean the gas, particularly at the small scale. If the biomass fuel or gasification process leads to the production of dust or ash, it may well be possible to remove it with a hot gas cyclone. If used in internal combustion engines (ICEs) or small turbines the gas may have to be cooled to intermediate or low temperatures due to temperature limitations in the fuel control system of the engine or turbine. Reducing the gas temperature will increase the volumetric heat value of the gas, but it will also increase the condensation of tars making the gas even less suitable. In these circumstances a gas cleaning system will be essential, possibly comprising cyclones, filters and wet scrubbers. Wet scrubbers are particularly effective as they capture the tars which are water soluble, collect the inert ash and mineral contaminants and reduce the gas temperature in a single operation. However, they produce a contaminated effluent stream with potential toxic and carcinogenic properties, the existence of which tends to undermine the clean image of biomass fuels and the concept of sustainability, especially at the smaller scale.

Gasifiers are classified according to how fuel and air are fed into the conversion plant in relation to each other.

Fixed-bed updraught   (or counter-current flow) gasifiers have combustion air blown into the reaction chamber from below while the fuel is fed in from above. The great advantage of updraught gasifiers is their suitability for a large range of fuel moisture contents and particle sizes, and fuels with a low slag melting point such as straw. For heat applications up to 10 MW$_{th}$, updraught gasifiers are most popular and because the gas leaves at relatively low temperatures, the thermal efficiency is high. This type of gasifier is not likely to be used for power generation due to the high tar production which would require extensive gas cleaning.

Fixed-bed downdraught   (or co-current flow) gasifiers have the fuel fed in from the top then undergo various processes as it moves downwards under gravity. Air is injected either into the middle section of the gasifier or from the top and flows in the same direction as the fuel. In their traditional design downdraught gasifiers are not suitable for fuels with a low ash melting point, such as straw. They are most popular for small-scale power generation but in order to operate properly, fuel moisture content and particle size have to be within narrow limits. In the 1980s many gasifiers of this design were installed in developing countries where the problem of fully unattended operation was not an issue due to cheap and abundant labour. Up-scaling of this design is difficult, and the maximum size is probably limited to about 3–4 MW_{th}. Since the 1990s, several projects have been executed in Europe to improve the downdraught gasifier, the latest research trend being the development of small scale, fully automatic units fuelled by one well defined biomass fuel type in order to reduce problems due to size and moisture content variations.

Fixed-bed cr oss-flow gasifiers are suitable for very small-scale applications (smaller than or equal to 10 kW_e), using charcoal as fuel. They operate at high temperatures up to 1500°C, offering the possibility of thermal tar cracking to a level satisfactory for subsequent direct use of the fuel gas in an ICE engine without the need for extensive and costly cleaning. Opportunities for up-scaling are limited though in Europe a project commenced in 1997 with the objective of developing a cross flow gasifier for cogeneration applications in the 500 kW_e to 2.5 MW_e range. The technical feasibility of this type of gasifier has been shown in a small laboratory gasifier and a 100 kW_{th} pilot plant is being designed, mainly for use in developing countries.

Fluidized-bed gasification of biomass was first implemented in the 1980s to meet the need for more efficient technologies for the utilization of low-grade fuels such as biomass. BFB and CFB gasification reactors operate under similar principles to comparable combustors but are at the large commercial scale.

There are two major reasons why small-scale biomass gasifiers are not commonly utilized and accepted at present for electricity generation:

- The large variation in the key parameters determining the quality of biomass fuels (moisture content, ash content, particle size, particle size distribution, chemical composition, bulk density, volatile matter content, etc.) causes extreme engine wear due to tar contamination, unreliablility and unstable operation due to slagging and variations in pressure drop.

- Small-scale gasification systems are characterized by minimal automatic measuring and control systems in order to keep the costs down. As a result the systems often show variable performance and need the continuous presence of experienced and expensive personnel.

To make a gasifier more reliable, a standardized fuel could be produced using a mixture of different biomass fuels by combining pre-process technologies and advanced briquetting. Cheap and reliable hardware and software from the automotive industry could be used to control gasifier and gas engine operation automatically. A fully automated biomass fixed-bed gasifier in combination with an engine/generator would be warranted.

Biomass integrated gasification combined cycle (BIGCC) is an advanced power cycle technology which has created recent interest at the medium scale. A gas turbine is fed on producer gas and the exhaust heat from the flue gas from the turbine is used to produce steam which is then fed into a turbine to generate power. Several demonstration projects have been designed between 7 and 30 MW$_e$ located mainly in Europe and the USA.

The advantages of advanced power cycles include increased conversion efficiency which results in reduced biomass feedstock consumption and thus lower operational costs, reduced environmental impacts and the possibility for co-firing biomass with fossil fuel and wastes. Investment costs for the first fully commercial plants were \$2300–2900/kW$_e$ in the power range 5–30 MW$_e$ for atmospheric IGCC, and \$2200–5900/kW$_e$ for plants in the power range 7–60 MW$_e$ for pressurized IGCC.

Pyrolysis

During the pyrolysis process biomass is heated either in the absence of air (i.e. indirectly), or by the partial combustion of some of the biomass in a restricted air or oxygen supply. The mix of products is extremely varied, consisting of gases, vapours, liquids, oils, solid char and ash, the composition and proportions depending on temperature, input composition, pre-treatment and air composition conditions. At temperatures of around 500°C and short reaction times of 0.5–2 s, pyrolysis oils are produced. At higher temperatures of 700–800°C, pyrolysis reactions produce a much higher proportion of gas, with correspondingly less liquid and solid products. The gas has a calorific value of 15–20 MJ/Nm^3, which is sufficient for many types of combustion engine and turbine without modification, though tars must be removed.

Pyrolysis liquid or bio-oil has a high heating value of 22–23 MJ/kg on a dry basis and about 17 MJ/kg as normally produced with a moisture content of about 25% (wet basis). It is composed of a very complex mixture of highly oxygenated hydrocarbons, which are chemically and physically unstable and so require special considerations during handling, storage and utilization. It will not mix with any hydrocarbon liquids. Density, viscosity, surface tension and heating value are key properties for combustion applications in boilers, furnaces and engines, but other characteristics such as char level, particle size and ash content also have a major effect. It is theoretically possible to upgrade crude pyrolysis oil into more familiar petroleum products, although the technology for this has yet to be used commercially. Pyrolysis is the favoured technology for annual crops and their residues because the moderate process temperatures are less likely to cause ash melting or softening, and the alkali metals are retained in the char.

A wide range of reactor configurations have been devised, such as fluidized bed, ablative, entrained flow, rotating cone, transported bed, molten salt and vacuum reactors at scales using from 20 to 650 kg/h of biomass fuel. Commercial operation is currently only being achieved using a CFB system.

A key advantage of producing liquids rather than gases is that the fuel production can be decoupled from power generation. Peak power provision is thus possible using stored bio-oil, and higher plant availability using intermediate fuel storage may also result. Several pilot and near-commercial projects running on different biomass fuels are being developed in Europe, the USA and Canada. Bio-oil has been successfully fired in diesel test engines, where it behaves similarly to diesel in terms of engine parameters, performance and emissions. Continuous runs of 10 h have been achieved on raw bio-oil without dilution or processing. The diesel engine may have to be modified to include a pilot injector ignition system if not already present because pyrolysis oil will not self-ignite. Diesel fuel is therefore needed for start-up, typically 5% of fuel in larger engines. No significant problems are foreseen with power generation up to 15 MW_e per engine. Pyrolysis oil has also been successfully used as a boiler fuel, in a modified dual-fuel diesel engine and in gas turbines, but these are more suitable for larger-scale applications.

Secondary energy conversion technologies

The gaseous, solid or liquid energy carrier produced during the primary biomass conversion step is converted into a more useful form of energy (space heating, hot water, electricity or process steam) in the secondary conversion step. Small-scale, electricity-generating technologies and also heat systems include:

- internal combustion engines
- steam turbines
- steam engines
- Stirling engines
- indirect-fired gas turbines
- direct-fired pressurized gas turbines
- micro-turbines
- fuel cells.

A summary of the electrical efficiencies and investment costs (including primary energy conversion technology) of these technologies is given in Table 10.1. The information is based on claims from manufacturers and data calculated from feasibility studies.

Internal combustion engines

Internal combustion engines (ICEs) are among the most widely used prime movers for biomass fuels. Electricity conversion efficiencies (fuel energy to

Table 10.1 Summary of the electrical efficiencies and investment costs for energy conversion technologies including primary and secondary

Technology	Efficiency range (% based on LHV)	Investment cost range ($/kW$_e$)
Internal combustion engine	25–30	800–1200
Steam turbine	15–35	1700–4800
Steam engine	10–25	700–2000[a]
Stirling engine	20–30	1000–4800
Indirectly fired gas turbine	15–24	3000–6100
Directly fired pressurized GT	25–30	1200–1600
Micro-turbine	20–30	1000–1300

[a] Investment cost for steam engine only

electricity) of 25–30% (LHV) make ICEs an economic option. Several types are commercially available, but those of most significance to stationary power applications are four-cycle spark ignition (Otto cycle) and compression ignition (Diesel cycle) engines. Diesel engines develop higher part-load efficiencies than Otto cycle engines because of leaner fuel/air ratios at reduced load.

The advantages of ICEs include ready availability over a wide size range, fast start-up, good part-load efficiency, reliability and long life. Gas cleaning for tar and particle removal is the key issue, as none of the current gas-cleaning systems safely meets the gas quality requirements for satisfactory ICE applications. Hence manufacturers are loath to maintain their warranties

Several projects have been evaluated linking biomass gasifiers with ICEs:

- In the UK a Fluidyne 30 kW$_e$/60 kW$_{th}$ downdraught gasifier (Figure 10.3) was connected to an internal combustion engine generator set (30 kW$_e$) and fuelled by short-rotation coppice willow. The basic idea of the system was to provide electricity and heat supply to a single farm, or to a co-operative of two or three small adjacent farms. The surplus heat produced was either used for wood chip drying or dumped. In energy terms the system was feasible, but mechanical problems were encountered, including bridging of the biomass fuel in the hopper, poor flow of the charcoal/ash through the bottom of the reduction zone, and inefficient operation of the gas-cleaning chain. The economic feasibility of the system was not evaluated but is probably high since there is no mass production of small-scale gasifiers.

- In Germany a fixed-bed gasifier fuelled by forest arisings was coupled to a spark ignition engine. The system had a capacity of 470 kW$_e$/1 MW$_{th}$ but economic operation of the plant was not possible owing to the high investment and maintenance costs.

- In Northern Ireland a downdraught gasifier with the capacity to fuel a dual-fuel gas/diesel engine driving a 100 kW$_e$ induction-type generator was tested. The engine was designed to operate on 90% gas, with the diesel fuel

Figure 10.3 Starting a batch Fluidyne downdraught gasifier from cold takes just 2–3 min before the ICE can be run on the gas

providing the necessary ignition. It was installed in a cogeneration configuration supplying 120 kW$_{th}$ for a central heating system and 80 kW$_{th}$ for drying the incoming short-rotation willow coppice fuel. Problems identified included fuel variation characteristics and gas clean-up. Modifications to fuel feed systems, hearth design and particulate and tar removal methods resulted in more consistent production of high-quality gas. The system was claimed to be installed for around \$800/kW$_e$ including connection to the national electricity grid, and at this cost had potential where a suitable match can be made between customer requirements for both heat and power.

- Under a US government funded Small Modular Biopower Project, Bechtel National Inc. undertook feasibility studies on a commercial KC gasifier (800 kW$_e$ and 1600 kW$_{th}$) and V-16 engine with a turbo-compressor and electronic ignition. The gasifier was tested satisfactorily utilizing a wide range of biomass fuels including rice husks, straw, corn cobs and switchgrass.

- The US company BG Technologies is marketing a small fixed-bed downdraught gasifiers coupled to an ICE/generator set in the size range 40–400 kW$_e$ for an investment cost of \$1000–1200/kW$_e$.

Steam turbines

Steam turbines have traditionally been used in large-scale power generation and cogeneration plants with a nominal capacity greater than 10 MW$_{th}$. The basic steam cycle is based on the closed Rankine cycle: water is heated, evaporated and superheated in a boiler. Energy contained in the steam is converted into rotational motion by expanding it through a turbine connected to an electric generator. The exhaust steam is passed through a condenser, where it is cooled and turned into liquid again to complete the cycle. System efficiencies vary between 15% and 35% according to the steam parameters. At smaller scales, steam turbine efficiencies decrease and become less than that of a steam engine at the same output. Under partial load the efficiency of a steam turbine decreases further.

Capital costs for a steam turbine unit range from $280/kW$_e$ at the 250 kW$_e$ scale to $130/kW$_e$ for 2 MW$_e$. Cost is the main obstacle for uptake even though it is a mature technology with no technical barriers, as turbines can withstand high pressures and temperatures. Other disadvantages include low efficiency at the small scale, poor part-load efficiency, and low suitability if the steam is of poor quality in terms of low pressure and high water content.

Several small-scale bioenergy steam turbine projects are being evaluated:

- In the USA Agrilectric Power Inc. is developing a power generation unit in which rice hulls are combusted and steam from the connected boiler is led to a turbine and generator. Three model sizes are planned of 500, 1500 and 5000 kW$_e$ with expected unit costs of $2900, $2700 and $1700/kW$_e$ respectively. The main markets will be in developing countries.

- Carbona Corporation conducted a feasibility study for a small modular cogeneration project that would usewoody biomass in an air-blown fixed-bed updraught gasifier of 1 MW$_e$ to produce steam for utilization in a back-pressure steam turbine generating power and providing heat. Calculated electrical efficiency was 18.6%, and the expected investment cost was $3880/kW$_e$.

- A system consisting of a fluidized-bed boiler coupled to either a back-pressure or condensing steam turbine, depending on the relative need for process heat or electricity, has been developed in the USA. The primary markets for the system include forest and agricultural processing facilities and remote area communities. Investment costs for the 500–5000 kW$_e$ plant size range are $3700–1600/kW$_e$ respectively for the condensing turbine system and $4800–2300/kW$_e$ for the back-pressure turbine system.

Steam engines

Steam engines are well proven for constant-speed operation in industrial environments but are only produced in small numbers, which makes them relatively expensive per unit. The efficiency of steam engines depends largely on the quality of the steam, requiring boilers with good steam pressures of at least 1 MPa and temperatures of at least 180°C. They are reliable, and have low maintenance costs. The principle of a steam engine with reciprocating pistons is similar to that of an ICE; the main difference is that in a steam engine the working fluid is steam rather than fuel combustion products. Most steam engines are double acting in that steam expands during both the forward and backward strokes of the pistons. As a result a steam engine is lighter and smaller than an internal combustion engine for the same output. Steam engines can be divided into traditional, modified (such as the Spilling engine) and those designs that use diesel engine components, thus reducing the production cost. The main market for steam engines using biomass would be as cogeneration units where quantities are too small for economic operation of a steam turbine.

Steam engines are commercially available in different sizes ranging from 8 to 1400 kW$_e$. Investment costs range from around \$7000/kW$_e$ for an 8 kW$_e$ engine to around \$270/kW$_e$ for 1270 kW$_e$. Compared with internal combustion engines, capital costs are too high and efficiencies too low for steam engines to be cost competitive. Advantages, however, include a proven, robust technology that can withstand constant operation in an industrial environment, the lack of internal contamination compared with an ICE or gas engine, and reasonably good efficiencies for small engines when used in connection with standard industrial boilers. Disadvantages include low efficiencies when used with low-pressure steam and the continued use of old designs and outdated technology, giving a poor image.

Stirling engines

Stirling engines are external combustion engines in that the fuel is combusted outside the engine and the heat transferred into the cylinder by a heat exchanger or via the cylinder wall (Figure 10.4). Any source of heat can be used provided that it is of sufficiently high temperature. The principle is that a working gas, enclosed by two pistons in a vessel, moves continuously back and forth between hot and cold spaces in a regenerator and is therefore continuously heated and cooled. The pistons are driven by the expansion of the gas caused by differences in gas volumes due to temperature differences. Since there is no contact between the gas or its contaminants and the moving parts of the engine, the lifetime is relatively long and maintenance intervals are large. The power output depends on the mean cycle pressure, swept volume of the pistons, speed, and temperature difference between hot and cold spaces.

Using any heat source, the air or helium gas sealed inside the expansion space under pressure of around 1 MPa is heated, and the expansion drives the piston. Cold gas is compressed in the compression space by the second piston, and heat produced by compression is removed by the cooling radiator and is available for cogeneration. The lower the cooled temperature, the higher the efficiency. The regenerator absorbs heat from the gas and then returns it later on in the cycle.

Typically, Stirling engines are single-acting as in Figure 10.4 with one common cylinder containing two reciprocating pistons. Various configurations

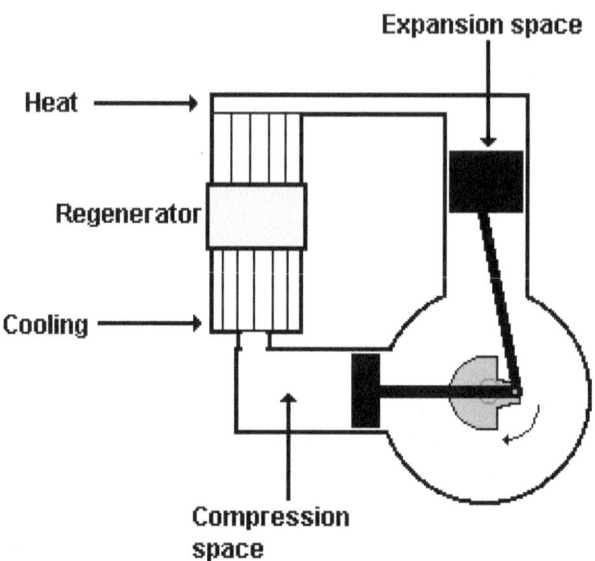

Figure 10.4   The Stirling engine uses a closed thermodynamic cycle that repeatedly expands and compresses a fixed mass of gas by heating and cooling it

exist, including a double-acting design with four cylinders and one piston in each: being mechanically more simple and compact this is more suitable for bigger engines. Even more simple designs are ones in which springs control the piston movements without a kinematic crank mechanism. During the late 19th and early 20th centuries several thousand Stirling engines with a maximum output of 4 kW were in operation in Europe and the USA, but they were later replaced by the more efficient and lighter internal combustion engine. Owing to the low emissions and noise levels, the Stirling engine was revived by Philips Electrical company in the 1930s but the need to find materials suitable for simple manufacture and operation under the high pressures involved was a limitation until more recently.

Stirling engines are now commercially available in the 0.5–150 kW$_e$ range, and a number of companies are working on their further development. Only a few engines have been introduced to the market, and commercially viability is unproven to date. The cost, for example, of a V160 Stirling engine is $4800/kW$_e$. However, if future sales of up to several thousands annually and mass production can be realized, expected costs will reduce to around $240/kW$_e$.

For Stirling engines specifically developed for biomass firing, overall efficiencies (fuel to shaft output) are around 30%. Other advantages include good partial load efficiency, low noise level, safe operation, low expected maintenance costs, suitability for a wide range of fuels and a long engine lifetime claimed to be 25,000 hours (the average life of a car engine, for comparison, is around 4000 hours). Only very small engines have been built to date, so experience with larger engines is lacking and there are few data on reliability and life times. High heat exchange temperatures may cause corrosion, and sealing of the engine is a technical problem yet to be fully resolved. The prospects for Stirling engines are good, but further development is required.

Several commercial projects involving Stirling engines are reaching the commercial stage:

- In Denmark two Stirling engines designed for direct combustion of biomass without any intermediate heat transfer system or particle separation have been operating, giving maximum electric power outputs of 36 kW$_e$ and 150 kW$_e$. Large-diameter tubes, which form a large, square combustion chamber, have been used in the heater, and the distance between fins is sufficiently large to avoid fouling. A screw feeds wood chips into a high-temperature revolving grate in the insulated combustion chamber, and the Stirling engine is mounted on the top. The system efficiency is 21% for the 36 kW$_e$ engine and 25% for the 150 kW$_e$ model. Engine maintenance (mainly replacement of piston and piston rod seals) is expected to be necessary after 10,000 h.

- In the UK the CRE Group Ltd in collaboration with Gamos Ltd designed a novel form of low-pressure nitrogen-charged Stirling engine of 150 kW$_e$/ 600 kW$_{th}$ outputs specifically for stationary applications. Previous Stirling engine development was directed towards minimizing engine weight and maximizing efficiency, which led to designs that required exotic working gases used under high pressures. The use of nitrogen at lower pressures (5 MPa) made it possible to avoid some of the difficulties of earlier Stirling engines.

- In the USA a 10 kW$_e$ Stirling engine was coupled to a small 1 MW$_{th}$ biomass atmospheric pressure fluidized-bed combustion unit utilizing pine chips. The SPS V160 Stirling engine was specifically designed for stationary use, low-cost production and long running time. The central problem for efficient operation was the fouling and plugging of the heat exchanger with fine particles from the combustion gas, which caused a decrease of heat transfer. Electrical efficiency of the system was 15–20%. Higher-efficiency heat exchangers and gas cleaning issues are yet to be resolved.

- The American STM Corporation developed a small modular biopower system based on their 25 kW$_e$ Stirling engine STM4-120 coupled to a vacuum updraught gasifier fired with sawdust. This 'BioStirling' system was equipped with an induction generator to produce electric power suitable for grid-connection. The cost was around $1600/kW$_e$.

- Sunpower Inc. undertook studies of a gasifier coupled to a Stirling engine with a capacity of 1–18 kW$_e$ for residential, small commercial and agricultural markets. The target high-volume system manufacturing cost was $1000/kW$_e$.

- The Whispergen Stirling engine developed in New Zealand is commercially available to provide hot water, space heating and electricity at the domestic scale. It is quiet, needs little maintenance and is fully automated. It switches on and off to maintain hot water temperatures and, if grid connected, can

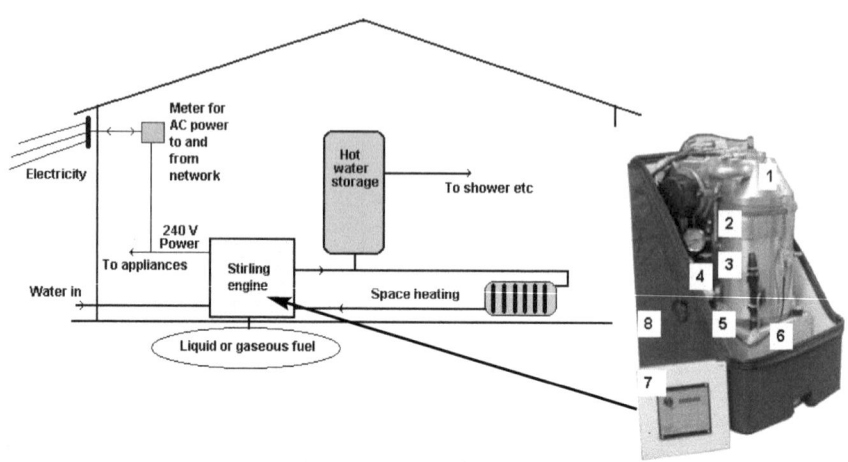

Figur e 10.5   The a.c. model of the Whispergen Stirling engine with dimensions of around 500 mm height and width can be used in dwellings either stand-alone or grid connected. 1 Combustion space. 2 Hot heat exchanger. 3 Cold heat exchanger. 4 Wobble yoke. 5 Sealed alternator. 6 Microprocessor. 7 Control panel. 8 Housing
Source: *Energy Wise News*, EECA

be set up to provide peak power to benefit from high electricity tariffs. A novel wobble yoke converts the reciprocating movement of the four pistons into a smooth rotary motion, which is used to power a generator. Natural gas is the normal fuel but this could easily be replaced by producer gas or biogas. A 200 W_e d.c. version has application for off-grid uses such as boats and mobile homes. An 800 W_e a.c. version, combined with a domestic gas-fired boiler and with simple controls, is suitable for grid connection and net metering (Figure 10.5). The heat-to-electricity ratio is 5:1, which is higher than for fuel cells. It is designed to fit into a kitchen or laundry with standard power plug, water supply and exit pipes just like a dishwasher. A protection device is built in so that, as for any net-metered device, for safety it ceases to supply power to the grid when a power cut has occurred. If incorporated with a condensing boiler to recover heat from the exhaust gases, when used in colder countries such as northern Europe where electricity is four times the value of gas, a payback period of 3 years from saved electricity purchases is claimed.

Indirect-fired gas turbines

Indirect-fired gas turbines are conventional gas turbines but with the combustion chamber replaced by a heat exchanger. Producer gas is burned in a close-coupled combustor-heat exchanger and the gas turbine operated with the resulting hot, clean air in a hot-air turbine cycle. Before the hot combustion gas enters the heat exchanger some cleaning is still necessary. The main research focus in the range 0.5–3 MW_e is on the heat exchanger system because all other components are already mass produced. In a Canadian project, specific investment costs were estimated to be $6100/kW$_e$ at the 530 kW$_e$ scale down to $4500/kW$_e$ at 972 kW$_e$.

Advantages include the use of mass-produced components, which can reduce costs, and high efficiency if the inlet temperature to the gas turbine can be increased to that of directly fired gas turbines. Disadvantages include the low efficiencies of standard metal heat exchangers, the large size of the heat exchanger, the need for a particle-cleaning system for the turbine to work reliably at high temperatures, and the heat exchanger's complicating turbine regulation to match load. Basically there are no technical barriers, but extensive testing has yet to be done to prove that the system works well.

Several projects are being evaluated:

- In Belgium a fluidized-bed gasifier designed for sawdust was coupled to an indirect-fired gas turbine through a high-temperature metallic air heater. The targeted electrical efficiency is 24%. To enhance power output and allow flexible power-to-heat ratios water injection into the air heater was included. Capacity of the gas turbine is 300 kW$_e$/1200 kW$_{th}$. Expected investment costs are $5000/kW$_e$; however, at the increased scale of 1.8 MW$_e$ a replicated design was claimed to cost $3000/kW$_e$.

- In the Netherlands small-scale stationary power generation under 500 kW$_e$ was demonstrated for cogeneration applications using an indirect-fired gas turbine with clean air as the working fluid. The combination of low temperatures and pressures enabled the turbine construction materials to be made from stainless steel instead of ceramics. A two-stage combustor was chosen as the most promising system coupled to the gas turbine. The system has an electrical efficiency of 20%. Investment costs were estimated to be $2600/kW$_e$, due mainly to the high cost of the heat exchanger, which is 50% of the total investment.

Direct-fired pressurized gas turbines

Direct-fired pressurized gas turbines utilize hot combustion gas from a pressurized downdraught gasifier. As the temperature of the hot gas is usually higher than the maximum allowable inlet temperature of the gas turbine, the gas is first mixed with cooler air. The operation of a gas turbine with biomass-derived gaseous fuels appears promising in the range of 5–20 MW$_e$ plant sizes, and several demonstration plants have been built in the USA, UK and Scandinavia.

Costs are expected to be around $1300/kW$_e$ for a complete plant or $600/kW$_e$ for just the gas turbine if in the 5–6 MW$_e$ range. Advantages are that no steam ducting system is required, thus reducing cost; that a standard gas turbine can be used; that relatively little erosion, corrosion or deposition occurs on the turbine blades, owing to effective hot gas clean-up; and that residual ash from the cyclone can be used as fertilizer.

Pressurized air-blown fluid bed gasification coupled to a gas turbine has been evaluated at the small scale:

- In North Carolina a 225 kW$_e$ system was tested with the gas passed through a hot gas clean-up system prior to injection into the combustor chamber of the turbine. Barriers to commercialization were the biomass-feeding system, hot gas clean-up problems and low conversion efficiency. The major engineering challenge is to design a turbine with a fuel and combustion system that will accept and burn the hot gas. When softwood and low sulphur content coal were tested, the part-load efficiency of the turbine was low.

Micro-turbines

Micro-turbines are similar in principle to gas turbines except that most designs incorporate a recuperator to recover part of the exhaust heat for preheating the incoming combustion air and hence give a higher efficiency of 20–30% LHV. By simply changing the fuel injector, a range of liquid and gaseous fuels can be used including biogas, producer gas, biodiesel and alcohols. The exhaust gas temperature is around 250°C, so heat recovery is possible for use either as heating or cooling in absorption chillers. Air is drawn through a compressor, mixed with gaseous or atomized fuel in the combustion chamber and ignited to power the turbine section, which drives the generator (Figure 10.6). Most manufacturers are pursuing a design in which the only moving part is the shaft on which the compressor, turbine and generator are mounted. The shaft, cast from aluminium and nickel alloy steel, is suspended, lubrication free, on a wedge of high-pressure air compressed during the first few revolutions at start-up, so maintenance is minimal and even at 120,000 rev/min there is little noise or vibration. For models with power outputs of 25–30 kW$_e$ electrical efficiency is 28–30%, fuel consumption around 430 MJ/h, and weight less than 150 kg; the unit is the size of a domestic refrigerator. The target cost is \$500/kW$_e$ though if mass produced as for automobile engines it could become far less than this.

Most operate at combustion temperatures below those where NO$_x$ form, so emissions are less than 9 ppmv. Initially the 30 kW$_e$ size was limited by the shaft thrust bearing, but several competing manufacturers are now developing engines in the 25–250 kW$_e$ range. If greater electrical output is needed, multiple units can be integrated; this also provides additional reliability. The high-frequency power that is generated is converted to grid-compatible power through power-conditioning electronics. For single-shaft machines a standard induction or synchronous generator could be used to avoid the need for any power-conditioning electronics. Net-metering and peak load matching are possible. Micro-turbines are a relatively new development, and therefore many of their repeated performance characteristics are estimates based on demonstration projects and laboratory testing.

The advantages of micro-turbines include a compact and lightweight design, good size range availability and low noise levels of less than 65 dbA at 10 m distance. They can be used in association with solid oxide fuel cells, the high-temperature exhaust gas at around 800°C replacing the combustor.

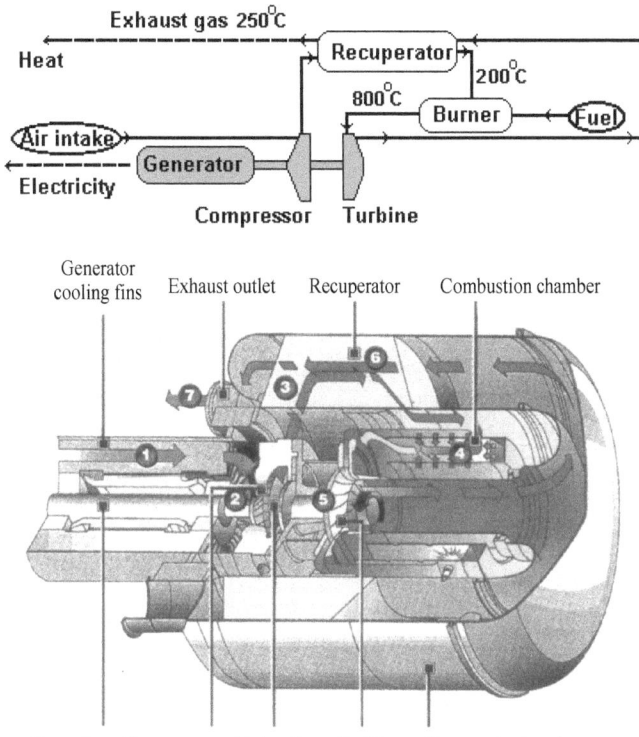

Figure 10.6 Components of the Capstone micro-turbogenerator, and schematic to show the preheating of the air
Source: Capstone Turbine Corporation

Reflective Energies is developing a small, reliable micro-turbine electric power plant, the Flex-Microturbine™. It will be able to run on fuel gases from gasification of biomass, biogas from landfills and animal wastes, and waste gases from petroleum and coal production operations. The investment cost of a Flex-Microturbine™ coupled to a simple downdraught gasifier including a small catalytic combustor will be around $1300/kW$_e$ for capacities between 27 kW$_e$ and 210 kW$_e$. A complete power plant system including a wood gasifier is expected to be available for about $1000/kW$_e$ within the next few years.

Fuel cells

Fuel cells were invented in the 1830s as a means of generating electricity through a catalytic chemical reaction, but only recently have they been considered as the 'new generation' energy conversion technology to meet the ever-increasing global electricity demand and concerns over climate change from fossil fuel use. These hopes for the technology are based on new materials that have been developed along with improved power electronic controls.

In a fuel cell, hydrogen, the energy carrier, is combined with oxygen to form electricity and water with no other emissions. Recent progress on fuel cell development has been rapid, and their market entry is imminent as the most likely future alternative to ICE for both vehicles and stationary applications. Fuel cells are not limited to 30% thermal efficiency by the Carnot cycle as is combustion of fuel in internal combustion engines. They avoid the need for mechanical devices, and can operate at high efficiencies at full or part load.

A fuel cell system consists of:

- a fuel reformer to generate hydrogen-rich gas from methanol, methane or other fuels in cases where hydrogen is not supplied directly from, for example, the electrolysis of water using solar energy

- a power section, where the electrochemical process occurs

- a power conditioner and inverter, where the direct current (d.c.) generated in the fuel cell is converted into alternating current (a.c.).

An anode and a cathode are in contact with an electrolyte, and the cell operates essentially by reverse electrolysis. When the anode and cathode are supplied with hydrogen fuel and oxygen (air) respectively, a voltage is generated between the two electrodes. When connected to complete the circuit, a current flows. Oxygen molecules are reduced to oxide ions at the cathode and hydrogen molecules are oxidized to H+ ions at the anode, which are then conducted through the electrolyte. Thus this electrochemical engine generates electric power (Figure 10.7), and water vapour is produced.

Each individual fuel cell produces between 0.5 and 0.9 V of d.c. electricity. So, just like cells in a lead–acid battery, several are linked together in series or parallel by a wire or ceramic conductor interconnector to form a stack and obtain more useful voltage and power outputs. The advantages of fuel cells include:

- high efficiency at all scales, even in the lower power ranges

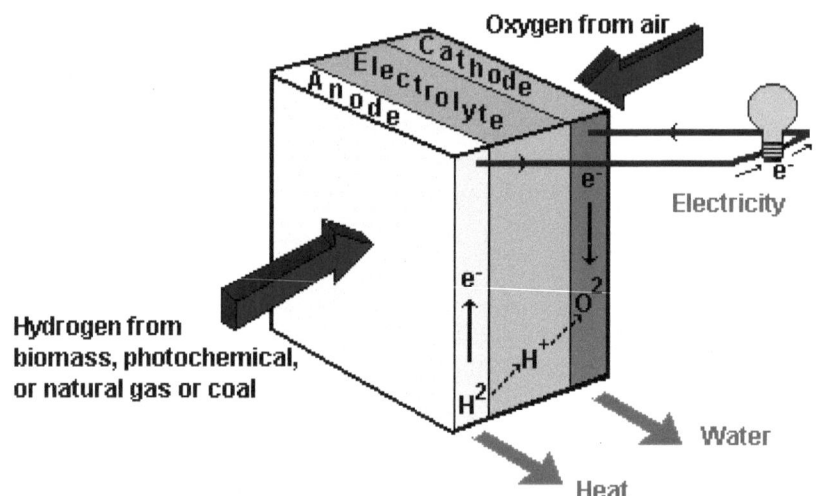

Figure 10.7 The operating principle of a fuel cell circuit to develop an electrical current

- a modular design, allowing a wide range of power ranges to be prodced
- very low or zero emissions (depending on the type of fuel used)
- low maintenance and high reliability
- no noise (as there are no moving parts).

Fuel cell technologies differ with respect to their electrochemical reactions, materials of construction, tolerance to contaminants, fuel flexibility and operational characteristics, which vary with the application. Stationary applications are very demanding on lifetime for economic viability, but less so on power-to-weight ratio of the system. For transport a shorter lifetime is acceptable, owing to the limited life of the vehicle, but the power-to-weight ratio and good performance are crucial. Transport applications set stringent performance and cost requirements. Several companies are working on the development of proton-exchange membrane fuel cells (PEMFC), including Ballard, IFC, Plug Power, Energy Partners, Siemens, De Nora, General Motors, Mazda and Toyota. Investments of millions of dollars have also been undertaken by Daimler Chrysler, Ford and General Motors, all aiming to have fuel cell vehicles ready by 2004. Daimler-Benz have stated that, if vehicles were powered by methanol derived from fossil fuels, there would be 30% less CO_2 emissions than for the most efficient ICE vehicle. If biomethanol were used there would be virtually no GHG emissions.

Both small and large power plants can be installed either in congested urban dwellings directly connected to the grid, or as stand-alone systems in relatively small or isolated villages, islands and utility centres. In the future, electricity generated may become cheaper than from other competing energy conversion systems owing to the flexibility of the fuel source, the doubling of the efficiency and the reduction in transmission and distribution costs.

Several types of fuel cell have been developed; the most common ones are described below. The type of fuel cell determines the operating temperature, the heat liberated during the process, and the cell's suitability for cogeneration applications. Low-temperature fuel cells generate heat only suitable for low-pressure steam and hot water applications. High-temperature fuel cells produce high-pressure steam that can be used in combined cycles and other cogeneration applications.

Proton-exchange membrane fuel cells

Proton-exchange membrane fuel cells (PEMFC) have a thin polymer electrolyte semi-permeable membrane that allows protons to pass but insulates the electrical contacts (Figure 10.8). General Electric developed the first PEMFC in the 1950s. PEMFCs have a low operating temperature of approximately 80°C, and the efficiency is 30–35%. Their electrical output can respond quickly, making them well suited for automobile applications and for providing power to residential and commercial buildings to meet ever changing demands. They are commercially available in the smaller size ranges under 10 kW$_e$ but at very high costs compared with an ICE, though this is expected to decrease sharply

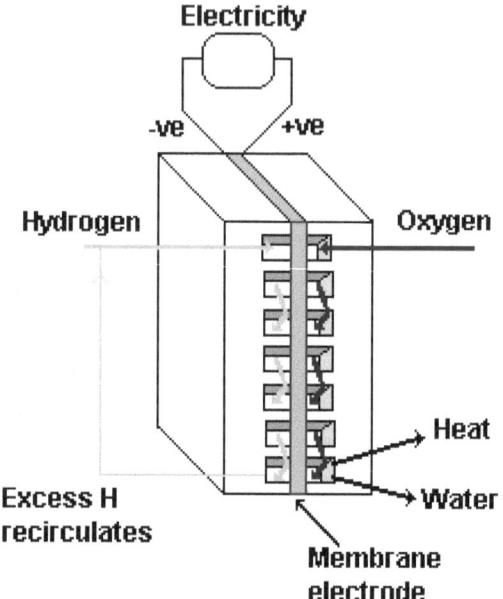

Figure 10.8 Proton-exchange membrane fuel cell stack, showing the flows of hydrogen and oxygen moving from cell to cell
Source: *Energy Wise News*, EECA

as mass production occurs in the automotive industry (Figure 10.9). At the moment Daimler Chrysler offers a fuel cell car engine at approximately $600/kW$_e$ and Ballard, who have run fuel-cell-powered buses in Vancouver and Chicago since 1993, is developing a 250 kW$_e$ stationary unit with a capital cost target of $1000–2000/kW$_e$. PEMFCs are likely to become the first mass-produced low-cost fuel cells in the marketplace owing to applications in the automotive industry.

Alkaline fuel cells

Alkaline fuel cells (AFC) have an operating temperature of 70–90°C with an efficiency of 30–40%. They have primarily been used by NASA on space missions, including Gemini from 1963 to 1966. The concept attracted renewed interest for mobile applications when potential efficiencies of up to 70% were claimed. AFCs are sensitive to the presence of CO_2, but less sensitive to CO. Therefore pure oxygen instead of air should be used as the oxidant, and this limits their applicability.

Direct methanol fuel cells

Direct methanol fuel cells (DMFC) use a polymer membrane electrolyte similar to the PEMFC, but the anode in the DMFC converts the methanol fuel directly and thus eliminates the need for a fuel reformer. Efficiencies of about 40% are expected for the low-temperature version, reaching up to about 50% for designs

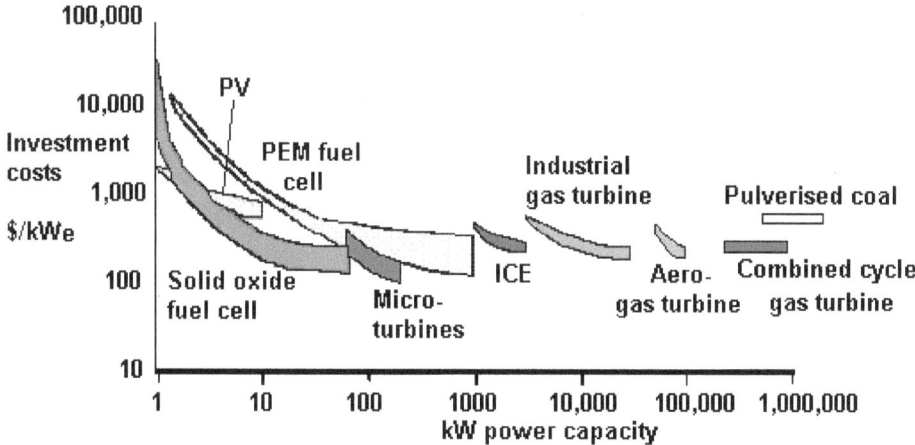

Figure 10.9 The costs of fuel cells and micro-turbines are expected to decline rapidly over the next decade or two, to compete with traditional power-generating sources at the smaller scale of less than 10 MW per unit and perhaps even overtake solar photovoltaic systems (although the cost of supplying the hydrogen fuel must be included whereas solar energy is free)

with operating temperatures above 100°C. All major PEMFC developers also have R&D programmes in DMFCs. The main technological challenges are the formulation of better anode catalysts and the improvement of the membranes and cathode catalysts in order to overcome cathode poisoning by migration of methanol from the anode to the cathode.

Phosphoric acid fuel cells

Phosphoric acid fuel cells (PAFC) use a phosphoric acid electrolyte and operate at approximately 200°C, giving an efficiency of 34–38%. Of all fuel cells the PAFC represents the most mature technology; it is commercially available and used in many applications including buses and stationary power units. For example, a 200 kW$_e$ ONSI PAFC operating on natural gas at the Australian Technology Park in Sydney generated 180 MWh in a year. It could also be used as a cogeneration unit, giving reliability averaging 96% based on current experience. Over 200 similar systems exist, mainly in the USA and Japan, with the size being similar to that of a shipping container. ONSI Corporation was initially selling this 200 kW$_e$ model for about $3000–4000/kW$_e$, but it is anticipated that the costs will soon approach $2000/kW$_e$ and, in the longer term, drop to $500–1500/kW$_e$.

Molten carbonate fuel cells

Molten carbonate fuel cells (MCFC) can be built from stainless steel and less exotic materials than the solid oxide type (below). An MCFC can accept carbon monoxide and tolerate carbon dioxide. It operates at 650°C and achieves a high efficiency of 45–55%. The MCFC is fuel-flexible and offers the best potential to be coupled with large biomass gasification processes. Stationary MCFC stacks of up to 250 kW$_e$ have been installed in several demonstration projects around the world, but they are not yet competitive as they cost around \$8000/kW$_e$.

Solid oxide fuel cells

Solid oxide fuel cells (SOFC) consist of monolithic, flat plate (planar) or tubular cells. A hard solid ceramic, zirconium oxide (zirconia), is doped with ions of the rare earth metals yttrium or scandium to maximize the capacity for oxide sites on the surface and to stabilize the molecular structure. Zirconia is a powder, which is mixed with the impurity, formed into a clay-like substance, and then extruded into the shape required, such as a 2 mm diameter tube, 100–150 mm long, which is easy to seal at the ends, before firing in a kiln.

In this type of SOFC the lanthanum manganite cathode is on the outside of the tube in contact with the air, and the nickel-based anode is on the inside where the fuel is introduced. The catalyst is heated directly so the fuel is reformed inside the device rather than needing a reformer (which requires its own energy source and so diminishes the net power output). The operating temperature of this fuel cell is around 1000°C at which the cubic molecular structure conducts best, though a design operating at 660°C is also under development. The efficiency is 45–47%; the cell is fuel-flexible, and stacks of 100 kW$_e$ have been demonstrated, though operational time has been limited to date.

Once proven it is expected that SOFCs will be geared towards distributed generation applications in order to become economically attractive because of their fragile nature for transport applications and the relatively high investment cost for the catalyst and materials.

A summary of fuel cell technologies and their characteristics is presented in Table 10.2. MCFCs and SOFCs are high-temperature fuel cells that are designed for continuous operation in larger power plants up to 2 MW$_e$ since less heat loss results than for smaller systems. SOFCs offer good potential for cogeneration markets, as they are efficient and flexible stationary power units. For smaller applications the low-temperature fuel cells PEMFC and PAFC are more suitable.

The types of fuel cell that seem to have good potential when linked with biomass fuels are PEMFC, PAFC, MCFC and SOFC. However, a novel fuel processor unit has been developed in combination with an AFC, and tested successfully using agricultural residues and wood charcoal. In addition, after extensive evaluation by the Electrotec Group of Industrial Research Limited, New Zealand, in partnership with the European fuel cell stack manufacturer ZeTek Power, AFC stacks were selected for research into distributed generation and back-up power sources in the range 1–100 kW$_e$.

The main barrier for commercialization of fuel cells is their high cost, as current stack technology is too expensive for most commercial applications. Significant R&D funding will be required to improve reformer design, reduce size and weight and develop manufacturing techniques in order to achieve the projected performance and cost reductions.

Table 10.2 A summary of fuel cell technologies and their characteristics

Type	Electrolyte	Anode	Cathode	Operating temperature (°C)	Electrical efficiency (%)	Applications
PEMFC	Sulfon-F-polymer Electrolyte Membrane	Platinum	Platinum	60–100	30–35	Transport; small-scale stationary; cogeneration
AFC	Potassium hydroxide	–	–	70–90	30–40	Transport; stationary
DMFC	Sulfon-F-polymer	–	–	70–100	40–50	Transport
PAFC	Phosphoric acid, H_3PO_4	Platinum	Platinum	190–210	34–38	Transport, small-scale stationary; cogeneration
MCFC	Molten carbonate, Li_2CO_3/K_2CO_3	Nickel	Nickel oxide	600–700	45–55	Large stationary; plant (electricity only)
SOFC	Solid oxide, ZrO_2/YO_2	Nickel	Sr-doped lanthanum manganite	660–1000	45–47	Large stationary plant (electricity only); small-scale cogeneration

Hydrogen is the most promising energy carrier for use in fuel cells because it is an energy-efficient low-polluting fuel. When hydrogen is combusted with air in a fuel cell the only products are water and a small amount of NO_x. Hydrogen has an oxidation rate around four orders of magnitude greater than CO, which has the same oxidation advantage over saturated hydrocarbons under similar conditions.

Hydrogen is found in many diverse compounds such as water, fossil fuels and biomass. At present it is produced for industrial use from fossil fuels such as natural gas, naphtha and inexpensive coal. In such a case the hydrogen is not renewable, and CO_2 is also emitted during the hydrogen production stage, similar in volume to that released during combustion of those same fuels.

Hydrogen can also be produced from the thermochemical gasification of biomass feedstocks. Typically it makes up about 6% by weight of dry biomass, which can be extracted by gasification to produce a gas containing around 10–25% hydrogen by volume. The use of hydrogen from biomass is attractive from an environmental point of view because the carbon cycle would be closed. The challenge is to overcome the economic barriers that current technologies present for converting biomass to hydrogen for use in clean, efficient energy-conversion devices.

Several new and promising concepts are under development to produce methanol or hydrogen from biomass. A Dutch modelling study compared the technical and economic performance of six methanol- and five hydrogen-producing systems at several scales. The most promising was ceramic membrane technology, which enables hydrogen to be separated from a mixed producer gas stream at high temperatures for direct use in a fuel cell. If biomass is available for $2–3/GJ, hydrogen can be produced for $6–8/GJ and methanol for $7–10/GJ.

Since producer gas coming from a biomass gasifier contains methane and carbon monoxide in addition to the hydrogen, the fuel cell electrolytes cannot use it directly so it must receive appropriate pre-treatment. The sensitivity of fuel cells to these fuel impurities is determined mainly by the operating temperature, the electrolyte and the catalyst. Sulphur oxides, metal vapours, tars and high CO levels can all be harmful. Table 10.3 summarizes the possible contaminants and toxin levels that would inhibit the operation of various fuel cell types. In addition, metal vapours and tars can inhibit SOFCs and MCFCs, but usually the levels are very low in gasified biomass. If bark, wood wastes or other kind of biomass were to be used as a source of hydrogen, removal of any metals as well as sulphur would be necessary.

Purification steps for the producer gas depend on the type of fuel cell, so the development of a small and efficient gas clean-up system would be a key step. Producer gas pre-treatment can involve adding steam at temperatures of around 600°C to convert the methane to carbon monoxide and hydrogen in a process known as **steam reforming**. At a lower temperature of 200–300°C steam reacts with the carbon monoxide to form hydrogen and carbon dioxide in the shift conversion reaction. Carbon dioxide and any sulphur compounds present can be removed. Methane reformers, shift reactors, CO_2 removal

Table 10.3 Contaminant and toxin levels that inhibit various fuel cell types

Type	CO	CH$_4$	CO$_2$ and H$_2$O	SO$_x$
PEMFC	Inhibits at >10 ppm	Diluent	Diluent	Unknown
PAFC	Inhibits at >0.5%	Diluent	Diluent	Inhibit at >50 ppm
MCFC	Used as fuel	Diluent	Diluent	Inhibit at >0.5 ppm
SOFC	Used as fuel	Used as fuel	Diluent	Inhibit at >1.0 ppm

systems and hydrogen purifiers are well-established technologies in the chemical process industries.

An alternative approach to the production of hydrogen from biomass is by fast pyrolysis to generate bio-oil followed by catalytic steam reforming of the oil or its fractions. This approach has the potential to be cost-competitive with other current commercial processes for hydrogen production.

There have been several studies linking fuel cells with biomass fuels:

- In one unique innovation, high-temperature fuel cells and gas turbines were integrated to boost electric generating efficiencies. The hot exhaust from the SOFC was used to drive the gas turbine. Energy recovered from the turbine exhaust was used in a recuperator to preheat the air from the turbine's compressor section before entering the fuel cell and the gas turbine. Any remaining heat energy from the turbine exhaust could be recovered for low-grade heating.

- The first biomass-based fuel cell power plant based on a small AFC module was demonstrated in 1999. The module consisted of 24 cells with a nominal power output of 350 W$_e$. The primary fuel input was producer gas obtained from gasification of wood char and agricultural residues. Power output using this feedstock showed potential as an efficient stand-alone power generator, but the generating cost is unknown.

- Testing in the USA of a pilot-scale air-blown fluidized-bed gasifier was linked with an MCFC when operated on gas derived from switchgrass using computer simulation techniques. The system showed that a 1.75 MW$_e$ power plant would operate at 46% efficiency; however, the effect of fuel gas quality on fuel cell performance and investment costs was not investigated.

- Several systems were compared for using switchgrass or woody biomass in downdraught gasifiers in conjunction with a 200 kW$_e$ fuel cell to generate electricity for on-farm installations: a Batelle indirect gasifier coupled with a PEMFC, a downdraught gasifier coupled with an MCFC, and a downdraught gasifier coupled with a PAFC. Cleaning the gas was included. The MCFC system with investment costs of $3200/kW$_e$ appeared to be more cost-effective than the PAFC system at $4200/kW$_e$ or the Batelle PEMFC system at $4450/kW$_e$. As well as the high capital costs of the fuel cell and gas preparation system, the primary barriers to development were low competing electricity tariffs and high operating costs. Automation of the system to minimize the operational costs is paramount for improving the return on investment for small power systems.

- A technical and economic comparison was made between a biomass gasification/gas turbine system and a biomass gasification/fuel cell system consisting of a Batelle gasifier, an MCFC and a steam turbine. Feeding the producer gas into the MCFC instead of the gas turbine allowed for a higher level of contaminants without risk. The fuel cell system also had a higher overall efficiency at 53% and a better economic performance at the larger scale of 1–2 MW_e.

Electricity generation system analysis at the small scale

There have been several studies of small-scale electricity and heat generation systems comparing different technologies. For example, the technical and economic performances of a variety of small-scale systems were assessed with the ECLIPSE process simulator program using short-rotation willow coppice, which gave the following results:

- At scales of around 5 MW_e, combustion plants have higher efficiencies than gasification/gas turbine systems in a simple cycle.

- Direct combustion for power generation was not feasible even up to 25 MW_e because of low efficiencies – ranging between 16% and 26% – and high capital costs, ranging from around $11,000/$kW_e$ for a 350 kW_e system to $3200/$kW_e$ at 4.75 MW_e. For cogeneration, sizes below 25 MW_e were considered more suitable.

- An atmospheric pressurized gasifier with ceramic filter hot gas clean-up coupled to a simple cycle gas turbine at power outputs of either 1.3 MW_e or 3.4 MW_e had low efficiencies of 13.9% and 16.0% respectively with investment costs of $8700/$kW_e$ and $7100/$kW_e$.

- Gasification integrated with a spark ignition engine/generator set resulted in moderate efficiencies ranging from 22.6% to 29.7% for plants between 150 kW_e and 2.27 MW_e with investment costs of $1900–3000/$kW_e$.

Another study examined gasifiers of various sizes in combination with a PAFC, and eight with an MCFC. One combination, for example, was a PAFC fuelled by an LPO atmospheric pressure oxygen-blown downdraught gasifier fed with dried SRC feedstock and with a wet scrubber for cold gas cleaning. Capacities were compared ranging from 15 to 500 kW_e, giving overall efficiencies in the range 13.7–16.6% and investment costs in the range $11,400–5200/$kW_e$. The same gasifier now linked with an MCFC gave overall higher efficiencies – between 24.8% and 28.3% – and lower investment costs in the range $7400–3200/$kW_e$. Given that the maximum size of an LPO gasifier is technically limited to 2 MW_{th}, the optimum plant size for a single downdraught gasifier integrated with fuel cells was between 300 and 500 kW_e. The most appropriate gasifier of those examined was the Koppers-Totzek entrained-flow gasifier, which was originally developed for coal gasification. It is commercially available, and has been assessed to produce gas with a low methane content from biomass so that reforming would not be necessary. This technology has potential if the lifetime and capital costs could be further improved.

Overall it was concluded that, at the very small scale of <1 MW_e, gasification plants integrated with MCFCs or gas engines were more efficient than the use of PAFCs. Combustion plants were the least efficient. However, until the MCFC technology is developed to the commercial stage, gas engines are the most appropriate for small-scale biomass systems.

Another study made comparisons between four different small-scale cogeneration technologies and drew these conclusions:

- The Stirling engine has good potential because of its relatively high efficiency and a wide spectrum of power ranges.

- To be efficient, the steam engine needs higher steam temperatures and pressures than those that can be achieved using standard small industrial boilers.

- Steam turbines are commercially available and well proven technology, but their efficiencies remain relatively low.

- Indirect-fired gas turbines have technical problems with high-temperature heat exchangers, which are still to be resolved.

- To eliminate the uncertainties associated with cost estimations for technologies not yet commercially available, the annual production expenses were calculated for a 3.12 MW_{th} capacity plant but excluding capital costs. The expenses were lowest for the Stirling engine, followed by the steam engine, the steam turbine and highest of all for the indirect-fired gas turbine.

- Cogeneration is important for the viability of small-scale combustion applications assuming there is a useful demand for the heat. It tends to increase the operating hours of the plant, but clean and robust technologies with high overall efficiencies still need to be developed. The lower limit for cogeneration plants is considered to be at a boiler capacity of 5 MW_{th} owing to the poor efficiencies of small-scale steam turbines and the higher investment costs of these systems.

- New small-scale cogeneration technologies have potential for the future. For example:

  - Few proven commercial Stirling engine systems are available at present, and investment costs for prototypes are high, but more systems are beginning to reach the market.

  - Newly developed cost-competitive steam engine designs with a screw-type motor should soon become available commercially.

- Organic Rankine cycles (ORC), with hydrocarbons or toluene as working fluids, operate at lower temperatures and pressures than conventional steam processes but are not yet economic for combustion plants smaller than 5 MW$_{th}$.

Another feasibility study was undertaken to evaluate the development of advanced small-scale power production systems based on biomass pyrolysis and gasification in the range between 5 and 60 MW$_e$ capacity (Table 10.4). It was concluded that, for peak load generation and less than approximately 2000 operating hours annually, a pyrolysis bio-oil-fired diesel power plant concept appeared the most promising in spite of having the lowest efficiency.

An overview study of gasification technologies for small-scale biomass cogeneration plant showed that, when integrated with steam turbines, efficiencies of 10–15% were obtained for capacities of 1–4 MW$_e$ with investment costs of $5500–2900/kW$_e$ respectively. For fixed-bed gasifiers integrated with a gas engine in the range 100 kW$_e$ to 2 MW$_e$ the system efficiency was 32–40% and investment costs were $5200–2300/kW$_e$. By way of comparison, gasification combined with either PEMFC and PAFC (indirect producer gas utilization) or MCFC and SOFC fuel cells (direct producer gas utilization) was expected to give efficiencies in the range 38–45%.

In summary, the operation of steam or gas turbines is considered to be relatively uneconomic in terms of operating, maintenance and capital costs for small-scale biomass plants (Figure 10.10). The preferred approach is to gasify

Table 10.4   Comparison of system efficiencies and costs for small-scale biomass projects

Primary system	Secondary system	System efficiency (%)	Investment cost ($/kW$_e$)
Atmospheric pressure fluidized-bed gasifier; dolomite tar cracker; gas clean-up	Integrated gas turbine combined cycle	37	4800
Pressurized fluidized-bed gasifier; gas clean-up	Steam-injected gas turbine engine	43	7100
Atmospheric pressure fluidized-bed gasifier; dolmite tar cracker; gas clean-up	Dual-fuel diesel engine	34	4600
Pyrolysis plant	Gas turbine	33	1600
Pyrolysis plant and bio-oil refinery	Diesel engine	30	700

or pyrolyse the biomass fuel to produce a gas with low calorific value that is subsequently combusted in a gas or diesel engine. Although they have fairly low efficiencies, their cheap cost due to mass production is hard to beat. Therefore, fast pyrolysis or atmospheric gasification with power generation by a modified ICE engine should remain competitive in the small-scale electricity generation market for some time. Model simulations have shown that fast pyrolysis has a market advantage in cases where the biomass resource is available some distance from the electricity market and where power supply is required at several remote sites, since the system can be decoupled. This means that the pyrolysis plant may be located near the biomass resource and the generator near the electricity market, thus reducing biomass transport costs. Alternatively, a single pyrolysis plant could be constructed to supply fuel for all generators at several remote sites, saving investment costs through scale economies.

This comparison is only indicative, and the broad ranges used show the uncertainty in the data. Published information is conflicting, and confusion arises between real project and estimated costs and between efficiency claims from laboratory tests and the usually considerably lower efficiencies achieved once systems are scaled up and installed at the commercial scale. Some economies of scale are evident, with the exception of internal combustion engines owing to their being mass produced at the smaller scale.

Gasification using a range of dry biomass feedstocks in conjunction with micro-turbines or fuel cells has the potential to generate electricity for rural

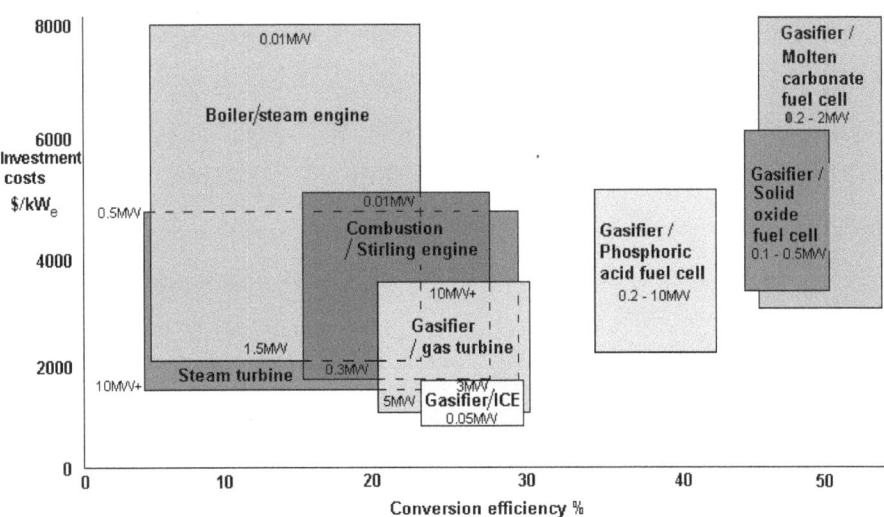

Figure 10.10 At the small scale, power generation technology systems cannot yet compete with the internal combustion engine, due either to their poorer fuel-to-electricity conversion efficiencies or higher unit costs or both (Note: it was assumed that the fuel cells were linked with an oxygen-blown gasifier and wet gas scrubber system)

community installations. Complete automation is essential for any small system to realize a return on investment. The Flex Microturbine™ appeared to be the most cost-effective system when compared at the 200 kWₑ level, being also competitive with gasifier and internal combustion engine systems in the longer term.

Case study 17

Small-scale bioenergy project, Taveuni Island, Fiji

This case study describes a project proposal for electrifying the Fijian island of Taveuni using not only biomass, but also hydro and solar PV. It typifies the problems of developing a small-scale biomass project in a developing country. The island population is around 14,000 but, with only a few diesel generation sets available, as in many other developing countries most of the people have no access to electricity. The project demonstrates a good example of how to integrate biomass with other renewable energy technologies. The options considered were as follows:

- **110 kW wood gasifier**   to meet peak loads and to provide heat as well as power using fuelwood supplied from old Fiji Tall coconut trees in the first instance and later from eucalyptus plantations to be grown nearby.

- **350 kW hydr o power**   station using a run-of-the-river scheme on the Naibili Creek near Somosomo with intake in the hills behind and a 4.1 km length of buried penstock pipe to the power station site at Vuniduva.

- **350 kW diesel engine**   to provide a stand-by supply for the short periods when the hydro turbine requires maintenance or on the rare occasions when the river flow is lower than design requirements in unusually dry periods.

- **solar systems**   using photovoltaic panels for the more remote domestic and small commercial applications not feasible to connect to the grid. In future micro-turbines or fuel cells may have application for such remote communities.

The objective was to develop an electricity generation and distribution supply network on Taveuni that would:

- provide a 24 hour power supply at an affordable tariff equivalent to that paid on the main Fijian islands

- give a reliable supply, urgently needed by the local hospital, hotel and food-processing industry

- be developed in four separate phases, beginning with the Somosomo region

- support existing industries and encourage them to expand

- encourage new industries to develop on the island

- contribute to the development of the island economy

- create employment opportunities

- improve the quality of life on Taveuni by improving health, education and communication services for the residents

- encourage further eco-tourism to develop.

Woody biomass gasification, small hydro power and diesel stand-by were identified as the mix of renewable energy technologies to best suit the available energy resources. If the distribution network is extended in the future to reach outlying communities, other power-generating technologies could be considered, including biomass combustion for steam generation.

Taveuni had not been electrified because in most previous feasibility studies only the commercial implications were considered, with little attention given to the social, cultural and environmental issues. In a proposed partnership between a developer, the community and the Government of Fiji, all these issues and benefits had to be taken into account during the analysis.

In most instances rural electrification is not commercially viable without some form of financial assistance. This case study not only details the true commercial costs involved in providing a 24 hour reliable power supply for the islanders of Taveuni but also indicates how a subsidized retail price of 11 c/kWh to match that paid by the mainland Fijians might best be achieved.

Local inputs into the project in the forms of fuelwood procurement, land rental income, maintenance of the power lines and connection of new customers should encourage local community interest in and direct involvement with the project. The substitution of diesel generation will help to reduce the drain on foreign reserves from oil imports.

Many old Fiji Tall coconut trees are due for replacement on the island. A scheme is in operation to encourage copra growers to upgrade their production stock with new hybrid trees. This initiative will provide a source of cheap fuelwood for the gasifier initially ($0.20/GJ delivered) which will later be supplemented by specially grown plantation fuelwood ($2.20/GJ delivered). Good soils and climate enable high growth rates to be achieved, averaging 30 t of harvestable biomass per hectare per year (based on the measurement of existing eucalyptus trees growing on the island). Around 20 ha of land will be needed initially to grow fuelwood to supply the gasifier to meet the anticipated load once the coconut wood resource depletes. An investigation of several sites together with discussions with landowners identified several options, but negotiations will be required to secure the land and gain access rights.

This new land use will not be in competititon with the fertile lowlands suitable for agricultural crop production, but will be sited further up the hillsides. The most ideal site identified was the Taveuni Coconut Centre, where 70 ha of surplus hillside land is available as it is unsuitable for coconut plantations. Good building and machinery facilities already exist, which could be shared. Land rental agreements would have to be negotiated. As the electricity supply system expands, a second gasifier will be needed to connect

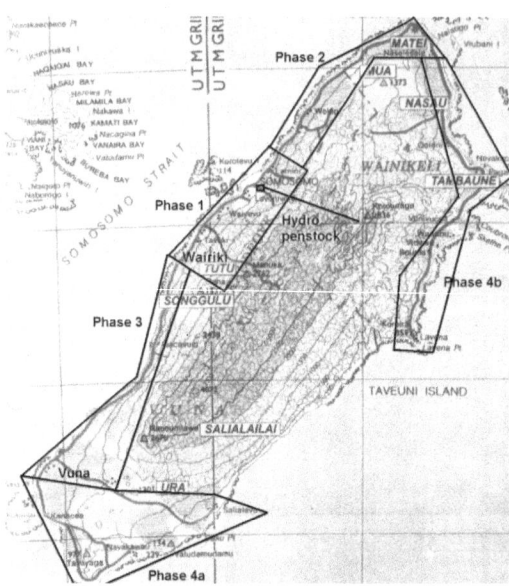

Figure 10.11   Taveuni, showing the main load centres and the four proposed phases of distribution line construction over the 40 km long island

to the grid to meet the increased peak demand. The Coconut Centre would be a possible site to build it, as it is close to the fuelwood supply and to the second phase major demand load around the Naselesele district (Figure 10.11).

Since this is government-owned land there should be no problems associated with the Agricultural Land Trust Act. A rental price will have to be negotiated for a shared-use arrangement to proceed. The first 6–8 ha of plantation area will need to be established in order to provide fuelwood for the gasifier once the coconut wood supply dwindles.

A 5000–8000 m³ plot is also required for the gasifier and turbine, diesel generating sets, diesel tank, fuelwood drying and storage areas, switchgear and administration building. Of the six sites considered, the preferred is close to the proposed hydro power station and in the central Somosomo location. If the owner is unwilling to sell, the alternative is to obtain freehold land at Waiyevu.

Phase 1

The main commercial and administration centre is situated around Somosomo and Waiyevu to the south. This area includes 550 households, the hospital (a new one soon to be built), police station, communications centre, government centre, the main resort, other tourism facilities and activities, the ice-making facility, several schools, and street lighting. A 13 km 11 kV three-phase backbone distribution line will be constructed between Lamini in the north to Wairiki in the south and will link with the existing buried cable network at Somosomo and the Waiyevu overhead network fed by the diesel generator. Single-phase high-voltage spur lines will also be required, with transformers to provide low-voltage mains to consumers.

Phase 2

A 20 km 11 kV backbone line will be constructed going north from Lamini to Naselesele including branch and spur lines. This will supply several small tourist facilities, the Taveuni Coconut Centre (where the second gasifier plant will be constructed) and the airport, as well as a further 300 households.

Phase 3

Phase 3 will extend the line south from Wairiki to Vuna, to enable a further 420 households to be connected and bring in some additional commercial load including the Tarte estate (with its steam engine generator) and the Vuna water supply. Approximately 22 km of 11 kV line will need to be constructed.

Phase 4

Phase 4, not considered in the current analysis, is the longer-term goal of extending the grid to enable another 700 or more households to be supplied with mains power, but it would require construction of a further 30 km or more of 11 kV line.

Power demand

It is well understood that it is not commercially viable to provide power to a rurally based and widespread community such as that on Taveuni, owing to the relatively high capital investment for distribution in proportion to the low revenue levels from sales of electricity. On Taveuni the problem is compounded by the peak power load, which at present occurs for only a short time each evening. Thus much of the installed generating capacity will lie idle for most of the time.

An analysis of the Somosomo region, based on current behaviour patterns, showed that if a mains 24 hour power supply were to be installed, then a base load of around 60 kW would occur 24 hours a day. A daytime load of approximately double this would occur between 7 am and 6 pm, with a slight peak mid morning. After 6 pm there would be a short and sudden peak load due to domestic demand, which would double the daytime load and last for only 3–4 hours (Figure 10.12). The existing communal power-generating schemes create this peak as they operate for restricted hours, usually between 6 and 10 pm, as determined by the local management committee. As Phases 2 and 3 come on stream the overall load will increase, particularly during the evening peak. In addition it has been assumed that there will be growth in demand from both new connections and also from greater use of the available power by those already connected. Once Phases 1, 2 and 3 are completed then a critical peak load of 350 kW$_e$ will result about 5 years later, at which stage further generation capacity will be required to prevent power outages occurring.

Reduction of the evening peak load should be attempted in order to delay the need to add expensive extra generation plant. This can possibly be achieved

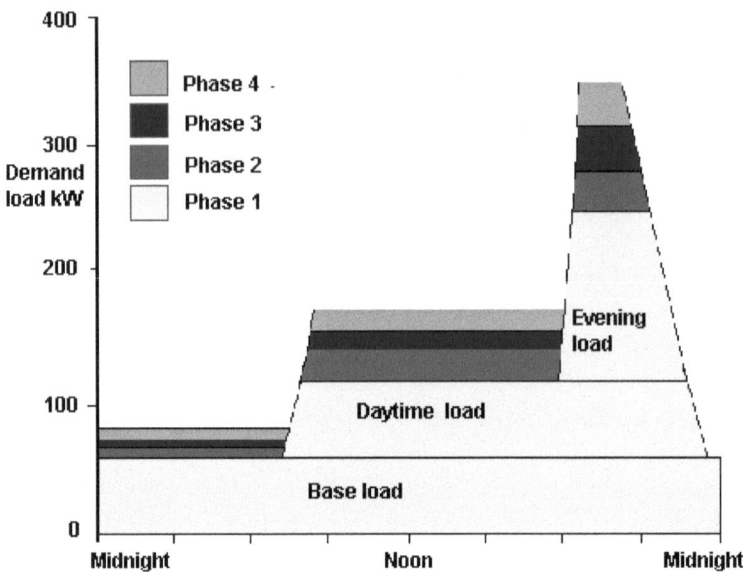

Figure 10.12   Assumed average daily load demand based on current uses of appliances by the islanders powered only by diesel generators and operating mainly only in the evening

by energy management techniques such as the use of low-wattage fluorescent light bulbs rather than the traditional incandescent type. In the future, energy management incentives may be a cheaper option than adding new generation capacity.

It was assumed that the hydro power plant will provide power for the base and daytime loads, with the wood gasifier being used daily to match the peak evening demand. The diesel generation will be used purely as a stand-by source. There will be ample total installed capacity for some time to meet the growing demand. However, to meet the peak load a second wood gasifier will be added to the system.

It is evident from the load diagram that surplus capacity will exist during the daytime (as well as the night-time). This provides the opportunity for industry to develop in this region, using the available power to give added value to existing commercial products. Construction of a kava-pounding plant and a cool room for storage of dalo to improve product quality and to give more flexibility to seek the higher market prices are examples. To encourage this daytime 'off-peak' demand for electricity in order to help flatten the peak load curve, the government could supply capital to assist local industry develop new enterprises.

It is not known how domestic behavioural patterns will change once 24 hour power becomes available. For example, ironing clothes is currently done in the evening since power from communal diesel-fired power schemes is only available at that time. Whether evening tasks will be carried out during the daytime instead, and by what proportion of the households, is difficult to determine. Behavioural changes will possibly produce a natural flattening of the peak load curve once 24 hour power becomes available.

It was assumed that connections to existing commercial and industry consumers will occur fairly rapidly once the mains distribution line is in place. Growth in domestic power demand was assumed to match the predicted population growth at 1.28% per annum. Growth in the commercial load was assumed to be 10% per year for the first 5 years after supply first became available, then slowing to 2% per year.

Power prices

The current annual average household income on Taveuni ranges from $3000 to $5000. This is above that earned on the main islands owing to the high prices paid for several of the main agricultural crops produced there and the income received from tourism activities. A detailed survey of the islanders conducted as part of the power project evaluation process showed that lower-income Fijian households spend around $6/month for the power generated from central diesel generation, where it is available, which they use for limited lighting, TV and refrigeration purposes. They each consume 40–50 kWh per month, and also purchase kerosene and benzene for additional lighting and many expensive batteries for portable appliances such as torches and radios. This expenditure would no longer be necessary if 24 hour power became available. Indian households tend to be wealthier and can afford more for their power. Many already own or have access to private diesel generation supplies and, in addition to the appliances possessed by a typical Fijian household, they also own a cooling fan, iron, washing machine, and several small kitchen appliances. Their monthly power bill amounts to around $20 and is equivalent to 22–25c/kWh if using 80–90 kWh/month as reported.

All survey respondents agreed that they badly needed a reliable 24 hour power supply. Many identified additional future domestic and commercial uses once it became available. For such a supply Fijian households are prepared to pay between $10 and $30 per month and Indian households between $30 and $40. An evaluation of the 'affordable power demand' for these low-income households estimated that a typical Fijian household could afford $10 per month and an Indian household $15. This confirms that even the poorest families will be able to afford electricity in sufficient quantities to produce a functioning household with the essential services of lights, refrigeration and TV/radio. All households would be better off than under their present energy supply cost arrangements.

Some respondents related their ability to pay to current consumption, while others realized that a 24 hour supply would cost more. How much a household is *willing to pay* was not clear from the survey, since many households had no clear idea of what their current expenditure was. However, it can be assumed that, once a 24 hour supply becomes available, most households will perceive the benefits more clearly and will soon seek a connection.

It was assumed that all commercial loads in the region will be connected to the network soon after power becomes available. Small shop owners, for example, currently pay around $50 per month for their power, so they could afford connection to gain better security of supply. It was also assumed that the government will lead by example and connect all their loads to the network as soon as practically feasible to do so. Existing diesel generating sets will probably be maintained in place by most of the owners initially as stand-by systems until confidence is gained in the reliability of the new network supply. In the longer term many diesel gensets will probably be sold as surplus to requirements.

Several prepayment options were evaluated including disposable cards, key pad systems and smart cards. The NPV differences for each system were marginal and so selection was based more on the social and cultural aspects of the Taveuni residents. A new, tamperproof and cheaper prepayment card system developed by a South African electronics company holds promise for these situations.

Capital investment costs

To undertake Phase 1 alone will require a capital investment of $2.1 million to construct the plant, approximately 70% being for the generation equipment and installation and 30% for the distribution, connection and metering of the supply network. The capital investment for the gasifier, based on a Dutch Kara downdraught model, was $647,000. The cost analysis assumed that:

- hydro power will provide approximately 80% of the total annual power demand

- the wood gasifier will supply 13% of the demand

- stand-by diesel generators will provide only 7%, being used only to ensure reliability of the system design during maintenance.

The retail price needs to be competitive with the existing diesel generation systems in order to drive participation, else little economic benefit would result from consumers' changing over.

The cost of generating and supplying power to the commercial and domestic centre around Somosomo, analysed over a 20 year period and with an internal rate of return of 10%, was $0.24/kWh excluding any taxation benefits. For a rural electrification project such as this the high cost of delivered power to consumers spread over a wide area with relatively low total power demand is not unexpected. Therefore, to make the project commercially viable it is necessary to obtain subsidies and other forms of government or international aid assistance.

Under the rural electrification extension policy of the Fijian government, it was assumed that the ownership of the distribution network requiring capital of $0.6 million would be subsidized by the Fijian government. For political reasons, in order to meet a domestic retail power tariff similar to that currently charged on the main island, a contribution of around $1 million to the total capital investment is required to give the desired 11c/kWh tariff for both domestic and commercial customers. This contribution could come from one or more of the following:

- taxation relief

- accelerated depreciation of the plant

- overseas aid such as World Bank development grants

- additional government support by way of a direct grant.

A project management group would need to be established and a project manager appointed to oversee the project. This role will be critical for a successful outcome to be achieved. The qualities required are experience in project management, a thorough knowledge of the renewable energy technologies, and a good understanding of the social and cultural issues of Taveuni residents. The project manager will have to supervise a small team of engineers and contract out the design and construction work to successful tenderers. He or she will also liaise with Government officials as appropriate, who will be formally represented on the management group. Once the project is completed, it will be imperative for the project manager to continue in the role for at least a further 12 months to ensure that the operation and maintenance of the plant are managed appropriately and that the Taveuni customers receive maximum benefit from the service provided.

A separate company would need to be established to manage and maintain the generating and distribution plant and to administer the sales and collect the revenue. A general manager will be appointed, to be based on Taveuni, and will manage a small staff of one administrator and three technical operators, one of whom will be trained to be the gasifier specialist. The distribution lines will be maintained under contract and the fuelwood purchased from local growers.

Chapter 11

The future potential for bioenergy and the barriers

Global trends are moving, albeit too slowly, towards sustainable production, waste minimization, reduced vehicle pollution, distributed electricity generation, conservation of native forests and greenhouse gas emission reductions. In addition, linked to sustainability are development and equity as sought by the people living in developing countries, 2 billion of whom have no access to electricity. Biomass has a role to play in each one of these environmental and social drivers.

New energy conversion technologies are being developed rapidly to help solve some of the world's problems relating to energy and the environment. Biomass is a suitable fuel to use in many instances, so several of these technologies have been discussed in earlier chapters. The key challenge is to encourage their uptake in countries such as China, India and South East Asia, which, owing to their rapid economic development, are consuming fossil fuels inefficiently at ever-increasing rates, resulting in greater local and global environmental problems. The western world created these problems in the first instance and is now seeking solutions, partly through developing new technologies. The aim should also be to ensure that these 'leapfrog technologies' are also made available to developing countries at prices they can afford.

Mobile cell phones are a good analogy. Land telephone lines were simply not constructed in many rural areas of Africa, Asia and South America as the cost of developing and maintaining the infrastructure was too high. However, go to Ghana, Uganda, Vietnam, Paraguay or Mongolia nowadays and mobile phones are commonplace. The need for the old-fashioned infrastructure has been overtaken by the new technology. So it is for electricity. Why build lines and poles when distributed power generation systems are just around the corner?

Biomass is already widely distributed, and so has good technical potential as an energy source for these systems. It will need to be better utilized than at present and managed in a sustainable manner in order to provide energy for people living in rural communities to have a more healthy and enjoyable quality of life. This will take time, effort, investment and political will to achieve. However, as is clear from the *IPCC Third Assessment Report* (2001), since sustainability, equity and development are all closely linked, biomass has a major part to play in determining the future of the planet and its people.

Greenhouse gas balances for bioenergy systems

The climate change problem is well understood:

- Society is highly dependent on energy.

- Fossil fuels are the cheapest sources of energy.

- Burning fossil fuels releases large volumes of greenhouse gases.

- Climate change will inevitably result, but to what degree is not fully predictable.

The solution is harder to define! We know the following:

- There is a finite amount of fossil fuel in the ground but reserves are not yet limiting. Other non-traditional resources such as oil shales and hydrates are yet to be fully developed, and scarcity of fossil fuels is not an issue.

- The available global fossil fuel energy reserves, particularly from coal and gas, will last for at least 600 years at current rates of extraction.

- Atmospheric concentrations for all greenhouse gases (GHG) continue to increase, and scientific evidence shows that this is due to anthropogenic activities, especially carbon dioxide emissions, which have risen from 280 ppm to 367 ppm since 1860, mainly as a result of fossil fuel use.

- Using all known fossil fuel reserves (excluding other resources) will release over five times the amount of carbon dioxide already emitted from all fossil fuels consumed since 1860, and will increase atmospheric CO_2 to unprecedented levels.

- Natural ecological systems are already adapting to climate change, and some are under threat. Human health will be affected and habitats threatened from sea level rise and possibly a greater frequency of extreme weather events.

- The Earth's temperature has increased over the last 100 years, and is continuing to rise. Ten of the hottest years have all occurred since 1980, and, based on tree ring records, the last few decades have been the warmest during at least the last 600 years.

- Insurance companies are becoming more concerned about the increasing frequency of extreme weather events and the cost of claims involved. Some have already withdrawn cover from islands in the South Pacific and Caribbean under particular threat from further cyclone damage.

- Several multinational companies, notably Shell, BP, Du Pont, General Motors, Ford and Siemens, have begun to reposition their business interests in preparation for a decarbonized world. Changes from their 'business as usual' approach include:

  making investments in renewable energy technologies and fuel cells

  instigating carbon trading within their internal business operations resigning from the US-based Global Climate Coalition, which strongly opposes climate change policies.

We do not know the following:

- What stabilization levels of GHG are needed to provide a stable climate and therefore an acceptable world to live in with minimal adaptation of ecosystems and food supplies? For carbon dioxide is it 350 ppm CO_2 or 750 ppm, or 1000 ppm?

- What mix of mitigation measures is needed to reach these stable levels and at what cost of investment in order to avoid the serious, and probably even more costly, alternative consequences of climate change?

- What incentives and regulations will be required to implement these measures sufficiently in order to meet the desired GHG reduction targets?

- How will agricultural emissions be measured and reduced over the next few decades, bearing in mind that one methane molecule from rice paddies or ruminant enteric fermentation has 21 times the global warming potential of a carbon dioxide molecule over a 100-year period, and that one molecule of nitrous oxide released from animal wastes and the use of nitrogenous fertilizers is 310 times as potent as CO_2?

- What are the true environmental costs of burning coal, oil and gas?

- What will be the future value of carbon once international trading begins in earnest, a 2000 World Bank estimate being around $30/tC ($8/tCO_2$)?

- What will be the price of future gas and coal supplies once carbon emission externality costs are added in, and hence what will be the increased cost of thermal electricity generation?

- What is the cost and long-term reliability of storing carbon underground or in the oceans? Will carbon forest sinks be accepted internationally as a legitimate and fair means of carbon sequestration? If so, how will they be monitored and measured?

- Will the Kyoto Protocol targets for the six major greenhouse gases be met by the ratifying countries, and when will the much higher targets needed for atmospheric stabilization of greenhouse gases take effect?

- Will developing countries agree to take a stake in the global climate change solutions by accepting the Kyoto flexible mechanisms, or will they continue to consider that it is not their problem and that such actions will block their economic development?

Regardless of the outcomes of the ongoing political deliberations, the science clearly shows that the problem will continue to increase under 'business as usual' scenarios. It should also be remembered that the reduction targets set at Kyoto in 1997, challenging to meet as they appear to be, were just a start. It is also of interest to note that the lifestyle of each American citizen currently results in approximately 7 t of carbon emissions being released each year, an Australian citizen 5 tC/y, a Dutch citizen 3 tC/y, and a Chinese or Indian citizen 0.6 tC/yr. Some climate change scientists claim that, to keep the climate in balance, average worldwide per capita emissions of only 0.5 tC/yr is the maximum allowable, which will indeed be a challenge based on our present-day lifestyles.

Strategies for carbon dioxide mitigation include reducing energy use, improving energy efficiency, using renewable energy sources, introducing lower carbon intensity fuels and nuclear power, as well as carbon capture, treatment and sequestration. Application of these options will impact on the potential role of bioenergy.

Reducing energy demand by seeking to change the behaviour of people would have significant societal impacts and therefore will only result from major government and community policy changes. Significant energy efficiency gains have been made, particularly during the mid 1970s to 1980s when the major driver was the oil price shocks leading to high energy prices and shortages of supply. Energy efficiency will have some impact on GHG emissions, but such measures by themselves will be insufficient to meet the necessary mitigation targets in order to stabilize atmospheric GHG at acceptable levels. So bioenergy and other forms of renewable energy will certainly have increasing roles to play.

Carbon sequestration, both physical and biological, is another option being closely evaluated as a method of reducing atmospheric increases. Carbon dioxide can be captured after combustion of fossil fuels, during transformation of the fuel, or directly from the atmosphere by biological sinks. It can be stored, possibly in perpetuity, in disused oil and gas wells and coal mines, in deep saline reservoirs, in oceans, or as solids such as mineral carbonates. These processes have been explored mainly for use in association with thermal power generation plants, but they are also relevant to other carbon dioxide producing industries such as cement and steel manufacturing.

Linking sequestration with bioenergy plants has additional benefits. As discussed in Chapter 1, biomass can be both a carbon sink, if the biomass is grown as a store of carbon, and also a carbon offset, if it is then used for bioenergy purposes to displace fossil fuels. In addition the carbon dioxide released at a bioenergy plant during combustion could also be captured and physically stored, as for a coal plant, so that the biomass then also becomes a cheap means of atmospheric CO_2 removal. In this way the next generation of biomass plantations grown to replace those harvested would not simply reuse the recycled carbon dioxide but would extract more from the atmosphere and hence lower the level.

Life cycle analysis techniques

When considering the reduction of GHG emissions arising from the use of bioenergy, it should be remembered that while biomass produces zero net carbon dioxide emissions when recycled between production and combustion, other parts of the life cycle do produce emissions. If the true effects of carbon emissions from energy use are to be evaluated accurately so that, for example, woody biomass can be compared with coal, then it is important that the emissions from all stages of the life cycle are assessed and included. For oil this 'well to wheel' analysis would need to include mining, processing, transportation and storage as well as the more widely understood emissions released during its combustion. For biomass it includes cultivating the land to grow the biomass, adding fertilizer, harvesting and transport, processing operations, constructing the conversion plant, and future decommissioning. Since biomass typically has a lower energy density (MJ/kg) compared with the more energy-concentrated fossil fuels, the delivery and conversion of this bulkier material require more machinery and larger structures per unit of energy produced, which in turn involves more embedded energy in their manufacture and construction. The elements to be analysed in detail in such a life cycle framework would need to include:

- whether the biomass was produced as the main product of a system such as SRF, or as a by-product such as wood processing residues, and what other by-products also need to be considered such that appropriate emissions and offsets can be allocated between them

- the factors and fluxes relating to biological carbon storage in forest sinks

- the balances between reforestation, afforestation and conservation of forests against utilization of resources for bioenergy

- whether bioenergy provides an irreversible mitigation effect by reducing CO_2 at its source, whereas afforestation and forest conservation are mitigation options subject to future management regimes[1]

- efficiencies of bioenergy conversion systems, which in many cases are low compared with fossil fuel systems, though technology improvements such as biomass integrated gasification combined cycle plants (BIGCC) have the potential to improve this

- leakage of carbon emissions, where biomass fuels simply provide a new energy source adding to the total energy consumption of 'business as usual' rather than displacing the use of fossil fuels to the extent anticipated

- the emission of other GHG emissions associated with both fossil fuels and bioenergy fuel chains, particularly methane and nitrous oxides.

The volumes of GHG emissions arising from the collection, transport and processing of biomass are dependent on a range of complex factors for specific projects, including:

- the system used to collect, transport and process the biomass feedstock materials including equipment used and types of biomass collected

- fuel consumption for the trucks and machinery used for each type of operation

- haul distances for supplying the biomass fuel to the conversion plant gate

- whether any electricity consumed comes from thermal, nuclear or hydro sources.

Life cycle analysis is a complex process, but there are some basic guidelines that have been developed through the International Energy Agency (IEA) and elsewhere. Where electricity is used, such as for processing chips or running conveyors, emissions are calculated based on the proportion of electricity generated from various sources (hydro, gas, coal, geothermal, biomass, etc.) and the emissions arising from each of these sources. The emissions can then be converted to CO_2 equivalence factors. Alternatively, emissions could be based on the marginal rate for the next new power station likely to be built, though this is usually less realistic. In theory the indirect energy and related emissions embodied in the machinery, buildings, equipment and fertilizers used should also be taken into account, but this becomes complex, and it is difficult to know where to stop the analysis.

Transporting biomass gives rise to a major component of the total GHG emissions. For example, an IEA assessment showed that transporting 2.9 million m^3 of forest arisings over a maximum haul distance of 120 km resulted in 26,400 tC being emitted, whereas the collection and chipping of the material produced only 8000 tC. To be thorough, the analysis should also include an allocation of GHG arising from the establishment and management of the forests prior to harvesting.

During the conversion of bioenergy several GHGs may be emitted depending on the fuel source, the type of technology and the plant efficiency. If conversion plants are operating at high efficiencies and achieving complete conversion of fuel to energy, then the emissions will be low or negligible. Combustion of coal gives similar values to biomass for methane and nitrous oxide emissions, but carbon dioxide emissions are far greater at approximately 220 gC/kWh generated compared with 0–10 gC/kWh for biomass.

Values from a life cycle analysis will vary depending on the technologies used and location, and also allowing for differences in extraction, processing, regional distribution and generation. Where biomass is substituted for coal or natural gas for electricity generation, carbon emissions can be reduced significantly (Table 11.1). More efficient biomass conversion systems can further improve emission reductions.

Not only does biomass have to compete with the other renewable energy

Table 11.1 Typical life cycle emissions for a range of conversion technologies for electricity generation (g/kWh)

Technology	C	SO_2	NO_x
Woody biomass gasification	5–10	0.05–0.10	0.5–0.6
Coal: pulverized IGCC	190–220	11.00–12.00	4.0–4.5
Natural gas: CCGT	90–120	0	0.5–0.6
Onshore wind farms	10–15	0.05–0.10	0.01–0.03
Decentralized PV	150–170	1.6–1.9	0.5–0.6

Table 11.2 A comparison of the power generating costs, carbon emissions and reduction costs for a range of technologies from the global perspective

Power plant	Energy source	Generating cost (c/kWh)	Emissions (gC/kWh)	Reduction costs ($/tC avoided)
Coal, pulverized; steam turbine	Coal	4.9[a]	229	Base case = 0
Integrated gasification combined cycle	Coal	3.6–6.0	190–198	−10 to 40
IGCC plus CO_2 capture and sequestration	Coal	5.4–7.8	20–28	50 to 150
Combined-cycle, gas turbine	Natural gas	4.9–6.9	103–122	0 to 156
CCGT plus CO_2 capture and sequestration	Natural gas	3.7–8.7	14–18	−23 to 185
Nuclear	Uranium	3.9–8.0	0	−38 to 135
Photovoltaic and solar thermal	Solar	8.7–40.0	0	175 to 1400
Hydro-turbine	Water	4.2–7.8	0	−31 to 127
Wind turbines	Wind	3.0–8.0	0	−82 to 135
Biomass integrated gasification CC	Dry biomass	2.8–7.6	0	−92 to 117[b]

[a]Based on average costs for projects in developed countries. The wide ranges acknowledge that each project is site specific, so that there are high degrees of uncertainty

[b]Biomass fuels as delivered ranged from $0/GJ for waste product on site requiring disposal costs to $4/GJ for purpose-grown energy crops

sources to gain a share of the electricity and heat markets, on a longer-term basis it must also compete with 'clean' technologies for fossil fuels in terms of $ investment per tonne of carbon emissions avoided. The *IPCC Third Assessment Report* compared a range of electricity generating technologies on a global basis, as summarized in Table 11.2. Details of the many complex assumptions used for this global assessment are not discussed here, but the point made is simple. It may under certain circumstances be cheaper from a national perspective, in terms of $/tC avoided, to replace say a coal-fired boiler with a bioenergy plant with virtually zero carbon emissions than to replace it with a cheaper combined-cycle gas-fired plant having some carbon emissions.

Social and other environmental issues

Developing a bioenergy project is not without its challenges in order to appease all the stakeholders involved:

- Environmental groups are sensitive to bioenergy projects, and will only accept use of 'sustainably produced' biomass.

- Biomass producers want good returns per hectare for growing the resource (which therefore has to compete with other land uses) or for collecting and storing it if produced as a waste (though as soon as a by-product material is seen to have a value it is no longer seen as a 'waste'!).

- Bioenergy plant developers want the security of long-term fuel supply contracts in place before proceeding.

- Power plant operators want quality fuel delivered all year round to an agreed prescribed set of standards and characteristics.

- Equipment manufacturers want to design their products to have improved thermal efficiencies, better controls and reliable feedstock handling to gain better returns and a fair share of the market. This often necessitates having a consistent biomass fuel supply on which to design the equipment.

- Financiers want to reduce the risks of investment by having heat or power purchase agreements in place, along with fuel contracts, and perhaps green pricing options.

- Competing markets want the biomass resource for mulch, pulp, fibre board, transport fuels, chemical feedstocks, etc.

- Communities, particularly in rural areas where most biomass plants will be located, want secure and long-term employment, independence, and some control of local resources.

Past assessments of potential energy projects have been based mainly on the economic return on investment. The 'triple bottom line' approach now being taken by many energy companies gives greater weighting to social and environmental issues. From the social perspective there can be little doubt that bioenergy projects protect existing employment, provide new jobs, give learning opportunities to transfer skills, introduce new skills, and provide training and educational opportunities. The trend towards independent power production using smaller-scale plants and embedded generation should result in a decline in urban drift once rural communities are able to develop and grow using the new sources of bioenergy available to them. This in turn will produce a sense of pride and independence, of particular importance to many indigenous or aboriginal communities who are struggling to maintain their cultural identities.

Local health benefits can also result, whether as a result of better wood stove designs for people living in rural areas of developing countries, or of reduced pollution from traffic emissions for those living in the centre of London, Tokyo or New York.

Bioenergy provides the opportunity for many people to have an improved quality of life. In developing countries this could result from having lights to read by at night or to be able to do homework. In wealthier countries it could be in terms of better levels of home heating or cooling. For individuals, communities and businesses who value achieving sustainability, the use of biomass as an energy source could help to meet their objectives of becoming self-sufficient and being recognized as environmentally aware and responsible. The IEA has an activity on *Socio-economic aspects of bioenergy systems* within its Bioenergy Agreement and more details can be found at www.eihp.hr/task29.htm.

Environmental benefits from biomass (other than reducing GHG emissions at the global level as cheaply as possible in terms of $/tC avoided) include reducing local emissions, using limited resources better, and improving and protecting the habitat and landscape (which growing short-rotation forests can do well if they are planned and designed carefully). Reducing waste disposal into landfills and waterways, avoiding the noise, maintenance and inconvenience of diesel generating sets, and minimizing the need for ugly power lines over the countryside are other benefits that bioenergy can provide.

There is a growing trend towards communities taking responsibility for their local environment. Private ownership of energy plants such as biogas plants in Denmark or bioenergy cogeneration plants in Sweden are good examples. The aim in Europe is for 100 communities (such as Vaxjo, Sweden, as described in Case Study 14) to become fossil fuel free in an endeavour to reduce environmental impacts and gain socio-economic benefits. This growing trend will be encouraged both by the development of new technologies such as fuel cells and micro-turbines, and by the moves away from traditional, government-owned, large-scale electricity and heat generating plants.

In standard economic feasibility studies a commercial price is usually assigned for the purchase of biomass fuels, which represents a significant cost factor of a project. However, small rural communities or households may be able to supply their own biomass feedstocks at much lower costs, especially in remote areas. Small-scale, biomass-fired electricity generation projects at the community or household level may therefore be feasible, especially when grid-connected power is not available or, if it is, embedded generation would boost the power quality in terms of outages and fluctuating voltages at the end of the distribution lines.

Distributed generation systems

These are smaller-scale generators that are connected to electricity distribution networks, and hence are termed 'distributed' in contrast to large-scale generators that are connected directly to the very high-voltage electricity transmission networks (Figure 11.1). They can also be referred to as decentralized, embedded or localized generation since the generation plant is normally located close to where most of the electricity is consumed, for example supplying electricity on-site, or to a neighbouring industry. Cogeneration and other types of small-scale generation such as fuel cells and photovoltaic solar cells are included in the definition as they are usually located near to the end user. Systems can be as small as a 3 kW$_e$ micro-cogeneration

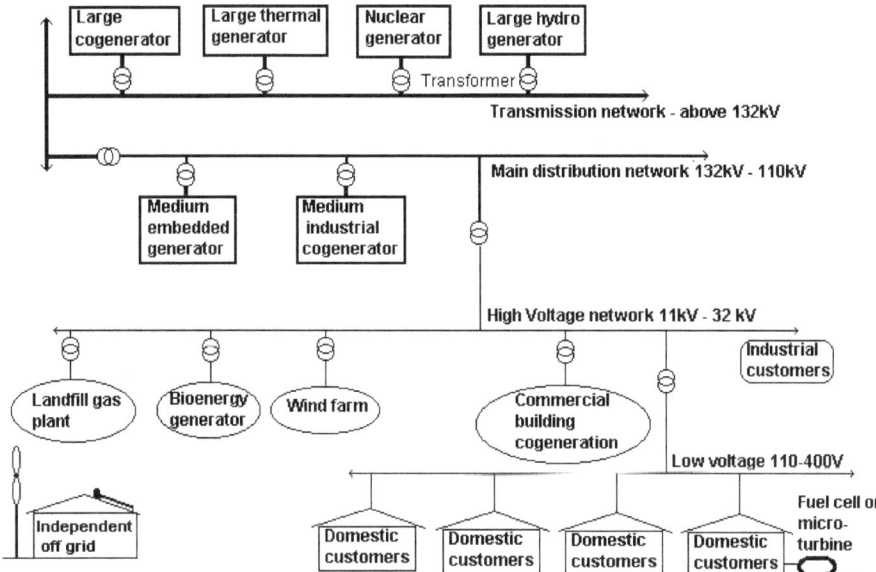

Figure 11.1 Large power stations are usually connected to the electricity transmission network whereas medium ones can be embedded in the local distribution network, as are small systems known as distributed generation

plant or fuel cell, or as large as a 450 MW$_e$ industrial system with the power generated used mainly on-site. Distributed generation tends to be environmentally friendly, flexible and efficient, and can be more cost-effective than traditional, centralized generation, especially in cases where network costs and losses are taken into account and/or where power lines will soon need upgrading to carry increased loads.

A wide range of small-scale bioenergy conversion systems are suitable for distributed generation at the community or household scale, as are other forms of renewable energy (Figure 11.2):

- Direct combustion for power production at this small scale is not always economically feasible since a small bioenergy plant can cost twice as much as a similar capacity wind farm so would need negative fuel costs to compete in regions where wind is an option.

- Cogeneration is more promising but even then the lower limit is around 5 MW$_{th}$ owing to poor electrical efficiencies and diseconomies of scale.

- Internal combustion engines are well proven whereas steam turbines have lower efficiencies at this scale with relatively high investment costs.

- Steam engines are proven in industrial operations, are reliable and have low maintenance costs but, because they are mainly manufactured only in small numbers, also have high investment costs.

- Stirling engines have many advantages at the small scale, such as good efficiencies, low noise levels, low maintenance and long lifetimes. They can be coupled to combustion and gasification systems or any other heat source, and their price should drop within a few years as they become more common.

- Indirect-fired gas turbines are mature, with system efficiencies of 20–24%, but when biomass-fired require a specifically designed heat exchange system, which still needs developing.

- Direct-fired, pressurized gas turbines coupled to gasification systems have potential for biomass fuels mainly in the 5–20 MW$_e$ range depending on the development of a small and efficient gas clean-up system.

- Micro-turbines between 25 and 250 kW$_e$ are compact, light, have low noise levels, can be fuelled by producer gas or biogas, and their currently high investment costs are expected to drop considerably.

Fuel cell designs under development also have the advantage of high system efficiencies, low noise levels, low emissions and good reliability, but their investment and operating costs will be crucial for market penetration, being too high for most commercial applications. It is expected that the price will drop considerably owing to developments occurring in the automotive industry. Until then, gasification and gas engine systems appear to be the best bioenergy option for distributed generation. Pyrolysis oil production coupled to a diesel

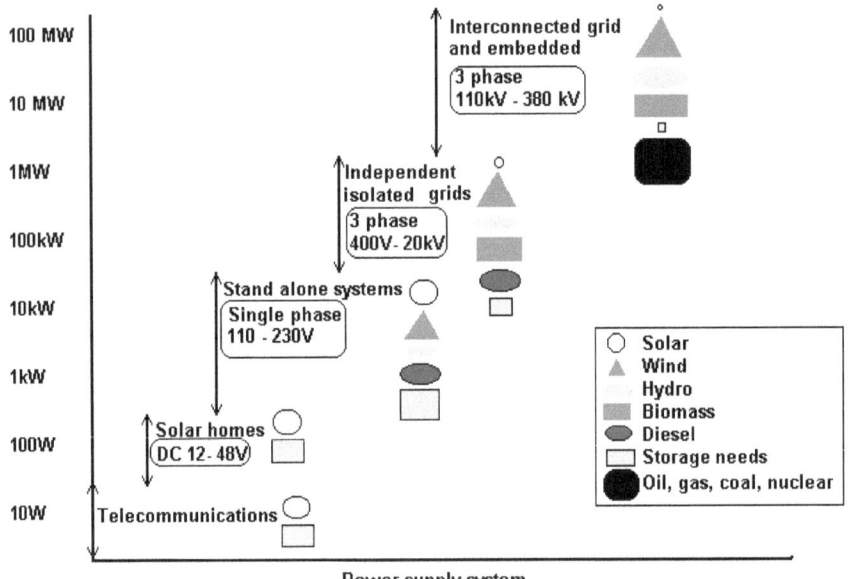

Power supply system

Figure 11.2   Different renewable energy technologies suit various scales of generation, with biomass around 10 kWe to 1 MW$_e$ suiting independent isolated grids, and up to 100 MW$_e$ for grid-connected embedded generation

engine or a gas turbine also has small-scale potential, especially for peak power plants.

Automation of small-scale distributed biomass power plants is important to realize an acceptable return on investment, although this may be less important in countries where labour is cheap. The development of a standardized biomass fuel such as pellets for such systems would be a useful step towards automation, but the high fuel pre-treatment costs would make such a system non-viable.

Barriers to implementation

The role that biomass will play in future consumer energy supplies depends on the many commercial opportunities that exist overcoming barriers to progress and commercial investment. Six questions need to be asked by bioenergy project developers, stakeholders and investors:

- What biomass resources are available or can be sustainably produced?

- What technology developments available now or in the near future will enable environmentally acceptable bioenergy products and services to be generated?

- What impacts will the increasing use of biomass have on the environment and on social issues such as employment, equity and development?

- What markets for bioenergy exist now or will be established in the future?

- What level of investment will be needed to establish the bioenergy industry?

- What is the level of risk from investing in such business?

A significant barrier is the popular concern that using some forms of biomass is non-sustainable. In some instances this view is correct. There are simply some sources of biomass that for a variety of reasons should never be used for energy purposes. The industry needs to clarify this issue immediately, since public concerns about the environmental impacts from using biomass as an energy source lead to a number of frequently asked questions:

- Will an increasing number of wood-fired heat and power plants lead to an incentive for investors and shareholders to support the cutting down of existing forests?

- Will stack emissions from municipal solid waste-to-energy plants, and also possibly from wood-fired biomass plants, contain toxic substances such as dioxins?

- Will planting large areas with fast-growing trees as energy forests reduce both water runoff and percolation into the groundwater, thereby affecting downstream users?

- Will soil nutrient levels be depleted by continually removing large quantities of biomass material such as crop residues from the land to supply nearby conversion plants?

- Will biodiversity be further threatened and agrichemical use increased if ever greater areas of monoculture crops are grown?

- Will genetically engineered trees and crops be developed specifically for use for biomass energy supplies?

- Will transport of large quantities of biomass to the power plants result in increased traffic congestion, noise, dust and road damage?

- Will the use of land for energy cropping reduce the area of land now used for food and fibre production so that scarcities will result?

- Will using waste for energy purposes reduce the desirable incentives to minimize and recycle waste materials if it is cheaper to burn it?

- Is biomass truly sustainable as well as renewable?

Many of these questions have been addressed in the preceding chapters. However, also to be considered in the debate must be any beneficial environmental impacts such as avoiding methane emissions from landfills, reducing odours from direct application of animal wastes to land rather than via a biogas plant, improving sewage treatment prior to discharge to waterways or oceans, obtaining dry salinity soil improvements, reducing the GHG emissions from fossil fuels, and planting of energy forests that encourage bird life and biodiversity.

Concerns by the public and environmental groups will continue to arise whenever a new bioenergy project is first mooted by the developer, often leading to prolonged debate over planning consents and at times leading to a ruling by the environmental courts. The development of good practice guidelines by a consensus process involving all potential stakeholders is the recommended approach that should be taken to avoid repeated conflicts.

The biomass industry association in the UK, British Biogen, in association with ETSU and the Department of Trade and Industry, recently developed a series of such guidelines following a lengthy but essential consultation process in each case with all stakeholders. Each guideline involved bringing together all the key organizations including municipal authorities, planning consultants, environmental groups, landowner associations, developers, researchers, equipment manufacturers, transport companies and local and regional governments. Through a series of meetings and circulation of draught documents, all the issues involved were thoroughly debated and, owing to good independent facilitation, eventually led to consensus solutions being reached.

The final version of each guideline was published, with each organization involved in the process showing 'ownership' of the document, and hence endorsing its contents, by having its logo clearly displayed. By this means all the relevant information is now available in one place to guide the local consenting municipality so that it should be easier to obtain resource and planning consents for future bioenergy projects, and the developers should be spared having to go through a full and costly consultation process each time.

Economic barriers

Many bioenergy projects are technically feasible but investments do not proceed because other forms of energy are more cost-competitive. High cost is the most significant barrier to achieving an increased uptake of biomass. Investors perceive the risks of bioenergy projects to be significant, and would rather invest in other renewable or conventional fossil fuel energy projects. Effective risk management could reduce these perceived barriers and thereby lead to an increase in the number of willing investors in bioenergy projects, which would result in more competitive financing opportunities.

Investors tend to seek a short payback period of 2–4 years, which favours conversion plants with low capital cost, albeit usually with a high fuel cost. Bioenergy plants usually have a higher capital cost compared with gas or coal plants, being around $1000/kW versus $700/kW. Increased depreciation rates would reduce this high capital cost barrier and help to encourage potential investors to favour bioenergy plants. For example, for tax depreciation assessments, boilers have a useful theoretical life of say 25 years, yet during that period it is quite likely the fuel supplies will change and obsolescence will therefore occur. A shorter life results for depreciation of the plant.

For small industrial investors in bioenergy, such as sawmillers or dairy farmers, the installation of energy facilities on the site may be the largest single investment that they will make, and the ability to raise the capital funds needed for the investment may be beyond their means. Such investors often have basic engineering skills so tend to purchase, modify and install second-hand plants that are often inefficient, require higher labour input, and involve more maintenance. Investment in new, efficient, low labour intensity plants would be a better economic option in the long term if capital was available. However, a lack of capital to finance projects results from poor understanding by bank managers of the project risks, leading to poor 'bankability'. Relatively high transaction costs are also commonly experienced for development of smaller-scale plant since it costs a similar amount to set up finance for a $2 million project of say 1 MW as it does for a 100 MW project costing $100 million.

Economic risks of biomass in the energy market are high owing to competitive costs from fossil fuels, hydro and other renewables such as geothermal and wind. Delivered biomass from forest residues for example can cost two to three times the $/GJ for coal, yet the savings on future environmental costs are not included. Hence bankers and financiers should be invited to become more involved during the project development process in order to fully understand the issues involved. The problems of low confidence in financial projections and possible insufficient debt service coverage could then be explored. Specialist lenders in renewable energy projects are now appearing, as are specialist insurance companies willing to offer cover for the many risks of a renewable energy project and its development.

Other economic barriers include deregulation of the power industry, which has made it more difficult for renewable energy projects to enter the market owing to a lower wholesale electricity price. Conversely, deregulation of the market means that independent power producers are now able to benefit from a bioenergy project where it is competitive.

Technical barriers

High costs always occur in the early stages of development of new technologies owing to technology risk, which can be partly mitigated by good design and by equipment manufacturers' warranties. Construction risk can be overcome by having a fixed contract price, and by buying insurance cover against delays and liability.

Lack of certainty of fuel supply at an acceptable cost is another significant barrier to achieving an increased uptake of bioenergy. Fuel supply risks from competing markets for the biomass (such as pulp chips, fibre boards, bark mulch and organic fertilizers) can be overcome by appropriate contracts and forward sales agreements. These can also cover variations in biomass quality as delivered to the conversion plant and long-term supply requirements to further reduce the risk. However, many biomass fuel suppliers are farmers, who are not used to long-term contracts. A fuel supply merchant could be contracted to transfer the risk and also to give incentives to the suppliers and growers to provide consistent fuel. Eventualities such as the death or bankruptcy of a supplier also need to be covered since access to the fuel by the project managers will still be required.

Biomass fuels, being of biological origin, are often bulky, have a high moisture content, and are usually of variable and unpredictable quality. Fuel standardization is needed, and techniques for fuel upgrading – for example by natural drying, pelletizing or briquetting – are advancing. For biodiesel, an international standard is in place. Fuel consistency may be achieved more easily with the production of dedicated energy crops than by using a variety of wood process residues or waste cooking oils, which vary daily. A bioenergy conversion plant is usually designed to suit a specific fuel, but it is likely that the mix of biomass available and its characteristics will change over time. This affects the design of the conversion plant and can shorten its economic life owing to obsolesence, as discussed above.

The physical handling of biomass fuels is a risk as it can be challenging to equipment designers and has even led to the failure of demonstration projects, particularly for bagasse. Fuel handling and processing equipment is often a difficult component to maintain and operate adequately. Overcoming potential problems by improved design is feasible, but then the higher cost for the better-quality plant becomes a barrier.

The lack of available information, and uncertainties over what are often deemed to be 'new technologies' by those making business decisions, or by those living in the proximity of a proposed conversion plant, create a barrier. The time available for company managers to learn more about possible bioenergy options is limited, and commitment is often lacking since energy inputs into many businesses are a small percentage of the total costs. Selection of plant for a specific project is therefore largely based on the effectiveness of the plant manufacturers' marketing.

Environmental barriers

Biomass use can be closely associated with health problems arising from open fires or from poorly designed bioenergy plants, which produce relatively high levels of particulate emissions. These can be overcome by proper installation of clean-burning combustors that meet modern air emission standards.

There is a lack of information available to potential bioenergy plant investors, and many rely on their own knowledge, which is often derived from magazines, from outdated publications or by word of mouth. Investors rarely seek and pay for quality advice. In addition relatively few senior business managers possess good information about their own processing plant, its energy requirements and the emissions. As a result many unsuitable bioenergy plants have been sold by unscrupulous or poorly informed equipment suppliers, so there is a need to research and publish information that will assist potential investors in bioenergy projects to make appropriate equipment selection.

For energy crops, monocultural production is deemed unacceptable by many environmental agencies, and there could be public rejection of growing SRF owing to changing landscape values and lack of biodiversity. Before planning consents for a bioenergy project can be obtained, an environmental impact assessment is often needed in which these issues can be addressed. It is sometimes worth considering planting a mix of species, not only for landscape benefits but also for added resistance to the spread of pests and diseases. In addition, continuous large-scale production of forest plantations and energy crops could reduce soil fertility levels, affect downstream water use, and lead to leaching of nutrients and increased use of agrichemicals.

The energy balance of bioenergy is not always considered to be favourable, though this is more the case for biofuels produced from annual energy crops than for woody biomass from perennial crops where the energy output is at least 10–20 times greater than the energy input. For ethanol fuels the promising development of enzymatic hydrolysis of ligno-cellulose has a more favourable energy balance than the use of sugar or cereal crops as feedstock. In addition the collection and transport of biomass often result in increased use of vehicles, higher exhaust air emissions and greater wear and tear on the roads infrastructure.

Social barriers

Land requirements for future energy crop and forest plantations will compete with land used for the traditional production of food and fibre products. Land use will ultimately depend on biomass crop yields as achievable on a sustainable basis, water availability, and the degree of conversion efficiency to usable fuels or energy services. As an example, for a 40% efficiency conversion plant fed with a forest energy crop yielding 15 odt/ha.yr, then 180 ha of energy plantation would be needed per MW_e of installed capacity when running the plant for 6000 hours per year.

There may be a future shortage of skilled workers for harvesting and collecting biomass. So although employment opportunities from greater bioenergy uptake are often quoted, it may not be easy to find willing workers for what can be somewhat arduous and repetitive work, in either developed or developing countries.

Policy barriers

Regulatory and fiscal barriers often exist, including an absence of effective markets such as green pricing to stimulate the biomass industry. Continued subsidies for fossil fuel energy, including incentives offered for further exploration, continued government R&D investment, and the fact that no value is applied to the associated externalities, together make it more difficult for bioenergy to enter the market.

Perception barriers

Biomass can have a poor image, particularly when used in out-of-date appliances, since it is viewed as a 'fuel of the past' because of its historically low efficiency and high atmospheric emissions. Many low-performing conversion technologies continue to operate, which adds to the poor image.

Overcoming the barriers

The removal of barriers to implementation is a challenge for developers and policy makers who wish to see more bioenergy projects up and running. Faster uptake could be made by offering a number of incentives.

Economic instruments

Carbon taxes imposed on a society would increase the cost of fossil fuels and therefore make biomass more competitive since, being carbon neutral, it would be exempt (as would other renewable energy systems as well as nuclear power). Climate change levies on electricity sales would also have the potential both to provide revenue and to create awareness if the revenue were used to encourage the use of biomass and other renewable projects. For example, in the UK a climate change levy was imposed on consumer electricity purchases by businesses from April 2001. Special discounts were made available to the large energy users. Renewable energy projects were exempted and this, together with the growing interest in selling 'green power', created demand for more bioenergy capacity. A $150 million fund from the levy was established to assist small and medium-sized businesses to cope better with the additional costs by providing advice on energy efficiency. Some of this revenue will also be used to undertake more research and promotion of renewable energy.

Carbon trading is also an incentive where renewable energy projects can be shown to displace fossil fuel use. Although trading has begun informally between organizations, and even within one or two large companies such as BP in order to gain experience, it is not yet known whether carbon emission trading will proceed internationally. If it does, some bioenergy projects will have additional value in terms of measurable carbon offsets.

Long-term feed-in tariffs are used in Germany to stimulate the renewable energy market, but elsewhere may need government-supported contracts to be attractive. Other grants and subsidies offered by governments could encourage further uptake of bioenergy projects, but these would need careful consideration as to the long-term reliance on them and how future removal would affect the industry.

Increased depreciation rates on plant and equipment for tax purposes would reduce the investment payback period and hence help to alleviate the capital investment and long payback period barriers that bioenergy plants currently face. Reduced excise taxes, especially if benefits can be shown to offset the loss in government revenue, may be applied to the use of fuels with a biofuel component, as for biodiesel in Germany and ethanol in France.

Non-economic instruments

Targets set by governments for new renewables can be successful in gaining increased uptake. Australia for example, with its 2% (9500 GWh/yr) new renewable energy obligation on electricity retailers by 2010, has made good progress, and most power companies soon identified commercial projects. It was then realized that bioenergy will be a major provider of this target, bagasse and forest residues being generally more cost-competitive than wind or solar under the Australian climatic conditions.

The non-fossil fuel obligation (NFFO) in the UK has been successful in encouraging investment in renewable energy projects and also in obtaining convergence in costs with fossil fuel projects. It has now been discontinued, and a feed-in tariff scheme is being implemented in its place. A limitation was that it did not encourage local development of plant and equipment, most being imported off the shelf in order to keep the project costs low. The UK experience was that significant investments of time and money were made in formulating project proposals, many of which did not receive support. Of those supported, a high proportion of projects did not proceed, either because planning consent constraints were not met or because the original cost assessments were too optimistic.

Green electricity markets have good potential. As well as selling the electricity privately or through the wholesale market, electricity generators and retailers are able to trade the renewable energy certificates issued after generation. Trading via an Internet-based market has begun in some countries. Advantages are that new project investors are able to estimate their green revenue in advance and so can cost this additional income into the project economic analysis, and then sell their rights into this market once the power is generated. The green certificate value can be capped by imposing a penalty that retailers will have to pay if they do not meet their green electricity targets as set by government.

Promotion of carbon sequestration has been achieved by the The International Carbon Sequestration Incentive Act of the US Senate, introduced on 27 July 2000, which rewards companies for voluntary environmental efforts on climate change mitigation. Eligible companies will receive an investment tax credit or access to low-interest loans and insurance options on carbon sequestration investments in other countries. Eligible projects must have a minimum length of 30 years, and receive funding at a rate of $5/tC avoided when verified as stored or sequestered. Up to 50% of the total project costs are met.

Education and access to information about the problem of GHG emissions may create greater awareness and encourage companies, communities and individuals to be prepared to act. This could involve behavioural change in the longer term and result in the immediate greater uptake of biomass and other renewables.

Case study 18

Western Power's integrated oil mallee eucalyptus project, Western Australia

Biofuels are not generally competitive with currently cheap fossil fuels, and therefore need some form of co-benefits to become a viable proposition. These could include:

- cogeneration of useful heat together with power

- reduced sulphur emissions from coal combustion by co-firing

- avoidance of the cost of waste disposal

- land treatment of municipal, agricultural or industrial effluents onto energy crops as sources of nutrients and irrigation

- improvement of degraded land such as mining spoil or saline soils

- a value on carbon sinks and offsets, giving an additional revenue stream.

The integrated oil mallee eucalyptus project is a good example of biomass used to produce a multi-benefit, multi-product business opportunity. In Australia, clearing the land of deep-rooted trees in low rainfall areas last century led to rising water tables, which later brought up natural deposits of salt from the subsoil to the surface, seriously affecting soil structure and plant growth. Not only has this become a problem for cropping and pasture lands but it also occurs in towns and cities such as Wagga Wagga in New South Wales, where trees were also cleared and where building foundations are now rapidly eroding as a result. The social effects are also serious, as many farmers are

Figure 11.3 Oil mallee short-rotation forest crops grown in strips across cereal land and ready for harvest after coppicing every 3–4 years

threatened with having to walk off their land within the next few years unless a solution can be found.

As a solution, cereal growers have recently planted oil mallee eucalyptus trees in strips across their degrading land, with the cereals planted in the alleys between (Figure 11.3). This attempts to drive down the water table through increasing the evapo-transpiration rates and thereby reducing the huge dryland salinity problem that is affecting millions of hectares across several states.

Many species of mallee exist, most of which coppice readily as a response to frequent forest fires. They grow well in dry land conditions, and can cope with fairly high salt levels. They are quick growing and have a high leaf oil content. The question then was what to do with the SRF mallee biomass. The farmer-owned Oil Mallee Company was formed in 1997 to commercialize the opportunities for the indigenous mallees for their oil content in the pharmaceutical and solvent markets and as feedstock for medium-density fibreboard. Several species were selected for planting after considerable research to identify which had the preferred oil chemical composition, in particular high levels of cineole. Those chosen included *Eucalyptus gratiae, E. horistes and E. kochii.*

Western Power Corporation, the state-owned electricity company, then became involved once the bioenergy value of the crop was also realized. They partnered with Enecon Pty Ltd, which holds the licence for a new technology developed by scientists at the CSIRO to manufacture carbon products and energy from woody biomass. The concept to develop a multi-product

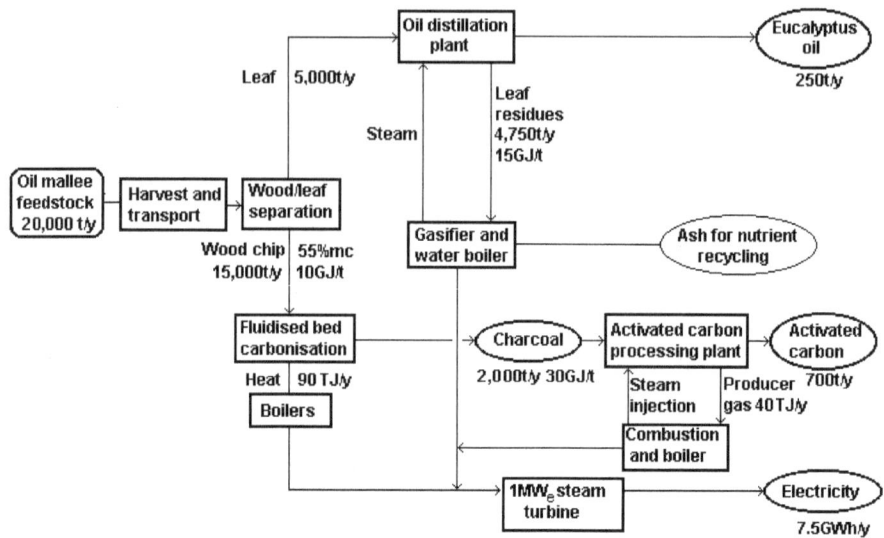

Figure 11.4   Integrated oil mallee process for multi-products of oil, activated carbon and electricity

processing plant was instigated (Figure 11.4). The original choice of mallee species mainly for their oil quality will now need to be reconsidered for future plantings since the wood yield and structure are also important, a greater proportion of the total revenue being likely to come from the activated carbon and green electricity sales than from the oil.

Following a detailed feasibility study including successful laboratory testing of oil mallee wood chips as a feedstock for activated carbon, a 1 MW$_e$ pilot plant was designed, and construction began in 2001. This will require feedstock from 800 to 1000 ha assuming a mean annual increment of 15–20 t green weight per hectare. A specialized prototype harvester was developed to cut the 2 million or so trees needed annually and chip them ready for transport to the plant as no suitable commercial machinery could be found. Delivering the chips for around $7–8/t remains the major challenge and possibly the weakest link in the production and processing chain. To extract the oil the biomass needs to be delivered to the plant green. The harvester therefore has to cut and chip one tree every 3–4 seconds and work 8 hours a day, 5 days a week, 50 weeks a year, at a forward speed of around 4–5 km/h. Maximizing truck payloads with either whole trees or with the wet chips depends on truck bin sizes available and local weight regulations. The options were to

- cut and chip in the field in one pass

- cut, transport whole trees to an intermediary station, chip, then transport to the plant

- cut, transport whole trees to the plant, and chip there.

When charcoal is produced traditionally, the volatiles are driven off and most of the energy content is therefore lost into the atmosphere as polluting gases. In this CSIRO system design the energy contained in the volatiles is utilized as recovered steam, and few energy losses occur. The decarbonization process using a fluidized bed is exothermic, so boiler tubes were incorporated into the bed to transfer some of the heat to the associated boiler to raise steam to drive the steam turbine. Also producer gas, mainly H_2 and CO formed during the carbon-activating process, is used in another boiler. Surplus steam is used to boost the steam pressure at the turbine, but other options for using this gas in the future include running gas engines or perhaps fuel cells if either prove to be more efficient in generating electricity from the gas than through the steam turbine route.

A full-scale 5–6 MW$_e$ plant is planned should the pilot plant prove successful, but greater transport distances would be involved and the delivered biomass costs in \$/GJ are therefore likely to be greater. This would need around 100,000 t of biomass per year harvested from mallee strips that cover perhaps 15% of the total land area to treat salinity. The plant would therefore be serviced by mallee grown in a total area of perhaps 1000 km^2 with an average transport distance of around 15–20 km to a central plant. Such a plant would produce approximately 1000 t of eucalyptus oil, which is valued at around \$1000/t, 30–40GWh/yr of electricity, which will be sold for around \$40/MWh as green electricity, and 10,000 t of charcoal valued around \$200/t for use in metallurgical processes, etc. To maximize the return on investment, value can be added to this charcoal by injecting it with steam at high temperatures to produce **activated char coal**, a powerful adsorbent used to clean up water, air and other gases in food industry and mining applications and valued at between \$600 and \$2000/t. Annual world demand is around 700,000 t and growing, but good-quality product is needed and this varies with the biomass feedstock and the process used.

It is clear that growing mallee for energy purposes alone would not be profitable. The value for the co-benefit of reducing soil salinity levels to maintain the productivity of the land has not been included in cost–benefit analyses but would be significant. What makes the investment worthwhile to the project developer and partners is the revenue streams for the charcoal or activated carbon and also for the eucalyptus oil, as well as for the electricity generated. Growing biomass to obtain several products and benefits, one of which is energy with its relatively low value, provides the key for identifying economic projects in the future.

Conclusions

Biomass continues to provide a significant amount of global consumer energy. Modern biomass is developing rapidly. Many new and improved bioenergy technologies are reaching the market and, in some cases, are successfully competing with fossil fuels even without government incentives. There are also many opportunities to improve traditional biomass practices to ensure that it is used sustainably, with low or zero emissions to the environment, and also to improve public health.

The use of biomass for conversion to energy is constrained principally by cost. Fossil fuels remain the fuel of choice, because useful energy can commonly be produced more cheaply from a fossil fuel conversion facility than from one fuelled by biomass. Biomass-fuelled energy facilities are likely to be most popular in the 5–100 MW$_{th}$ range, though many heat projects will be below 5 MW$_{th}$. The global trend towards distributed electricity generation will provide further opportunities for bioenergy at the smaller scale, providing future employment in rural areas in both developed and developing countries. The use of biofuels for transport will continue to grow slowly, but only with government support as petroleum fuels will remain comparatively cheaper for some time. The opportunity for biofuels to provide hydrogen for fuel cells is a particularly exciting option.

Over the next decade, as the carbon dioxide mitigation benefits of biomass become better understood by investors and carbon emissions trading begins, there is likely to be a significant increase in the total installed capacity of biomass-fuelled plants, including cogeneration facilities for heating and cooling. Increases in the cost of disposal of wastes and residues, driven by growing environmental concerns, will drive the need to find alternative options. Waste-to-energy projects not only avoid the cost of disposal but will provide useful and valuable outputs in the form of heat and electricity.

Many bioenergy facilities operate with uncertain future fuel supplies because low-value biomass fuels currently available could become potential feedstocks for newly developing markets. Wastes and residues available for conversion to energy are of mixed quality, which causes problems for plant designers and operators since fuels outside the design specifications may have to be utilized at times. Improving fuel quality and security of supply by long-term contracts with fuel suppliers will reduce these risks. Potential investors in bioenergy facilities will be cautious about projects with a life cycle of 15–20 years where there is uncertain fuel availability at no fixed prices, and hence the project is exposed to market risks. By comparison fossil fuel plants often have a shorter payback period.

Purpose-grown energy crops are unlikely to become economic within the next decade unless they can show multi-products and benefits, are used as a hedge against fuel supply risks, or where a policy is introduced that dramatically improves their economics.

Many potential investors in bioenergy projects do not have a good understanding of the technical, social and environmental issues they should address. Ready access to information on plant design and technology, fuel supply, fuel quality issues and conversion plant economics will assist their understanding. This will enable their objective to be met of installing good-quality plant of appropriate design at the right price in order to gain a good return on investment from meeting energy demand with biomass fuels.

Visionary board members of companies considering the future of their energy supplies, together with those dealing in business investment opportunities, should put the following premise to their colleagues:

- if we think that there may be something in climate change concerns
- if we think there is a strengthening public opinion that supports such environmental issues together with moves towards waste minimization and
- if we think governments might actually be serious about meeting their international obligations to reduce GHG emissions, then,
- if we are a responsible company, we should be asking the following questions:
- can we afford to ignore climate change?
- can we capture any win/win, no regrets, energy/cost-saving opportunities?
- can we benefit from climate change opportunities by investing in bioenergy, and as a result gain future revenue from carbon sinks and carbon offsets?
- can we use the opportunity of using bioenergy to enhance our company's public image?
- can we benefit from new technology innovations relating to energy use that are rapidly coming on stream, such as micro-turbines, fuel cells and distributed generation?

For many commercial organizations, bioenergy has the potential to produce new business opportunities. If carefully implemented using sustainable biomass supplies to displace fossil fuels in meeting energy demands, then *biomass can be forever!*

Notes

[1] See the IPCC report *Land Use, Land Use Change and Forests* (2000) for a detailed discussion of this issue.

References and bibliography

General papers and Chapter 1

D'Apote S L (1998) IEA biomass energy analysis and projections. In: *Biomass Energy: Data, Analysis and Trends, Proceedings of OECD/IEA Conference*, 23–24 March. OECD/IEA, Paris, France.

EECA/CAE (1996) *New and Emerging Renewable Energy Opportunities*. Energy Efficiency and Conservation Authority, Wellington, New Zealand.

ESD (1996) *Energy for the Future: Meeting the Challenge. TERES II (The Renewable Energy Study)*, Report and CD-ROM prepared for DG XVII, European Commission. ESD Ltd, Corsham, Wiltshire, UK.

FAO (1997) *Regional Study on Wood Energy Today and Tomorrow in Asia*. FAO Field Document no. 50, Bangkok.

Grassi G (1998) Modern bioenergy in the European Union. In: *Sustainable Agriculture for Food, Energy and Industry*, (eds N El Basram, R H Behl and B Prochnow), pp 829–836. James & James (Science Publishers) Ltd, London.

Hall D O (1998a) *Analyse the Needs for Fuelwood Plantations*. Shell International Petroleum Company and World Wide Fund for Nature.

Hall D O (1998b) The role of bioenergy in developing countries. In: *Proceedings of the 10th European Conference on Biomass for Energy and Industry, Wurzburg, Germany*, pp 52–55. C.A.R.M.E.N., Rimpar, Germany.

Hall D O and Rosillo-Calle F (1998) *Biomass Resources Other Than Wood*. World Energy Council, London.

Hall D O and Scrase J I (1998) Will biomass be the environmentally friendly fuel of the future? *Biomass and Bioenergy*, **15**, 357–367.

IEA Bioenergy (1998) *The Role of Bioenergy in Greenhouse Gas Mitigation*. Position Paper Task 25, IEA Bioenergy: T25: 1998: 02, http://www.forestresearch.co.nz/ieabioenergy.

IEA (1998, 1999 and 2000) *IEA Bioenergy Annual Reports*, (ed. J Tustin), IEA Bioenergy Secretariat, Forest Research, Rotorua, New Zealand.

Interlaboratory Working Group (1997) *Scenarios of US Carbon Reductions: Potential Impacts on Energy-Efficient and Low-Carbon Technologies by 2010 and Beyond*. Lawrence Berkeley National Laboratory and Oak Ridge National Laboratory (LBNL-40533 and ORNL-444).

IPCC (2000) *Third Assessment Report, Intergovernmental Panel on Climate Change*. www.ipcc.ch.

Jakobsen H H, Houmoller S and Pedersen L T (1998) *Technologies for Small Scale Wood-Fuelled Combined Heat and Power Systems*. International Energy Agency, Bioenergy Agreement, Report T13 Combustion 02.

Moomaw W, Serchuk A, Unruh G, Sawin J and Sverrison F (1999) Renewable energy in a carbon limited world. In: *Advances in Solar Energy 3* (ed. D Y Goswami), pp 68–137. American Solar Energy Society, Boulder, CO, USA.

Moore T (1998) Electrification and global sustainability. *EPRI Journal*, January/February, 43–52.

Nakicenovic N, Grübler A and McDonald A (2000) *Global Energy Perspectives*, International Institute for Applied Systems Analysis (IIASA) and World Energy Council (WEC). Cambridge University Press, Cambridge, UK.

Overend R and Costello R (1998) Bioenergy in North America: an overview of liquid biofuels, electricity and heat. In: *Proceedings of the 10th European Conference, Biomass for Energy and Industry, Wurzburg*, pp 59–61. C.A.R.M.E.N., Rimpar, Germany.

Porrit J (2000) *Playing Safe: Science and the Environment*. Thames & Hudson, London.

Sulilatu W F (1998) *Co-combustion of Biofuels*. International Energy Agency, Bioenergy Agreement, Report T13 Combustion 06.

SVEBIO (1998) *Environmental and Energy Policies in Sweden and the Effects on Bioenergy Development*. Swedish Bioenergy Association, www.svebio.se/environment/env_contents.html.

Watson R T (1999) The climate change debate: where are we? In: *Proceedings of the 6th Sustainable Energy Forum Conference, Sustainable Cities, Auckland*, pp 125–132. SEF, Wellington.

WEC (2000) *Energy for Tomorrow's World: Acting Now!* World Energy Council, Warwick St, London. http://www.worldenergy.org.

World Bank (1996) *Rural Energy and Development: Improving Energy Supplies for Two Billion People*. World Bank Industry and Energy Department, Report no. 1512. GLB, Washington DC.

Chapter 2. The woody biomass resource

Gifford J, Hooper G, Nicholas I, Hall P, Li J, Senelwa K, Sims R and Clemens T (2000) 'Woody biomass as a sustainable energy source'. In: *Proceedings of the Renewable Energy Research Showcase Seminar 'The role of renewable energy technologies in New Zealand's future'*, March, Paper 3. Massey University, Palmerston North.

Gifford J, Cox B and Sims R E H (2001) *Woody Biomass in New Zealand*. www.eeca.govt.nz.

IPCC (2000) Special report on *Land Use, Land Use Change, and Forests*. Summary available at www.ipcc.ch.

Larsson S, Melin G and Rosenqvist H (1998) Commercial harvest of willow wood chips in Sweden. In: *Proceedings of the 10th European Conference, Biomass for Energy and Industry, Wurzburg*, pp 200–203. C.A.R.M.E.N., Rimpar, Germany.

Riddell-Black D M, Sims R E H, Roygard J, Clothier B, Green S and Edwards R (1996) 'Water and nutrient use by fuelwood species: preliminary results from an intensively monitored lysimeter study'. In: *Proceedings of the 9th European Bioenergy Conference, Copenhagen*, Vol 1, pp 775–780. Pergamon, Oxford, UK.

Senelwa K. and Sims R E H (1998) 'Determination of relative yield indices for short rotation forestry species grown in New Zealand'. In: *Sustainable Agriculture for Food, Energy and Industry*, (eds N El Bassam, R H Behl and B Prochnow), pp 773–779. James & James (Science Publishers) Ltd, London.

Sims R E H and Riddell-Black D M (1996) The practical and economic feasibility of fuelwood production and use in a municipal sewage management system. In: *Proceedings of the 9th European Bioenergy Conference, Copenhagen*, pp 769–774. Pergamon, Oxford, UK.

Chapter 3. The non-woody biomass resource

Dixon T (1998) *Cogeneration from Bagasse*. Report, Sugar Research Institute, Mackay, Queensland, Australia.

El Bassam N (1998) *Renewable Energy: Potential Energy Crops for Europe and the Mediterranean Region*. FAO Regional Office for Europe, REU Technical Series 46.

El Bassam N, Graef M and Jakob K (1998) 'Sustainable energy supply for communities from biomass'. In: *Sustainable Agriculture for Food, Energy and Industry*, (eds N El Bassam, R H Behl and B Prochnow), pp 837–843. James & James (Science Publishers) Ltd, London.

Veenendal R, Jorgensen U and Foster C (1997) European energy crops overview project: synthesis report. *Biomass and Bioenergy*, summer/autumn special issue.

Chapter 4. The supply chain: harvesting, transport and processing

Browne A J, Palmer M, Hunter H A and Boyd J (1996) *Transport and Supply Logistics of Biomass Fuels. Vol 1: Supply Chain Options for Biomass Fuels*, ETSU report B/W2/00399/REP/1. University of Westminster, UK.

CEC (1999) *Evaluation of Biomass-to-Ethanol*. Californian Energy Commission Report P500–99–011, August.

Deboys R (1996) *Harvesting and Comminution of Short Rotation Coppice*, ETSU report B/W2/00262/REP. Forestry Commission, Technical Development Branch.

Dornburg V and Faaij A (2000) System analysis of biomass energy system efficiencies and economics in relation to scale. In: *Proceedings of the 1st World Conference on Biomass for Energy and Industry, Seville*, pp 847–850. James & James (Science Publishers) Ltd, London, UK.

ETSU (1990) *Wood Fuel Supply Strategies*, Aberdeen University, Contract report ETSU B 1176–P1. ETSU, AEA Technology, Harwell, UK.

FCA (1997) *Baling Forest Residues in the United Kingdom*. Forest Contractors Association, Aberdeen, UK.

Forestry Commission (1994) *First Field Evaluations of Short Rotation Coppice Harvesters*, Technical Development Branch Report 11/94. Forestry Commission, Dumfries, UK.

Hall P, Sims R E H and Gigler J K (2001) Delivery systems of forest arisings for energy production in New Zealand. *Biomass and Bioenergy*, in press.

Lowe H T, Sims R E H and Maiava T (1994) Evaluation of a low cost method for drying fuelwood from short rotation tree crop for small scale industry. In: *Proceedings of the 8th European Biomass Conference, Vienna*, Vol 1, pp 461–467. Pergamon, Oxford, UK.

Nellist M E (1997) Storage and drying of arable coppice. *Aspects of Applied Biology* **49** , 1–11.

Silsoe Research Institute (1997) *Drying and Storage of Short Rotation Coppice*, ETSU report B/W2/00391/00/00. AEA Technology, Harwell, UK.

Sims R E H and Culshaw D (1998) Fuel mix supply reliability for biomass-fired heat and power plants. In: *Proceedings of the 10th European Conference on Biomass for Energy and Industry, Wurzburg*, pp 188–191. C.A.R.M.E.N., Rimpar, Germany.

Sims R E H, Lowe H T and Maiava T (1994) All year round harvesting of short rotation coppice eucalyptus. In: *Proceedings of the 8th European Biomass Conference, Vienna*, Vol 1, pp 507–514. Pergamon, Oxford, UK.

Chapter 5. Thermochemical conversion by combustion and the steam cycle

European Commission (1998) *New Solutions In Energy Supply*: Addressing the constraints for successful replication of demonstration technologies for co-combustion of biomass/waste, DIS/1743/98-NL. Energie, Brussels.

ETSU (1996) *Energy from Waste*. Department of Trade and Industry, London.

FEC Ltd (1990) *Forestry Waste Firing of Industrial Boilers*. ETSU Report B1178. AEA Technology, Harwell, UK.

Van Doorn J, Bruyn P and Vermeij P (1996) Combined combustion of biomass, municipal sewage sludge and coal in an atmospheric fluidized bed installation. In: *Proceedings of the 9th European Biomass Conference for Energy and the Environment, Copenhagen*, pp 1007–1012. Pergamon, Oxford, UK.

Chapter 6. Thermochemical conversion by gasification and pyrolysis

Bridgwater A V (1998) The status of fast pyrolysis of biomass in Europe. In: *Bioenergy 98, Proceedings of the 10th European Bioenergy Conference, Wurzburg*, pp 268–271. C.A.R.M.E.N., Rimpar, Germany.

Bridgwater A.V (2001) Towards the bio-refinery: fast pyrolysis of biomass. *Renewable Energy World*, **4**(1), 66–83.

Joseph S (1997) EDL/BEST low cost biomass gasifiers for heat and power. In: *Proceedings of the Biomass Taskforce Symposium, Canberra*, October (ed. S Schuck). Bioenergy Australia, Sydney, Australia.

OECD (1998) *Projected Costs of Generating Electricity: Update 1998*. OECD, Paris, France.

Overend R P (1998) Biomass gasification: a growing business. *Renewable Energy World*, November, **1**(3), 26–31.

Pitcher K (2000) Turning willow into megawatts. *Renewable Energy World*, **3**(6), 35–45.

Senelwa K and Sims R E H (1999) Fuel characteristics of short rotation forest biomass. *Biomass and Bioenergy*, **17** , 127–140.

Senelwa K (1997) *The air gasification of woody biomass from short rotation forests*. PhD thesis, Massey University, Palmerston North, New Zealand.

US Department of Energy (2000) *Annual Energy Outlook*, DOE/EIA-0554(2000). US Department of Energy, Washington DC.

Chapter 7. Biochemical conversion of wet biomass

Biogas Technology (2001) Solution in search of its problem; a study of small scale rural technology in India. www.he.gu.se/dot.

Bridle T and Skrypski-Mantele A (1997) The ENERSLUDGE process: an energy and cost efficient sludge management system. In: *Proceedings of the Biomass Taskforce Symposium, Canberra*, October (ed. S Schuck). Bioenergy Australia, Sydney, Australia.

British Biogen (1997) *Anaerobic Digestion of Farm and Food Processing Residues*, Good Practice Guidelines. British Biogen, London.

ETSU (1996) *Energy from Waste*. Department of Trade and Industry, London.

New Zealand Standards (1987) Codes of Practice: *The Production and Use of Biogas, Farm Scale Operation. Part I Production of Biogas. Part II Uses of Biogas*. New Zealand Standards Association, Wellington, New Zealand.

Sims R E H and Richards K (1990) Anaerobic digestion of crops and farm wastes in the United Kingdom. *Agriculture, Ecosystems and Environment*, **30** , 89–95.

Chapter 8. Cogeneration of combined heat and power

Johnson U (2000) 104 MW Cogen block for Vaxjo. *Cogeneration and On-site Power Production*, **3**, 52–57.

Langreck J (2000) Cogen-absorption plants for refrigeration purposes and turbine air inlet cooling. *Cogeneration and On-site Power Production*, **2**, 46–49.

Lofstedt R E (1995) *The Use of Biomass Energy in a Regional Context: The Case of Vaxjo Energi AB*. Swedish Policy Report. Vaxjo Energi AB, Vaxjo, Sweden.

Pogoreutz M. (2000) Economical and technological comparison of small-scale CHP on the basis of biomass. In: *Proceedings of the 1st World Conference on Biomass for Energy and Industry, Seville*, pp 821–824. James & James (Science Publishers) Ltd, London.

Sinclair, Knight and Merz (1999) *Assessment of Cogeneration for the Australian Cogeneration Association*. http://www.cogen.com.au/va/cogen/.

Chapter 9. Biofuels for transport from the biochemical conversion of biomass

Austrian Biofuels Institute (1997) *Biodiesel Development Status Worldwide*. Report to the International Energy Agency, Paris. Austrian Biofuels Institute, Vienna.

Beer T, Grant T, Brown R, Edwards J, Nelson P, Watson H and Williams D (2000) *Life-Cycle Emission Analysis of Alternative Fuels for Heavy Vehicles*. CSIRO, Australia.

Calais P and Sims R E H (2000) A comparison of life-cycle emissions of liquid biofuels and liquid and gaseous fossil fuels in the transport sector. In: *Proceedings of the 37th Annual Conference of the Australia and New Zealand Solar Energy Society, Brisbane,* December. ACN 006 824 148. ANZSES, Maroubra, Australia (available only on CD-ROM).

Howell S A and Weber J A (1995) US Biodiesel overview. In: *Proceedings of the 2nd Biomass Conference of the Americas, Portland,* pp 840–848. NREL/CP-200–8098. National Renewable Energy Laboratory, Golden, CO.

IEA (1993) Vegetable oils as transport fuels. In: *Proceedings of the International Energy Agency Seminar, Pisa, May* (ed. G Caserta). IEA, Paris.

Korbitz W (1998) Biodiesel: from the field to the fast lane. *Renewable Energy World*, **1**(3), 32–37.

Larson E D and Jin H (1999a) Biomass conversion to Fischer-Tropsch liquids: preliminary energy balances. In: *Proceedings of the 4th Biomass Conference of the Americas, Oakland,* pp 843–853. Pergamon, Oxford, UK.

Larson E D and Jin H (1999b) A preliminary assessment of biomass conversion to Fischer-Tropsch cooking fuels for rural China. In: *Proceedings of the 4th Biomass Conference of the Americas, Oakland,* pp 855–863. Pergamon, Oxford, UK.

Moreira J R and Goldemberg J (1999) The alcohol program. *Energy Policy*, **27**, 229–245.

Overend R P and Costello R (1999) Bioenergy in North America: an overview of liquid biofuels, electricity and heat. In: *Proceedings of the 10th European Conference, Biomass for Energy and Industry, Wurzburg,* pp 59–61 C.A.R.M.E.N., Rimpar, Germany.

Reed T B (1993) An overview of the current status of biodiesel. In: *Proceedings of the 1st Biomass Conference of the Americas*, Vol II, pp 797–814. Burlington. National Renewable Energy Laboratory, Golden, CO, USA.

Saller G, Funk G and Krumm W (1998) Process chain analysis for production of methanol from wood. In: *Proceedings of the 10th European Conference on Biomass for Energy and Industry, Wurzburg,* pp 131–133. C.A.R.M.E.N., Rimpar, Germany.

Scharmer K (1998) 'Biodiesel from set-aside land'. In: *Sustainable Agriculture for Food, Energy and Industry,* (eds N El Bassam, R H Behl and B Prochnow), pp 844–848. James & James (Science Publishers) Ltd, London.

Schindlbauer H (1995) Standardization and analysis of biodiesel. In: *Proceedings of the First International Conference on Biodiesel, Vienna,* pp 72–75. FICHTE/Technical University of Vienna, Austria.

Sheehan J, Camobreco V, Duffield J, Graboski M and Shapouri H (1998) *An Overview of Biodiesel and Petroleum Diesel Life Cycles*. National Renewable Energy Laboratory, Golden, CO, USA.

Sims R E H (1995) The biodiesel research programme of New Zealand. In: *Proceedings of the 2nd Biomass Conference of the Americas, Portland,* pp 849–858. NREL/CP-200–8098, National Renewable Energy Laboratory, Golden, CO.

Sims R E H (1985) Tallow esters as an alternative diesel fuel. *Transactions, American Society of Agricultural Engineers*, 28(3), 716–721.

Sims R E H (1990) Tallow esters and vegetable oils as alternative diesel fuels. *Solar and Wind Technology*, 7(1), 31–36.

Sims R E H, Ritchie W R and Thompson M C (1990) Tallow esters and vegetable oil esters. In: *Proceedings of the 1st World Renewable Energy Congress, Reading,* Vol 3, pp 2128–2141. Pergamon, Oxford, UK.

Sims R E H, Ritchie W R and Chadwick A J (1990) Turbocharging of an agricultural tractor engine. *Journal of Agricultural Engineering Research*, **47**, 177–186.

Wideman B A (1989) Fundamental research to develop a standard for fuel quality of rapeseed oils and esters. In: *Proceedings of the 5th European Biomass Conference, Lisbon*, pp 9–13. Elsevier, London.

Worgetter M (1993) Biodiesel in Austria. In: *Proceedings of Seminar on Vegetable Oils as Transport Fuels, Pisa*. International Energy Agency Bioenergy Agreement Tasks VIII and X, pp 45–50.

Chapter 10. Small-scale bioenergy projects: present and future

Bauen A (2000) Sustainable heat and electricity supply from gasification based biomass fuel cycles: the case of Sweden and the UK. In: *Proceedings of World Renewable Energy Congress VI (WREC2000)*, pp 1381–1384. Elsevier, London.

Bowman L and Lane N W (1999) Micro-scale biomass power. In: *Proceedings of the 4th Biomass Conference of the Americas, Oakland*, pp 1445–1451. National Renewable Energy Laboratory, Golden, CO, USA.

Carlsen H (1996) Stirling engines for biomass: state of the art with focus on results from Danish projects. In: *Bioenergy 1996: Proceedings of the 9th European Bioenergy Conference, Copenhagen*, pp 278–283. Pergamon, Oxford, UK.

Cowburn D A, Gale J and Dando R (1997) Biomass combined heat and power based on a Stirling cycle engine. *Aspects of Applied Biology*, **49**, 443–448.

Craig J D and Purvis C R (1999) A small scale biomass fuelled gas turbine engine. *Transactions of the ASME*, **121**, 64–67.

Czernik S, French R, Feik C and Chornet E (1999) Hydrogen from biomass via fast pyrolysis/catalytic steam reforming process. In: *Proceedings of the 1999 US DOE Hydrogen Program Review*, NREL/CP-570–26938. National Renewable Energy Laboratory, Golden, CO, USA.

Czernik S, French R, Feik C and Chornet E (2000) Production of hydrogen from biomass-derived liquids. In: *Proceedings of the US DOE Hydrogen Program Review*, NREL/CP-570–28890. National Renewable Energy Laboratory, Golden, CO, USA.

Dawson W M, Forbes G and McCracken A R (1996) Small scale gasification of short rotation coppice willow for electricity generation. In: *Proceedings of the 9th European Bioenergy Conference, Copenhagen*, pp 1289–1294. Pergamon, Oxford, UK.

Dawson W M (1997) The potential of small scale biomass electricity systems. *Aspects of Applied Biology*, **49**, 423–428.

De Ruyck J, Allard G and Maniatis K (1996) An externally fired evaporative gas turbine cycle for small scale biomass gasification. In: *Proceedings of the 9th European Bioenergy Conference, Copenhagen*, pp. 260–265. Pergamon, Oxford, UK.

DOE (1999) Small modular biopower projects. www.eren.doe.gov/bioenergy_initiative/pdfs/small_modular.pdf.

Föger K (2000) Fuel cells for sustainable energy conversion. In: *Proceedings of the 38th Annual Conference of the Australian and New Zealand Solar Energy Society, Brisbane*, pp 35–42. ACN 006 824 148. ANZSES, Marouba, Australia.

Hasler P and Nussbaumer T (1999) Gas cleaning for IC engine applications from fixed bed biomass gasification. *Biomass and Bioenergy*, **16**, 385–395.

Heinzel A, Formanski B, Ledjeff-Hey K and Schaumberg K (1996) Fuel cells for electrical energy from gasified biomass. In: *Bioenergy 96: Proceedings of the 9th European Bioenergy Conference, Copenhagen*, pp 1462–1467. Pergamon, Oxford, UK.

Hoogers G and Potters L. (1999) Fuel cells: their potential as the ultimate clean source of power. *Renewable Energy World*, **2**(1), 51–57.

Houmann-Jakobsen H, Houmoller S and Pedersen L T (1998) *Technologies for Small Scale Wood-Fuelled Combined Heat and Power Systems*, Report for IEA Bioenergy

Agreement Task XIII Biomass Utilization. DK-Teknik, Energy & Environment, Soborg, Denmark.

Kiros Y, Myren C, Schwartz S, Sampathrajan A and Ramanathan M (1999) Electrode R&D, stack design and performance of biomass-based alkaline fuel cell module. *International Journal of Hydrogen Energy*, **24**, 549–564.

Knoef H A M, Wagenaar B M and Reumerman P (1998) Indirectly fired gas turbine for rural electricity production from biomass. In: *Proceedings of the 10th European Bioenergy Conference, Wurzburg*, pp 1334–1337. C.A.R.M.E.N., Rimpar, Germany.

Koppel T (2000) *Powering the Future*. John Wiley, Toronto, Canada.

Lobachyov K V and Richter H J (1998) An advanced integrated biomass gasification and molten fuel cell power system. *Energy Conversion Management*, **39** (16–18), 1931–1943.

McIlveen-Wright D R, Williams B C and McMullan J T (2000) Wood gasification integrated with fuel cells. *Renewable Energy*, **19**, 223–228.

Munster E (1996) 40 kW Stirling engine powered by wood chips. In: *Proceedings of the 9th European Bioenergy Conference, Copenhagen*, pp 1280–1282. Pergamon, Oxford, UK.

Obernberger I (1998) Decentralized biomass combustion: state of the art and future development. *Biomass and Bioenergy*, **14** (1), 33–56.

Prabhu E and Tiangco V (1999) The Flex-Microturbine™ for biomass gases: a progress report. In: *Proceedings of the 4th Biomass Conference of the Americas, Oakland, CA*, pp 1439–1444. National Renewable Energy Laboratory, Golden, CO, USA.

Schmidt D D and Gunderson J R (2000) Opportunities for hydrogen: an analysis of the application of biomass gasification to farming operations using micro-turbines and fuel cells. In: *Proceedings of the Hydrogen Program Review*, NREL/CP-570–28890. National Renewable Energy Laboratory, Golden, CO, USA.

Smeenk J, Steinfeld G, Brown R C, Simpkins E and Dawson M R (2001) Evaluation of an integrated biomass gasification/fuel cell power plant. www.cvrcd.org/research/papers/evalution_paper.

Twidell J (1998) Biomass energy: technology update. *Renewable Energy World*, **1** (3), 38–39.

Wang D, Czernik S and Chornet E (1998) Production of hydrogen from biomass by catalytic steam reforming of fast pyrolysis oils. *Energy & Fuels*, **12**, 19–24.

Chapter 11. Future potential for bioenergy and the barriers

Chegwidden A (2000) Mallee eucalyptus for electricity, activated carbon and oil – a new regional industry. In: *Proceedings, Bioenergy Australia 2000, Gold Coast*. Bioenergy Australia, Sydney, Australia.

ETSU (1996) Good Practice Guidelines: *Short rotation coppice for energy production*; *Wood and forest products*; *Anaerobic digestion of farm and food processing residues*. ETSU, AEA Technology, Harwell, UK.

Faaij A, Hamelinck C, Larson E and Kreutz T (2000) Production of methanol and hydrogen from biomass via advanced conversion concepts: preliminary results. In: *Proceedings of the 1st World Conference on Biomass for Energy and Industry, Seville*, pp 683–686. James & James (Science Publishers) Ltd, London.

Stucley C (2000) Carbon products and energy from wood. In: *Proceedings, First Conference, Bioenergy Australia 2000, Gold Coast*. Bioenergy Australia, Sydney, Australia.

Index

agricultural residues   52, 154
animal wastes   52, 61, 172, 263, 296
ash content   19, 21, 42, 73, 87, 107, 108, 113, 167, 212, 252

bagasse   52, 53–57, 107, 197
bagasse and MGW cogeneration   69
BIGCC   158
biodiesel   215, 218–225, 262
biogas   60, 62, 64, 168, 169, 173, 178–179, 184, 187, 197, 214, 241, 246, 262
bio-oil   160, 164, 216
biorefinery   114, 121, 163, 236
black liquor   133, 200
bulk density   13–14, 21, 42

carbon offset   24, 37, 287, 301
carbon sink   24, 25–27, 37, 287, 302
chemical composition   4, 10, 42
climate change   8, 23, 153, 217, 263, 300
co-firing   23, 37, 52, 108, 112, 116, 134, 163, 201
cogeneration   17, 49, 51, 55, 192, 196, 248, 256, 265, 272, 273, 306
co-incineration   117
combined-cycle   105, 155, 156, 198, 242, 290
combustion 7, 104, 113, 192, 210, 246, 247
   characteristics   19
comminution   81, 83, 91
condensing turbine   128, 196
contracts   209
cooling   197, 201, 262, 292, 306
coppice   34, 38, 80, 89, 93, 97, 255, 272

densification   14–15
developing countries   5, 35, 81, 106, 115, 157, 169, 181, 197, 242, 251, 256, 276, 284, 306,
digester   172, 173, 181, 183, 184, 186
   design   174–175
distributed generation   156, 292–295, 306

emissions   115
energy balance   84, 146, 299
energy crops   28, 38, 44, 52, 59, 75, 193, 216, 235, 306
energy density   16, 17, 218
energy ratio   38, 75, 153, 221
ethanol   214, 233

forest arisings   28, 29, 42, 43, 44, 45, 46, 85, 98, 101, 107, 209
fuel cells   181, 192, 214, 235, 246, 253, 262, 263, 294, 305
fuel supply   70, 99, 157, 291, 298
   contract   18, 73,
fuelwood supply   43

gas engine   179, 182, 185, 188, 192,
gas turbine   55, 92, 105, 120, 144, 151–152, 201, 206, 253, 260–262, 272, 290
gasification   168, 246, 249–252, 274
gasifier   276
greenhouse gas (GHG)   168, 181, 197, 199, 201, 250, 221, 244, 284, 285
   emissions   181

heat pump   197
higher heat value (HHV)   16, 18

hydrogen 270
hydrolysis 236–238

incineration 116, 129, 131, 142, 146, 181
industrial heating 51
internal combustion engine 225, 250, 253

kiln drying 49, 50

landfill gas 63, 65–66, 71, 130, 135, 168, 182, 190–193, 193, 194, 197, 263, 296
laws of energy 8, 196
life cycle analysis 287–290
liquid fuel 104, 156, 160, 214
lower heat value (LHV) 16, 18, 20

methanol 145, 240, 265
micro-turbines 246, 253, 262
moisture content 13, 15–16, 21, 22, 42, 50, 84, 87, 93, 102, 105, 107, 113, 135, 188, 207, 251
municipal green waste 34, 41, 146, 173, 181
municipal solid waste 34, 62, 106, 116, 121, 129, 161, 169, 182, 190, 193, 235

nutrients 33, 179, 184, 188, 207, 295, 302

particulate 116, 138, 142, 159, 212, 221, 229, 242, 255, 299
pass-out steam turbines 128, 204
pellets 14, 58, 85, 107, 295

photosynthesis 3, 57
planning consent 174, 296, 301
producer gas 144, 246, 262, 270
pyrolysis 160–163, 168, 241, 246, 252–253, 274

scale of conversion plant 6
scrubbing 177, 185, 190, 192, 241
social issues 277, 282, 290–292
socio-economic aspects 292
solar energy 2, 59
steam turbine 54, 70, 87, 90, 104, 122, 126–129, 134, 136, 139, 153, 199, 201, 209, 211, 248, 253, 256, 290, 305,
Stirling engines 246, 253, 257–260
straw 52, 57, 67, 89, 96, 106, 107, 108, 173, 197, 224

tallow 220, 226
technical potential 9, 39, 57, 61, 284
traditional biomass 4, 28, 37

vegetable oils 214, 220, 227
vehicles 192, 181, 214, 239, 264, 265
vehicle biofuels 179, 214, 306
volatiles 105, 114, 121, 137, 145, 146, 180, 187, 305

waste-to-energy 129, 141, 163, 168, 306
wood process residue 28, 43, 45, 98, 133, 207
wood stove 106